THE CHALLENGE OF
THE 21st CENTURY

THE CHALLENGE OF THE 21ST CENTURY

Managing Technology and Ourselves in a Shrinking World

HAROLD A. LINSTONE

Portland State University

with

IAN I. MITROFF

University of Southern California

FOREWORD BY IDA HOOS

State University
of New York
Press

Published by
State University of New York Press, Albany

Production by Susan Geraghty
Marketing by Bernadette LaManna

Printed in the United States of America

For information, address State University of New York Press,
State University Plaza, Albany, N.Y., 12246

Library of Congress Cataloging-in-Publication Data

Linstone, Harold A.
     The challenge of the 21st century : managing technology and
ourselves in a shrinking world / by Harold A. Linstone with Ian
I. Mitroff ; foreword by Ida Hoos.
          p.   cm.
     Includes bibliographical references and index.
     ISBN 0–7914–1949–5 (alk. paper). — ISBN 0–7914–1950–9 (pbk. :
alk. paper)
     1. Technological innovations—Management. 2. Technology—Social
apsects. 3. Information society. 4. Twenty-first century.
I. Mitroff, Ian I. II. Title. III. Title: Challenge of the twenty
-first century.
HD45.L5284   1994
658.5'14—dc20   93–296698
                                                            CIP

10 9 8 7 6 5 4 3 2 1

For our grandchildren,
who will grasp the challenge.

CONTENTS

LIST OF FIGURES

LIST OF TABLES

FOREWORD

by Ida Hoos

The 21st century was born prematurely not long after World War II. The Bomb, the Pill, the Box, and the Tube, in synergy, caused profound changes in every facet of human experience and everywhere on earth. Here was the dawning of a "brave new world," the shining wave of the future, with optimism rampant and clichés galore. The smashing of the atom would release cheap energy in abundance; meters would become as obsolete as ear trumpets. The Pill would accomplish that which Malthus foresaw as essential for survival of life on this planet, namely, population control and by means far kindlier than his dire prediction of famine, war, and disease. Computers would do the dirty, heavy, and dangerous tasks in factory and field. In the office, paperwork would be reduced, streamlined; the "white plague" would be over. And everywhere humans would be forever freed from dull, stultifying routine. It even became fashionable in academic circles to ponder the implications of leisure for the masses! Television would bring education and, yes, culture and enlightenment into every home. Illiteracy would be vanquished; even the most remote hamlets of Africa and India would receive instruction via satellite. Telecommunication would link all peoples everywhere. In this global village the dream of one world realized at last, "E pluribus unum" would take on new meaning. Now all men would be neighbors, perhaps even brothers.

Reckoning by sociological time, we were living in the age of technology, its epiphany science and technology. Is it any wonder that the precepts of management science, panoplied in the raiment of rationality, attracted wide acceptance? How the techniques migrated from the war front to outer space and thence to the inner city is a tale often told. What needs to be emphasized as relevant here are certain distillates: the notion that problems, animal, vegetable, mineral, technological, sociological, and human, can be *managed,* that there is a *rational methodology* for this management, and that *cost-effectiveness* is its avatar.

Thus, in the political arena from county to Congress, here was the basis for resource allocation and budgeting. In academic circles from kindergarten to college, decisions about curriculum, student/faculty ratios, library acquisitions, fee structure—in short, everything—were tailored to meet this criterion. Matters of medicaments, drug research and testing, food additives, provision of health care—all were governed by this "rational" stricture, namely, to be cost-effective. The environmental impacts of everything from cow manure (as a source of methane and, hence, renewable energy) to the disposal of nuclear wastes were assessed through the methodology of "risk analysis," this having become the accepted way to calculate cost-effectiveness in the future tense.

Technology on all fronts advanced so swiftly and with such pervasive effects that they seemed déjà vu almost before they occurred. The denouement was subtle; by the time the year 1984 came around, its symbolic meaning was all but lost. In the novel, Winston Smith was surrounded by signs warning him that Big Brother was watching. For us, this was an all-too-familiar phenomenon, data banks, dossiers and electronic surveillance devices having rendered privacy a quaint anachronism. Orwell's once-dreaded "thought control" occurs whenever one turns on the Tube. Subscribers to cable television even pay for the privilege of being brainwashed by having their prejudices and preferences astutely and purposefully programmed. This is how we choose up sides in war, tote up defeat or victory, select shampoo, or elect presidential aspirants.

Television has done its job all too well as a teaching device. Influential and beguiling, it has effectively taught violence, inculcated deplorable values, and has corroded the capacity for healthy familial and interpersonal relations. For many persons of all ages, life has become a spectator sport, the screen having sapped independent thought. "Educational television," oxymoronic though it may sound, could be and sometimes is a useful pedagogical tool. But there are better and more cost-effective methods of teaching-learning. For the already underprivileged youngsters, children's television is a long step toward addiction. "Sesame Street" legitimizes television viewing because it is purported to teach reading skills. Ironically, the school records of two generations of Sesame Streeters show that there is little carryover from learning one's ABCs to the acquisition of reading and comprehension skills. These are now at such a low point that America's schools have had to lower their requirements so as to grant diplomas to near-illiterate graduates.

By the seventies the computer had metamorphosed from the pampered behemoth that needed an air-conditioned, dust-free environment to a laptop genie, paradoxically servant to man and master of his fate. As to its effects at work, at home, and in every aspect of our lives, we were like the six blind men of Indostan who went to see the elephant: we were all right and we were all wrong.

Automation took over many heavy and hazardous tasks, to be sure. But whole categories of occupations disappeared. With more and more skills built into the machines, work became splintered and routinized. Our technological era created machines that can think and people who cannot! Stress resulting from alienation and dehumanization have become predominant occupational hazards and a key problem in industrial relations. Factories have come to look like offices and offices like factories. The volume of paperwork follows Parkinson's Law: it has expanded to exceed capacity. We cannot keep up without computers, nor can we keep up with the work they generate. Thanks to all kinds of unanticipated glitches and bugs, not least of which are maliciously destructive viruses, we have no notion of the way the Box has become Pandora's!

As to population control, what the Pill failed to accomplish directly, it has done in ways diverse, diffuse, and difficult to document. It has contributed to a revolution in which mores and morals have undergone fundamental change. Suspiciously concomitant with the loosening of traditional constraints has been the spread of AIDS (acquired immunodeficiency syndrome), which has caused untold suffering and claimed the lives of thousands. At risk are young and old, male and female, weak and strong, homosexual and heterosexual. And there is no medication or cure in sight. In the end, Malthus is still right.

The Bomb is, in the vernacular, not all it was cracked up to be. The smashed atom has not yielded cheap and abundant energy. In fact, from uranium mine to spent rods, the fuel cycle has been costly and fraught with dangers, not all of them fully assessed or acknowledged. In fact, zeal to protect the *image* of safety has so far outweighed concern for public safety that information is often skewed and unreliable. Invocation of the Freedom of Information Act and reports by "whistle-blowers" indicate how far "official" information is from truth. Only time will tell that, since the effects of radiation are insidious, long-lasting, and not fully understood. Even the results of such recent and much-publicized events as the disaster at Chernobyl have been deliberately miscalculated and grossly skewed—to allay well-warranted public fears. How unreliable then are the models contrived to predict the risks posed by nuclear wastes with a half life of centuries to come!

Dropping the Bomb on Hiroshima in 1946 did not, as military strategists proclaimed, put an end to war. Quite the contrary, nuclear capability has become the means by which every present and future petty tyrant, whether in emirate, sheikhdom, shogunate, or revolutionary faction, can hold hostage the very survival of life on this planet. There is considerable known trafficking in fissionable materials, and nuclear capability is a big ticket item in the international arms market. Even without resort to bomb dropping, the continued manufacture of weapons-grade material produces quantities of plutonium, a substance so-named because of its hellishly toxic properties. Who is to say what historic, economic, or political circumstances will breed terrorists likely to use this formidable nuclear blackmail?

As the end of the 20th century draws near, Professor Linstone reviews the modes of thought that have made the fin de siècle what it is. Drawing on theoretical sophistication and vast experience, he analyzes the shortcomings and pitfalls of systemic approaches that may have had iatragenic effects in the long run. Arguing persuasively for a new multi-perspective paradigm, one that encourages unbounded thinking and ethical action, he presents us with *The Challenge of the 21st Century: Managing Technology and Ourselves in a Shrinking World.*

PREFACE

Historian Elting Morison has given us an intriguing account of the fate of the USS *Wampanoag*, a 4200 ton "advanced technology" destroyer built for the U.S. Navy and commissioned in 1868. She had sails and a steam engine and was fast (exceeding seventeen knots). Sea trials proved her to be a magnificent technical achievement—ahead of ships in any navy at that time.

In 1869 all naval steamships were scrutinized by a board of naval officers. The mood of the board is documented. The steam vessel, said the board, was not a school of seamanship for officers or men:

> Lounging through the watches of a steamer, or acting as firemen and coal heavers, will not produce in a seaman that combination of boldness, strength, and skill which characterized the American sailor of an elder day; and the habitual exercise by an officer of a command, the execution of which is not under his own eye, is a poor substitute for the school of observation, promptness and command found only on the deck of a sailing vessel.[1]

The board examined the *Wampanoag* and developed a bill of particulars leading it to the conclusion that the ship was "a sad and signal failure" and could not be made acceptable. The country was in a state of peace, and the board opposed building iron-clads, needed in war, to avoid unnecessary alarm. There was a large supply of timber in the Navy Yards "which the interests of economy demand should be utilized." They noted the familiarity of the workmen with wooden ship building and their dependence on it for a livelihood.

The ship was laid up for a year and soon sold by the Navy. Morison ponders this strange turn of events:

> Now it must be obvious that the members of this Naval Board were stupid. They had, on its technical merits, a bad case, and they made it worse by the way they tried to argue it . . . [But after a time] I began to be aware of a growing sense of dis-

ease ... Could it be that these stupid officers were right? I recalled the sagacious judgment of Sherlock Holmes. The great detective, you will remember, withheld the facts in the incident of the lighthouse and the trained cormorant because, as he said, it was a case for which the world was not yet fully prepared. Was this also the case with the *Wampanoag*?

What these officers were saying was that the *Wampanoag* was a destructive energy in their society. Setting the extraordinary force of her engines against the weight of their way of life, they had a sudden insight into the nature of machinery. They perceived that a machine, any machine, if left to itself, tends to establish its own conditions, to create its own environment and draw men into it. Since a machine, any machine, is designed to do only a part of what a whole man can do, it tends to wear down those parts of a man that are not included in the design ...

I don't happen to admire their solution, but I respect their awareness that they had a problem ... [It] is not primarily engineering or scientific in character. It's simply human.[2]

A decision that appeared "stupid" from one point of view suddenly became reasonable when seen from another.

In a very different context, Richard M. Cyert, coauthor of the well-known book *A Behavioral Theory of the Firm*, observed after he became president of Carnegie-Mellon University:

As a professor of organization theory and management, I used to wonder about the practical value of these academic fields. For the last eight years, I've had some first-hand experience finding out ... And I've concluded that the study of management makes a useful, but only a limited, contribution to the practising manager.

Organization theory hasn't provided me any framework to judge possible appointees. The theory hasn't even been very useful in developing new organizational structures ... Finally, theory doesn't shed much light on how a manager should get information about how his organization operates.[3]

Both revelations make the point that most highly trained academics and technologists have only a partial view of the world they deal with. Essentially, their analytic, data- and model-based approach to problems is one-dimensional in a three-dimensional world. It is a lesson also learned by the authors, Harold Linstone, who received his doctorate in mathematics, and Ian Mitroff, who obtained his in engineering science.

Linstone spent the first half of his professional life in the aerospace industry. He started an operations research group at Hughes Aircraft Company (actually a high-tech electronics manufacturer) and went on to become associate director of corporate planning at the Lockheed Corporation. He learned at first hand the place of systems analysis in the corporate decision-making process. In particular, it became clear that analyses offered an adequate basis for decision on very minor problems but formed only one essential element in the really important decision questions. In these latter situations, no amount of clever cost-benefit analysis could suffice, and very different ways of looking at the same problem or issue were required.

In doing analyses of future customer needs, he also learned valuable lessons. The primary client, the Department of Defense, bought not necessarily what it needed according to our "rational" analysis but what it was comfortable with. The top client priorities differed from those Linstone's team developed. Top dollar items, big firepower and big vehicles (tanks, ships, and aircraft) were more crucial to the customer than communications equipment that ranked high on the priority needs list Linstone's group derived. Such experiences led to a sense of dis-ease akin to that of Elting Morison. The team had been naive in its reliance on analysis, on data and modeling. It was not looking at the real world where organizations and individuals play vital parts not captured in the idealized, "objective," systems analysis approach.

For example, there may be good reasons to do things that appear not to be cost-effective. A company may undertake a research and development program to keep its superb engineering team together, knowing that it cannot make a profit on the project. Engineers want to be on the leading edge of technology, and the preservation of that talent may be more important for the long-term future of the organization than the near-term profits. A military organization may buy certain equipment partly to maintain a high-morale force in peacetime, a difficult task. Both examples illustrate human factors often neglected by the dazzling computer models.

Mitroff studied National Aeronautics and Space Administration (NASA) scientists, specifically those involved with the Apollo lunar landing program. His many interviews showed that they themselves had serious doubts about the existence of "unbiased observers." Indeed, they were very human, at times highly subjective and irrational. For example, they might hold on to pet hypotheses in the face of overwhelming contradictory data.

Two academics also played important roles in encouraging the authors in the direction of the approach taken in this book. The first was Harvard's Graham Allison, who, in 1971, wrote a study of the Cuban missile crisis, *Essence of Decision*, using three points of view. Each provided unique insights that could not be obtained with the others. The second was the University of California's West Churchman, a systems philosopher who has inspired the authors and many others with books such as *The Design of Inquiring Systems*, *Challenge to Reason*, and *Thought and Wisdom*. Indeed, he continues to inspire them.

We still may wax nostalgic when we remind ourselves of the beautiful, elegant, and satisfying results achieved with the paradigms of science and engineering with which we grew up. But we must now face complex systems where everything interacts with everything, where human and technical factors must both be fully appreciated and ethics means much more than logic and scientific rationality. This effort represents but one step along the path of the unbounded systems thinking that prepares us for the opportunities and dangers in the global village of tomorrow.

Finally, let us turn to the questions the reader should ask the authors: First, what is the content of the book? There are five parts. Part 1 (chap. 1) introduces us to two systems of very different scales, both typifying the need for new thinking in managing our technology and ourselves. Part 2 (chaps. 2–5) examines the first, the 1989 Alaska oil spill, from various points of view. Part 3 (chap. 6) presents the essential features of the multiple perspective approach. In part 4 (chaps. 7–11) the concept is applied to the second system, a developed society in the evolving global village. Finally, part 5 (chap. 12) draws implications from the multiple perspective exploration of these two cases. Although the cases are anchored in the American experience, the approach itself, and many of the implications, are clearly relevant to complex systems in many other settings.

Second, who will find the book of value? As noted earlier, Linstone observed its use in a natural, informal way by senior executives in large, high-tech corporations. He has been asked to present management seminars and talks on the concept to professional groups in the United States, Japan, Germany, Korea, Mexico, and elsewhere. Under the auspices of the National Science Foundation, he has used it to develop insights on technology assessments in the United States and on regional development in China. His graduate students have applied it to problems involving organizations such as Tektronix, Portland General Electric, US West Communications, American and German automobile manufacturers, Bonneville Power Administration, a health sciences university nursing program, and an urban perinatal health care program.

The book is designed to sensitize any thoughtful reader to the importance of sweeping in multiple perspectives with distinct paradigms when confronted with messy, complex problems. These may involve private or public sector decision making, local or global scale, near-term or long-term issues. Breaking out of the single-perspective constraint poses a challenge not to be underestimated. As an aid to the reader, each page, beginning with chapter 2, features a box in the upper right-hand corner that identifies the perspective(s) being discussed.

The discussion should help corporate planners and managers, engineering administrators and policy analysts. It can wean students in science, engineering, and business administration from exclusive concern with the technical perspective. Their training often involves a kind of brainwashing that focuses on analytic tools, that is, models and data. The traditional textbook problems—clear problem definition, single set of assumptions, definite solution—leave the students quite unprepared for the real, messy world. By opening up their thinking, they become better practitioners. Engineers gain a deeper insight on technological risk treatment; students in management programs gain deeper understanding of the organizational operating environment they will face.

The two systems that form the substance of the book, the Alaska oil spill and an advanced society in the evolving global village, should be of interest to all who are concerned about their rapidly changing world. It is hoped that the discussion will help in bursting the chains of the old mind set with its obsolescing assumptions and suggest the new issues that must be addressed. In this way the reader dips at least a few toes into the turbulent sea that is shaping the new century.

PART 1

Introduction

CHAPTER 1

Hedgehogs or Foxes?

The mature individual is the individual who can hold conflicting
world views together at the same time, and act, and live and that his
or her life is enriched by that capability—not weakened by it . . . To
be able to see the world globally, which you are going to have to be
able to do, and to see it as a world of unique individuals . . . That is
complexity, that is really complexity!

C. W. Churchman

The test of a first-rate intelligence is the ability to hold two
opposed ideas in the mind at the same time, and still retain the
ability to function . . . One should, for example, be able to see that
things are hopeless and yet be determined to make them otherwise.

F. Scott Fitzgerald

We begin our exploration with two vignettes, one looking back
and the other looking ahead.

MIDNIGHT AT BLIGH REEF

At 12:04 A.M. on March 24, 1989, in calm seas, the tanker *Exxon
Valdez* struck Bligh Reef in Prince William Sound, Alaska. Eight
of the eleven cargo tanks were punctured, and the result was the
largest oil spill in U.S. history. Of the 53 million gallons (or 1.26
million barrels) of crude oil carried, 21 percent spilled (10.8 million gallons). Almost all of the spill (10.1 million gallons)
occurred in the first five hours after the accident. The oil spread
over 3,000 square miles in Prince William Sound and the Gulf of
Alaska. The ship was under the command of Captain Joseph
Hazelwood. A sobriety test was given to him ten hours after the
accident and showed a .06 percent blood alcohol level, above the
USCG limit of .04 percent, but below the state's .10 percent limit.

Built in San Diego in 1986, the *Exxon Valdez* is 987 feet long and has a deadweight of 213,755 tons. The spill was by no means the largest in the world to date. In 1967 the *Torrey Canyon* dumped three times as much (30 million gallons) off the British coast and in 1978 the *Amoco Cadiz* spilled six times as much crude oil off Brittany (68 million gallons).

Captain Hazelwood notified the Coast Guard 22 minutes after the grounding. The on-scene coordinator (USCG) notified the National Response Center, the State of Alaska, and Alyeska, the petroleum service company responsible for pipeline and Valdez port operations, within one hour of the accident. Alyeska's response was delayed more than twelve hours after notification— far beyond the five hours stipulated in its contingency plan. Within twenty-four hours, the *Exxon Baton Rouge* was positioned alongside the *Exxon Valdez* to transfer the nearly 80 percent of oil still in the tanks of the grounded ship. On the second day Exxon Shipping Company assumed responsibility for the cleanup. Estimates of the shoreline contaminated by oil ranged from 730 to 1245 miles. Figures for sea bird deaths climbed from 28,000 in 1989 to 90,000–270,000 in 1990 to 580,000 in 1991. Estimates of sea otter casualties ranged from 872 in 1989 to 5,500 in 1991. There was a loss of at least $12 million in herring fishery, while 30 percent of the salmon spawning grounds were threatened. Three years after the accident, the governmental Exxon Valdez Oil Spill Trustee Council reported continuing damage as the oil worked itself through the fish-spawning and animal breeding cycles. Hydrocarbons have accumulated in the bile and blood of river otters, which eat mussels that are still contaminated with oil in the Sound. Bird nesting sites continue to be disrupted by oil, high abnormality rates in fish eggs have been observed, and the social structure of killer whale pods is breaking down. Even the migrations of salmon and birds have been affected.[1]

The cost of the spill to Exxon from March 1989 through September 1990 was $2.2 billion.[2] This figure may be compared to the $1.3 billion cost of the Three Mile Island nuclear accident. *In neither case were there any human fatalities.** At the end of the

---

*It is interesting to note that the Bhopal, India, chemical accident at the Union Carbide plant, with estimates of 2,500 to 4,000 fatalities and possibly 200,000 non-fatal injuries (many of them severe), was settled by the company for only $470 million in damages paid to victims. However, this settlement was subsequently disputed.

1989 cleanup effort, Exxon executives said it was time to celebrate the successful "treatment" of the shoreline. Otto Harrison, the company's cleanup chief, pronounced the entire shoreline of the Sound to be "environmentally stable," almost free of oil, and posing no further danger to fish and wildlife. However, Exxon announced that it would have a 450-person crew in the area after the pullout to monitor and to study developments. Alaska's Governor Cowper disagreed with Exxon's assessment: "Obviously the beaches are not clean, and we are not satisfied with the condition of the beaches and the water." The Coast Guard's Admiral Yost observed that the results of Exxon's laudable summer cleanup project did not measure up to the efforts. It was evident that the corporation, already faced with more than 140 law suits, was likely to be billed for additional cleanup work that the state would perform.

A year after the spill, in March 1990, many wildlife species, such as otters, porpoises, and bald eagles had returned to Prince William Sound. But tarry oil layers were still found on some beaches, and oil washed out from the beaches was found to have been redeposited. Early estimates of oil removed in Prince William Sound were 4–12 percent by man and 30 percent by natural marine processes.[3] Subsequently, the National Oceanic and Atmospheric Administration (NOAA) sampled 600 sites and claimed that the ferocious waves of winter had removed 50 percent of the oil buried along the shores and up to 75 percent of the surface oil. But one hundred miles of beaches remained tainted, oil could be seen on the water surface, dead sea otters were still occasionally washed up on shore, and Eleanor Island waters were too dirty to fish for shrimp.[4]

Exxon returned in April 1990 for more cleanup with up to 150 workers. There was no more massive high–pressure steam cleaning of rocks as in the first summer. Instead, specific spots of heavy oil were tilled, washed, or dug up. Nitrogen and phosphorus fertilizer were sprayed to promote the growth of microorganisms that can break down oil.[5] By the end of the second summer the heavily oiled beaches were down to four miles, but oil was still buried in gravel and sand at about 800 sites. A NOAA survey during the summer concluded that oil could remain a decade or more before disappearing, noting that "the same forces that buried it can unbury it."[6] The oil spill coordinator for the state of Alaska estimated that 60 percent of the oil remained in Prince

William Sound two years after the accident. However, by 1993, both government and Exxon scientists agreed that the remaining oil appears to pose no significant environmental threat.[7]

Environmental damage assessment was hindered for three years by legal maneuvers to block the release of data relevant to pending law suits. Much uncertainty remains as to the long-term effects of the original spill. The full biological, ecological, and social impacts are still unknown. There was a record catch of pink salmon in the sound in 1990, but studies have raised concerns about their long-term reproductive vitality.[8] A 1992 overall assessment of Alaska's wildlife by federal and state agencies indicated that the total population of sea otters was 250,000, "healthy and increasing," that of bald eagles was above 35,000 and healthy. Large land mammal and bird populations were stable or increasing, as were those of some sea mammals, such as sea otters, beluga whales, and ringed and ice seals. However, other sea mammals—steller sea lions, harbor and northern fur seals, and humpback whales—were declining precipitously. Four years after the spill, there is still sharp disagreement between government and Exxon scientists about the fate of sea birds, specifically murres, and wild pink salmon.[9] In August 1993, the local fishermen staged a demonstration by blockading the Valdez Narrows to protest their continuing low catch of pink salmon and Exxon's refusal to meet with them.

As a postscript, it should be noted that, in October 1992, another oil tanker, the British Petroleum-chartered *Kenai*, carrying 35 million gallons of Alaskan crude, developed steering trouble 16 miles from Bligh Reef and came perilously close to running aground on Middle Rock in Prince William Sound. Recent disastrous tanker accidents off the coasts of Spain, Scotland, and Sumatra underscore the global nature of the pattern. As of March 1993, five of the world's fifteen largest oil spills have occurred in the four years since the *Exxon Valdez* incident.[10]

TWO FORCES THAT ARE TRANSFORMING OUR WORLD

As we speed toward the 21st century, there is much talk about "a new world order." The cold war is over and the "knowledge society" is beckoning. The two factors that undoubtedly will have a decisive impact on life in the next quarter-century are population and technology. The impact of their explosive growth will be pervasive and affect all passengers on spaceship Earth in obvious and subtle ways, raising unprecedented challenges to system management capability.

Population

While the earth's land area has remained virtually fixed, population is multiplying as never before. At the time of Christ, the global population was only about 200 million. By 1950 it reached 2.5 billion, and this figure doubled to 5 billion in 1987. In 1900, there were 16 cities with more than 1 million inhabitants; in 1990, there were 276. The World Health Organization estimates that there are 100 million acts of sexual intercourse every day, with 910,000 conceptions.[11] With 40 percent to 60 percent of all pregnancies unplanned in the U.S., the global rate is undoubtedly even higher.[12] From 1987 to 2000, 1.3 billion additional human beings will rub elbows with one another, raising the world's population to 6.3 billion. The net annual population growth currently is 92 million. In the year 2000, there will be at least 21 cities with 10 million or more population; 18 of these will be in the developing countries.

The projected 2025 total of 8.5 billion* means that the population has multiplied by a factor of more than forty in two millenia. More than 62 percent of the people will live in crowded urban areas. In 2025, China and India together will have a population of 3.2 billion, equal to the entire world population in 1940.** While the developed countries in 1990 averaged 1.8 children per woman, the figure for Africa was 6.1. In Rwanda, the number of births per woman in 1989 was 8.5. Women bear many children to improve their status. As a Rwandan women's advocate observed, "The more children you have, the more stable the marriage. The children become your strength against your husband—they will fight for you if he tries to hurt you."[13]

By 2020, Kenya's population will have jumped from 23 million to 79 million; Nigeria's, from 112 million to 274 million.

---

*Based on past forecast accuracies, a range between 7.2 and 9.8 billion reflects the uncertainty.

**The Chinese government has recognized the severity of the problem and instituted Draconian family planning measures to reverse its disastrous population growth. It claims that the expected number of children per woman has declined from 2.5 to 1.9 in the period 1988 to 1992.

Even if birth control were to be practiced widely, the status of women in the Third World raised significantly, and old-age security provided, the impact would not be apparent for a long time. United Nations population estimates for 2050 range from a low of 8.5 billion to a high of 12.5 billion.[14] The pressure on the natural environment and resources as well as on social institutions of the global population growth is crushing. In 1992, 60 percent of the world's people are living in coastal areas; in fact, 65 percent of all cities with 2.5 million or more population are on coasts. The pressure on the marine environment is obvious.

There is a growing gap between rich and poor. The fastest population growth is occurring in the poorest, the slowest in the wealthiest nations. While the population in the wealthy areas is aging, that in the poor areas is becoming younger. In 2000, one-third of the world will be teen-agers, mostly in Asia, Africa, and Latin America. This growth pattern hardly seems to be touched by the simultaneous death due to malnutrition and disease of 35,000 children daily (the estimate of the United Nations). Within the United States we observe a similar pattern. The income of the top 5 percent of the population has risen in the last decade from 15 times to 22.5 times that of those at the poverty level.[15] The family income of the wealthiest 5 percent rose 23 percent while that of the poorest 25 percent declined 6 percent.

The U.S. population has increased from 151 million in 1950 to 250 million in 1990. About one million immigrants arrive each year, with 300,000 of them estimated to be illegal (90 percent of the illegals being Mexican). One estimate anticipates at least 15 million immigrants, legal and illegal, arriving during the 1990s, with another 30 million by 2020.[16] By 2030, three minorities—Hispanic, Black, and Asian—will jointly constitute the majority population. In the fast-growing Hispanic population, only 26 percent are fluent in English and only half complete high school.[17] One example:

> At age 32 [Andre] Sherman [of Baltimore] acknowledges six children whose ages range from 8 to 16 years. They were born to four different women, none of whom he married . . . Each of the mothers receives welfare payments to help raise the children, whom Mr. Sherman now says he fathered impulsively and recklessly. He learned at age 16 that he had first impregnated a girl . . . In rapid succession [he] fathered more children. At one point 12 years ago, three women were simultaneously pregnant

with his children. "I was wild like all kids," he offered by way of explanation.[18]

For poor teenage girls in grim ghettos, pregnancy gives their seemingly hopeless lives purpose and meaning.

Technology

The overarching technology of our time is that of information—its gathering, processing, storing, movement, and display. Major advances are occurring at a dizzying pace in areas such as microelectronics, optoelectronics, artificial intelligence, computer architecture, and networking. From computerized medical diagnosis to learning games, from professional work stations to desktop publishing, from computer-integrated manufacturing to environmental simulation, from simple robots to intelligent machines that move, sense, and respond to environmental changes, new applications are propelling us deeper and deeper into the information age. Even genetic engineering can be considered in the context of information technology, as it deals with biological levels of information.

Nineteenth century technology gave us machines that were extensions of the human limbs so that we could move faster and lift larger loads; information technology creates extensions of the human brain, nervous system, eyes, ears, and mouth. Indeed, the distinction between living and nonliving systems is becoming blurred, as shown by the current evolution of biochips and genetic manipulation.

Information differs strikingly from matter and energy. If we give you physical goods, you gain them and we lose them. If we give you information, you gain it but we do not lose it. The production of information uses very little energy in comparison with the production of material. Matter and energy are bounded on a finite earth; information is not. Thus the focus on information appears uniquely suited to the future on this planet.

Progress is accelerating in other fields as well. Trains moving at 250 MPH to 300 MPH, as well as new supersonic aircraft, are on the horizon. The creation of new materials (composites, ceramics, design at the atomic and molecular scale) and energy conversion systems (photovoltaics, fuel cells, nuclear fusion) should provide substitutes for problematic resources. Planetary engineering projects will become feasible: routine terrestrial mapping using satellites, education systems for Third World areas involving satellite transmission, and possibly even large-scale amelioration of adverse environmental changes.

The Interaction

The combination of population growth and technology is shrinking the earth to a kind of global megalopolis. Economies, enterprises, television-borne culture, starvation, and environmental concerns transcend national boundaries. Information technology brings distant events into our living rooms instantaneously. Perhaps the most striking example is the Apollo manned lunar landing seen "live" on home television screens. Air transportation brings people and goods easily and quickly from one continent to another. The synergism resulting from striking advances in both information and transportation technologies truly transforms the relation between the human being and territory in an anthropological sense. The world becomes one's neighborhood.

At the same time, the very same combination of more people and more technology is creating unprecedented waste, a stinking mess that befouls air, land, and sea. Both rich and poor are degrading the environment: the rich by overproduction and overconsumption, the poor by overpopulation (even in the face of underconsumption). The impact of industrial processes, such as chemical effluents (as in the Rhine River) and nuclear radiation (as in Chernobyl), moves effortlessly from one country to another. The burning of fossil fuels produces massive amounts of carbon dioxide that create the global greenhouse effect, which may raise the surface temperature and sea level significantly in the 21st century. The use of chlorofluorocarbons destroys the stratospheric ozone layer protecting life on the earth. Humans are decimating forests from the United States and the Amazon Basin to Borneo and Nepal.

The rapidly expanding global demand for energy and materials is inexorably increasing the possibilities for catastrophic accidents. The proliferation of biological, chemical, and nuclear weapons and high-tech delivery systems is increasing the opportunities for terrorist leaders and "crazy states" to cause human catastrophes on an unprecedented scale. The frustrated and seething masses in the poor world are easily inflamed by tribal feuds or fundamentalist crusades into fanatic engines of destruction. *For the first time in history, human-induced crises such as modern wars and environmental disasters have the potential to rival natural disasters in their scope and magnitude.*

So many new products are created that we cannot keep up with studying them to determine their toxic effects. Of 48,000 chemicals listed by the EPA, little is known about the toxic effects of 38,000 and fewer than 1,000 have been tested for acute effects.[19] A 1984 National Academy of Sciences study found that adequate information on potential health hazards existed for only 18 percent of the 1,815 pharmaceuticals studied, 10 percent of the 3,350 pesticide ingredients, and 11 percent of the other commercial chemicals considered.[20]

Often the impacts of a technology are not apparent for decades. The building of the Welland Canal to connect the St. Lawrence River to the Great Lakes in 1829 made it possible for lampreys to bypass the natural barrier of Niagara Falls and gain entry into Lake Erie. It took 110 years for them to decimate fishing in the upper Great Lakes. Fortunately, by 1955 a selective chemical poison was found that killed the larvae of lampreys, and by 1962 the threat was overcome.[21] There is always the hope that new knowledge will reverse environmental threats. But it is just as possible that we may face devastation before the threat is effectively countered, or perhaps before it is even widely recognized. The time factor also comes into play with population growth. If the world's population is to be stabilized, how many decades (or centuries) will it take?

Of most concern is the threat to freedom posed by the combination of exploding population and technology. Consider Los Angeles as a primary example. It has been rapidly transformed from a "laid-back," decentralized community to a traffic-grid-locked metropolis. Initially, traffic lights sprouted, then "freeways" mushroomed. Today, clogged streets and freeways often create a nightmare of immobility. Two freedoms appear to conflict: It is difficult to imagine the accommodation of the freedom of growth without increasing the need for controls and thus reducing the freedom of the individual. How can democracy continue to spread and offer prized freedom to individuals?

The combination of population and technology is giving rise to new forces, which will impact everyone in the crowded global megalopolis of the 21st century. Technology now permits unprecedented centralization *and* decentralization simultaneously. It is creating a single global system while at the same time, through networking, permitting greater local autonomy.

Must we redefine terms like *security, freedom,* and *democracy?* Is a reformulation of the concept of 'growth' in order? Will islands of wealth continue to be surrounded by a roiling sea of poverty? Will the wealthy rely on technology to protect themselves from being swamped by the waves of restless poor, the "barbarians at the gate"? Or are we smart enough to exploit it to enhance the quality of life for all?

NEW CENTURY—NEW THINKING

The two problems we have considered seem very different. One is local, the other global; one deals with the past, the other with the future. But they have vital aspects in common. Both draw in human beings as well as technology; both are complex. It is the thrust of this book that such problems must be viewed from several perspectives. No one way of "seeing" them suffices to give an adequate understanding.

In aerial reconnaissance we often find that a puzzling object on the ground can be identified from the air once we have a sequence of photos, each taken from a different position. Looking at one picture, we cannot identify the building. But several photos from distinct angles enable us to see the shadows thrown by the building and provide vital added information. We can do even better by stereoscopic photography, that is, calling on the human ability to "see" in three dimensions. Our two eyes do not merely see slightly different images; the difference between the two enables the brain to compute dimensions invisible to both eyes. The total is more than the sum of its two parts.

In this book, we shall use three very different kinds of perspective to illuminate complex systems: the technical or analytic (T), the organizational or institutional (O), and the personal or individual (P). Each views a system through a different lens. Most importantly, each perspective provides insights not obtainable with the others. Together, they give us a deeper understanding of complexity. Once again, the total is more than the sum of its parts.

A prime example of the importance of using multiple perspectives is the case of human-caused crises. In virtually every major crisis, there has been simultaneous interaction or breakdown of technology, organizations, *and* individuals.

The idea of using several perspectives in examining a problem or concept is very old. Indeed, one finds a curious preference for three perspectives throughout human history. In Egypt, Ra, Amon, and Ptah form three aspects of one supreme and triune (three-in-one) deity. In Judaism, God is viewed as nature (Malhuyot), as history (Zihronot), and as revelation (Shoferot). In Christianity we have the Holy Trinity: God the Father, God the Son Jesus, and God the Holy Ghost.

Historians have also found multiple perspectives valuable. The distinguished French historian Fernand Braudel divided his work on *The Mediterranean and the Mediterranean World in the Age of Philip II* into three parts:

1. the timeless history of humans and their interactions with the physical, inanimate environment
2. the social history of groups and groupings that generate forces leading, for example, to wars
3. history on the scale of individuals

Correspondingly, he saw historical time as "geographical time, social time, and individual time."[22]

We want to make very clear, however, that viewing the past from different perspectives in no way means denial of the actuality of the past, as has been done by certain "revisionist" historians, for example, those who deny the Holocaust ever occurred.

Max Weber, the great German sociologist, proposed three kinds of legitimacy: rational, traditional, and charismatic. Sigmund Freud used three quasi-archeological layers in order to understand human complexity: the professional, the political, and the personal. He found the first to be the most current and accessible, the third to be the deepest, least current, and least accessible. Jürgen Habermas saw man in three relationships: man to outer nature (or technical), man to man (or societal), and man to inner nature (or the self).

In 1959 C. P. Snow gave a lecture at Cambridge University entitled "The Two Cultures." In this widely discussed talk, he decried the gulf of misunderstanding between the scientific and nonscientific cultures. He saw this polarization as "a sheer loss to us all." He also drew a distinction between the individual and the social "condition," noting that the individual becomes fully human only in the context of a social setting.[23]

Political scientist Graham T. Allison is the most immediate link to the concept used in this book.[24] He introduced three "models" to examine the Cuban missile crisis: (1) the rational actor, (2) the organizational process, and (3) the bureaucratic politics. Allison's models were subsequently used by historians James W. Davidson and Mark H. Lytle to examine the decision to drop the atomic bomb in their 1982 book *After the Fact: The Art of Historical Detection.*[25]

Archilochus, a Greek poet, said, "The fox knows many things, the hedgehog knows one big thing." In his book *The Hedgehog and the Fox* Isaiah Berlin used this quotation as the basis for his examination of the fundamental difference between the single- and multiple perspective modes of thinking.

> There exists a great chasm between those, on one side, who relate everything to a single central vision, one system less or more coherent or articulate, in terms of which they understand, think and feel—a single universal organizing principle in terms of which alone all that they are and say has significance—and, on the other side, those who pursue many ends . . . ; these last lead lives, perform acts, and entertain ideas that are centrifugal rather than centripetal . . . moving on many levels . . . The first belongs to the hedgehogs, the second to the foxes.[26]

For complex systems we want the foxes rather than the hedgehogs. It is our hope that the reader will come to feel at ease with multiple perspectives and recognize their usefulness in dealing effectively with his and her own challenges.

In part 2, we shall use the technical, organizational, and personal perspectives to view the Alaska oil spill. The anatomy of this case is instructive because it is indicative of technology management challenges that we will confront around the globe with increasing frequency.

As noted in the Preface, the box in the top right-hand corner will serve to remind the reader which perspective or perspective linkage is under discussion.

PART 2

The Alaska Oil Spill

CHAPTER 2

Technical Perspectives

It is not considered likely that we can move to the point of
guaranteeing containment and recovery at sea.

American Petroleum Institute Report, June 14, 1989

PHYSICAL HAZARDS AND RISKS

Physical hazards have always been a feature of life. Until the 20th
century, the primary hazards were natural ones such as storms
and earthquakes. In addition, living organisms such as viruses,
animals, and fellow human beings posed major threats. In modern times a new set of threats has been added: major fires, explosions, chemical and oil spills, drugs, pollution, and radioactive
material, to name just a few. The combination of an ever exploding global population and ever-more-powerful technologies has
become deadly. Population growth not only multiplies the need
for energy and materials, which industry is striving to fill, but also
places many more people in harm's way. Thus, we should not be
surprised to find in the 21st century a growing number of industrial accidents with catastrophic consequences. The Three Mile
Island and Chernobyl nuclear accidents and the chemical explosion at Bhopal, India, give us a foretaste of the intensifying challenge to the management of technology.

Analyses of such hazards and their consequences has traditionally involved engineers and scientists who have calculated
probabilities of equipment failure and sought "objective" measures to quantify human life and "acceptable risks" for a society.
Traditional forms of analysis fail to capture vital aspects that are
inherent in such problems.

They fail to recognize that the determination of risk constitutes an inherently ill-structured problem. All such disasters
involve a series of complex relationships among humans, private

17

and/or public organizations, as well as the technologies themselves. Given a particular hazard, different parties see inherently different risks. A study of such accidents, and of technologically induced risks generally, indicates the importance of viewing them from multiple perspectives to understand how and why they occurred and what can be done to prevent or minimize them in the future.

In this chapter, we examine the 1989 Alaska oil spill from a technical or analytic perspective. Specifically, we focus on the engineering aspect and leave aside, except in passing, other important technical perspectives such as the economic one. In the following chapters of part 2, we introduce the organizational and personal perspectives, then draw distinct implications from each type and integrate them. The reader may integrate the perspectives in his or her own way and arrive at different conclusions.*

THE STATE OF THE TECHNOLOGY

Alaska can consider itself lucky if the cleanup efforts do not compound the original damage and create a double disaster. That was the misfortune of the *Torrey Canyon* oil spill. Some 2 million gallons of detergents were used to treat an estimated 13,000 tons of oil on Cornish coasts, and another 0.5 million gallons were sprayed at sea. Scientists found that the detergents did much more harm to shellfish than the original oil spilled. In addition, some of the aromatic hydrocarbons used to dissolve the detergents and to aid in mixing the oil also caused much damage to wildlife.[1]

Oil spill cleanup methods can be divided into two kinds and both were used in this case: *(a)* oil containment and recovery and *(b)* oil degradation and removal. *The most striking revelation of the T perspective is the inadequate state of the available methods.* In their report to the president, Samuel Skinner, secretary of transportation, and William Reilly, environmental protection agency administrator, concluded that "oil spill cleanup procedures and technologies are primitive."[2] General Accounting Office data imply that no more than 10 to 15 percent of oil lost in a major spill is ever recovered by human effort. The reasons are apparent:

---

*An early version of this work was prepared for the State of Alaska Oil Spill Commission at its request in September 1989. However, this material should not be construed as reflecting the opinions or recommendations of the Commission.

- *Manual removal along the shore* is labor intensive and inefficient. No evidence for the primitive state-of-the-art is more compelling than the pictures on television news of thousands of Exxon workers in the summer of 1989 wiping off oiled rocks on the beaches with paper towels. Steam cleaning was also widely used with mixed results.

- Lawrence Rawl, chairman of Exxon, admitted that "with a large spill like this one, you can't get *booms* around it."[3] The Coast Guard was initially concerned that oil collected in booms around the tanker could give off fumes that might have formed a dangerous gas bubble. Boom was finally put in place at 11 A.M. on March 25. The fishermen helping with the booms complained about "the low quality boom . . . [that] continually broke, fractured, and pulled apart as the oil gushed."[4]

- *Skimmers* were used with some success in conjunction with the booms, "but long periods of inactivity resulted when they became disabled . . . only about 10 percent of the designed recovery rate could be achieved." Clogging was a continuing problem. Temporary storage of the recovered oil on storage vessels was slow because of difficulties in pumping the heavy, greasy material.[5] It should be noted parenthetically that booms and skimmers have restricted effectiveness at best, being most suitable with current velocities of less than one knot and waves of less than two meters. Dredges with large oil pumping and storage facilities may be useful, but they have not been subjected to adequate investigation. A Russian dredge has been designed and built specifically for oil spill recovery.[6]

- *Dispersants* may work when there is some water turbulence, but very little testing has been done in the last decade. There is much uncertainty about their effectiveness and the possible harm they may do to fish. No clear governmental policy directives on their use were in effect prior to the accident, and the resulting confusion delayed decisions. At the height of the crisis (at 3 P.M. on the first day) a trial application was authorized. Three more trials were undertaken on the second and third days. Nothing of any consequence was accomplished and "the issue of dispersant use remains in dispute."[7]

- *Burning* in-situ may be suitable in calm-sea conditions and was tried, but there was disagreement between Exxon and the state of Alaska about its effectiveness, and nothing of any significance was accomplished. Burning produces toxic chemicals, including carcinogens and acid rain.
- The spill triggered some experiments with *bioremediation*. This process involves the use of microbes to biodegrade spilled hydrocarbon molecules. The concept appears environmentally attractive and relatively inexpensive, but much research remains to be done.

The shutdown of Exxon operations for the winter due to severe temperatures, wind and wave conditions, and reduced daylight, indicates that any similar oil spills occurring during these months would have to rely almost entirely on natural processes. Exxon reports that many of the cleanup ships are "at great risk in bad weather."[8] Presumably tanker operations would be halted and the pipeline flow slowed or stopped in the event of winter storms.

Prevention of oil spills has focused on improving the design of tankers, for example, use of double hulls. There is dispute about their effectiveness. They add about $15-20 million to the tanker construction cost. A Coast Guard study claims that more than half the 11 million gallons might not have been spilled if the ship had had a double hull. But the ship would be more vulnerable to leakage in case of a collision. Charles DiBona, president of the American Petroleum Institute, insists that the *Exxon Valdez* accident would have meant penetration of a double hull.[9] Vice Admiral Clyde Lusk believes that double hulls would make tankers less stable.[10] It is also conceivable that vapors can fill the space between the two hulls and cause a ship explosion.[11] The Alaska Oil Spill Commission found the arguments against double hulls to be "without merit."[12] Another promising design is the MARPOL ship with the hydrostatic vacuum feature, which follows the International Maritime Consultative Organizations's 1978 protocol on reducing marine pollution.

IMPACT ANALYSIS

It is characteristic of systemic disasters that the impact is difficult to measure. Even definitions present serious difficulties. Foremost in this case are the questions: What constitutes "cleanup"? How clean is "clean"? Exxon's Otto Harrison favored "environmentally stable" and ADEC's Dennis Kelso talked about "treatment."

How does one measure loss when the effects are not only economic but also ecological and human? Costs have been estimated in both monetary and energy terms. The National Research Council has determined cleanup costs of $1,500 to $38,000 per ton of oil spilled and $10,000 to $90,000 per ton of natural resource damage.[13]

The effort to calculate the economic and ecological costs in terms of total energy uses a measure known as "emergy." The term refers to the energy required, both directly and indirectly, to generate matter and/or energy. Such analysis finds that the largest single amount of ecological damage from the spill was the natural resource emergy value of zooplankton. But the value of the total natural resource emergy loss was far smaller than that of the economic system, with its major component the human services involved in the cleanup operations. The emergy of the actual oil spilled was less than 2 percent of the total emergy loss from the spill. As a fraction of annual emergy use, the spill was small for Alaska as a whole (about 1 percent) but catastrophic for Prince William Sound (330 percent to 490 percent).[14]

Interestingly, both of these approaches cast doubt on the cost-benefit value of available spill prevention technology such as new tanker designs.

A TRULY COMPLEX SYSTEM

The 21st century will feature increasingly powerful technology, creating an ever-larger potential for accidents that have unprecedented impacts. Besides the obvious effects of the well-publicized industrial accidents noted earlier, we must also be concerned about more subtle, long-term problems. Stratospheric ozone depletion by chlorofluoromethanes, nitrogen oxide, and carbon dioxide, as well as the rise in atmospheric carbon dioxide level due to fossil fuel burning, may lead to catastrophes. We are thus forced to examine complex industry-based systems in a new light. One such group comprises systems characterized by the combination of *(a)* very low likelihood of disastrous failure and *(b)* catastrophic consequence if such failure does occur (see the comments on probability in chap. 10). Supertankers have transformed the oil shipping system into just such a type.

In complex systems virtually everything interacts with everything. In our case, oil shipping from Alaska has connections at one level with the Alaskan economy, the Alaskan ecology, Alaskan lifestyle, oil prices in the U.S., the U.S. economy, U.S. Mideast policy, the global air and ocean environment, the oil industry, and alternative energy development. At another level, we must deal with the actors directly involved—Exxon, Alyeska, the state of Alaska, the federal government (Department of Transportation, Environmental Protection Agency, NOAA, Department of Justice, Department of the Interior), Valdez and other communities, the Coast Guard, the fishing industry, the insurance industry, United Nations International Maritime (Consultative) Organization, and environmental groups. On a third level the system includes *(a)* on the high seas: the ship itself, radio communications, and weather, *(b)* in Prince William Sound, the terminal, other ships, the Vessel Traffic System, shipping lanes, and ice.

Marine accidents have involved an astounding array of factors—radar image being misinterpreted, supertankers negotiating channels only two feet deeper than they are, tugboats blocking radio channels by playing music, monumental storms, captains playing "chicken" in sea lanes with forty ships about, a frying pan causing a grease fire that destroyed a luxury liner in hours, and in this case a captain with a history of alcoholism and a revoked driver's license in charge of a supertanker. The appalling condition of many tankers in operation is another factor. Recently, British Petroleum examined a Greek-owned, Maltese-registered tanker and found seventy-three deficiencies, including twenty-year-old navigation charts, unlicensed engineers, inoperative anti-collision radar, a hopeless fire-control plan, and manuals written in Serbo-Croatian for a Greek-Filipino crew. A Shell Petroleum Report estimates that 20 percent of the world's tanker fleet is suitable only for "the scrapyard."[15]

There is a multitude of ways that a series of very low likelihood events can interact to create an unexpected, disastrous system failure. In the case of the *Exxon Valdez*, we can list many such events. If any one of them had not occurred, the spill might well have been averted or minimized. Examples:

- If Captain Hazelwood had stayed on the bridge and not turned over control of the ship to Third Mate Cousins at 11:50 P.M.

- If the ship had kept its speed down to permit safe movement through the small ice floes (broken off from Columbia Glacier) present in the traffic channel (its increasing speed made departure from the ice-strewn channel necessary)
- If Helmsman Kagan had quickly followed the simple turn command given by Cousins
- If the Coast Guard had been able to monitor the ship's movement through Prince William Sound
- if Alyeska had been in the status prescribed by its own contingency plan
- if Exxon had followed its existing policy and dismissed Captain Hazelwood after the first drink he had subsequent to his alcohol rehabilitation (suggested by Lawrence Rawl, chairman of Exxon, in an interview).[16]

The pattern is typical for complex systems. Unfortunately, it is still not widely understood. The pattern obviously makes the establishment of liability exceedingly difficult. Each of the accused parties or stakeholders can convincingly spread the blame.

It is easy to neglect *interactions* among the many possible subsets of a system and to miss important feedback loops (see chap. 12). It is useful to distinguish between two kinds of interactions, *simple* ones and *intricate* ones.[17] An example of simple interactions is the production assembly line. There are many interactions, but they occur in sequence (fig. 2.1a). If there is a breakdown in step C, it is easy to ascertain the effect, as every step beyond C stops. And we can go back step by step to check B, then A for the cause.

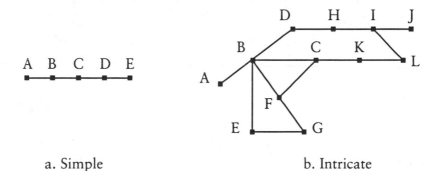

a. Simple   b. Intricate

FIGURE 2.1
Interactions

In intricate interactions, the situation is messier (fig. 2.1b). A failure in *C* could be due to a failure in *A, B, E, F,* or *G*. A failure in *B* immobilizes every component other than *A*. Such common-mode relations are far more difficult to manage than a linear one as in figure 2.1a. Another source of intricate interactiveness is physical proximity. A failure in one component located near another may cause failure in a second unrelated one, for example, a spark jumping from one to the other. Table 2.1 summarizes the many significant differences between simple and intricate interaction systems in industry.

Another important feature of systems is the coupling. *Tight coupling* denotes a lack of buffer or slack between two elements. *Loose coupling* correspondingly indicates lack of tight control. Tight coupling means close supervision and rapid response to changed requirements. Loose coupling means some local autonomy, ability to make spur-of-the-moment adjustments and substitutions. Table 2.2 summarizes the differences. Examples of the four combinations of interactions and couplings abound:

1. simple interactions plus tight coupling: dams, some drug processing plants
2. simple interactions plus loose coupling: post office, most manufacturing

TABLE 2.1
Comparison of Interactions in Systems

	System Interactions	
	Simple	*Intricate*
Equipment	Spread out	Tightly spaced
Common-mode connections	Few	Many
Subsystems	Segregated	Interconnected
Parts substitutability	Easy	Limited
Feedback loops	Few	Many (often subtle)
Controls	Single purpose	Multiple, interacting
Information	Direct, on-line	Indirect or inferential
Understanding	Extensive (expected interactions)	Limited (surprising, unplanned interactions)

TABLE 2.2
Comparison of Couplings in Systems

	System Coupling	
	Loose	*Tight*
Process delays	Possible	Not possible
Order of sequences	Changeable	Invariant
Alternative methods	Available	Not available
Slack in resources	Possible	Little slack allowed
Buffers and redundancies	Can be improvised	Must be designed in
Substitutions	Available	Must be designed in

Source: C. E. Perrow, *Normal Accidents* (New York: Basic Books, 1984), pp. 88, 96.

3. intricate interactions plus tight coupling: nuclear power plant, space mission
4. intricate interactions plus loose coupling: universities, R&D facilities

It seems obvious that systems of type 1 are best managed in a centralized manner while those of type 4 operate best in a decentralized way. For type 2 either centralization or decentralization usually works, but type 3 presents a real dilemma. For tightly coupled systems, centralization is desirable, but for intricate interactions decentralization is preferable. As we move into the 21st century, type 3 must be expected to become a more common occurrence. Fortunately, the information technology of the new century should be able to deal with simultaneous centralization and decentralization as never before (see chap. 7).

Examples of these couplings and interactions can be found in the marine transport system:

- Tight coupling: the very restricted maneuverability of modern tankers due to the large size combined with single screws and modest engines—it takes more than twenty minutes to stop a 250,000 tonner doing sixteen knots. Often ships operate with minimal clearance between their hulls and the channel bottom. Owners create a tight coupling of another kind by pressuring captains to maintain tight schedules. Another example is the traditional authoritarian hierarchy on board ship.

- Loose coupling: the tenuous connection between insurance rates and a shipper's operational performance, the loose enforcement of regulations by the underbudgeted and understaffed Coast Guard.

- Simple or expected interactions: the visible and planned operations that comprise oil shipping, such as the terminal-tanker oil transfer relation, the tanker navigation-designated shipping lane relation, and interactions in the boiler subsystem resulting in breakdowns.

- Intricate or subtle interactions: the connection between tank cleaning and gas vapor explosions, liquified natural gas leakage and vapor cloud flammability, detergent use and ecological damage.

The marine transport system evidently has aspects of *(a)* both tight and loose coupling, as well as *(b)* both uncomplicated/expected and intricate/unexpected interactions. A critical question is, How should the system be modified or redesigned by altering the interactions and coupling to improve its operation?

We hasten to point out that even elimination of unexpected interactions would not mean that the system is *fail-safe*. By this we refer to an ability to design the system so that catastrophic consequences cannot occur. In the case of oil shipping, this means designing the shipping system so that no large spills can occur. Engineers traditionally aim for fail-safe design and this is a sound approach for relatively simple systems such as bridges and buildings. However, complex human-machine systems cannot be made fail-safe, no matter how much redundancy and control is built into them. Anticipation that some "solution," some combination of preventive steps, can eliminate the possibility of serious tanker accidents, is not realistic.

A more reasonable objective is to make the complex system *safe-fail*. This approach trades avoidance of failure for survival of failure. It minimizes the cost of failure rather than the likelihood of failure. This is, incidentally, the design principle of advanced living systems, including human beings. In the absence of effective cleanup technology, is the system safe-fail?

AN ERROR-INDUCING SYSTEM

A curious feature of the marine transport system is that it is an error-inducing system. In such a system, the configuration of its components induces errors and defeats attempts at error reduction.[18] As such it contrasts with the air transport system, which is *safety-reinforcing*. Table 2.3 displays some key distinctions.*

Nearly half of the 3,250 transoceanic tankers are registered under five "flags of convenience" (Panama, Liberia, Greece, Bahamas, and Malta). They are characterized by particularly weak registration standards. Furthermore, 26 percent of the tanker fleet is at least twenty years old.[20] The poor safety record of ships—15 percent of the world's ships have some kind of collision each year[21]—thus becomes less puzzling: it is an integral characteristic of the system.

In an error-inducing system some aspects are too loosely coupled and others are too tightly coupled; some interactions are too simplistic, others too intricate. Increased electronic gear and automation are characteristic of the new tankers, and the technology is assumed to reduce human error. But the effect can be perverse: it easily leads to more carelessness and a willingness to take risks previously avoided. In such a system, the more complicated the equipment, the more likely it is to be out of order or operated improperly. In the past, improved instrumentation provided "greater economical efficiency and certainly greater ease, but the risk per ship would seem to remain constant," according to a captain who was a director of Shell.[22] The combination of *(a)* nonoccurrence of crises over a period of years and *(b)* the existence of contingency plans and equipment (ignoring partial dismantlement and current inoperable status) creates great confidence that nothing can happen. Significantly, the Institute of London Underwriters refused to insure 85 percent of the ships it inspected in 1992.[23]

It is typical of error-inducing systems that operator error is a prominently given explanation for an accident. But this argument is misleading. For example, workers coping with exhaustion due

---

*In March 1990, a Northwest Airlines (NWA) cockpit crew was arrested, terminated, and had its FAA licenses promptly revoked for drinking within twelve hours (NWA rule) and eight hours (FAA rule) prior to flight time, although the affected flight from Fargo to Minneapolis had proceeded normally. Only very recently has the Coast Guard mandated that ship captains be given a breath test for alcohol one hour before leaving port.

TABLE 2.3
Comparison of Safety-Reinforcing and Error-Inducing Systems

Air Transport	*Marine Transport*
Co-pilot shares responsibility, teamwork	Authoritarian captain, little sharing of responsibility
Moderate productivity pressure: captains cancel flights	Severe productivity pressure: owners force tight schedules
Ground controller shares responsibility, ATC mandatory: ATC can override captains	No equivalent of ground control: Vessel Traffic System (VTS) is only advisory
Federal presence large: tough standards and enforcement;* FAA has central responsibility	Federal presence minor: lax standards and enforcement (U.S. ranks only fourteenth in ship safety); VTS only small part of USCG duties; no FAA equivalent
Strong international cooperation	Weak international cooperation: "flags of convenience" provide weak regulation; toothless U.N. International Maritime Organization
Attractive work conditions: strict limits on work hours	Debilitating work conditions: overwork common
Neutral physical environment: storms avoidable; alternate airfields and delays acceptable	Hostile physical environment: storms not avoidable; alternate ports unacceptable
Accidents get high visibility: extensive media coverage	Accidents get low visibility: unless vast environmental side effects
Victims of accidents identifiable: airlines carry people whose support and business they need	Victims of accidents anonymous (examples: foreign seamen, fishermen, wildlife); no significant customer effect at the gas pump, even with sharp price rise after disaster

*The FAA record is by no means perfect. A recent General Accounting Office study reported that, in 1990, FAA inspectors failed to make the required three annual inspections on 36% of all U.S. airliners. The 1991–93 handling of the Boeing 757 wake turbulence danger reflects FAA's hesitancy to act on safety when industry profits would be adversely affected.[19]

to excessive work hours and routing short cuts made to avoid the anger of superiors in the home office in case of late arrival may easily make navigation decisions resulting in catastrophic accidents. Yet it would be totally inaccurate to simply state the cause of such an accident as "human error." Rather, it is a system error, that is, an error of the whole system, not of any one of its parts.

IMPLICATIONS

Working from the T perspective, we find that the oil shipping system can be improved in at least three ways:

- heightened prevention—better understanding of system coupling and interactions to institute changes that will make the system less error-inducing and more safety-reinforcing
- more effective response—upgrading of crisis management plans and procedures
- more effective response—major effort directed toward the development of improved cleanup technology

In the words of the industry's own task force report, "the frustrations in trying to contain and recover oil at sea and in attempting shoreline cleanup indicate a need for new technology and thus increased need for research."[24]

One example of a new cleanup approach under laboratory study in 1992 involves the use of hollow microscopic glass beads coated with titanium dioxide pigment. The coating is a semiconductor that, when exposed to sunlight, energizes oxidizing chemical reactions. The coated beads oxidize hydrocarbons adhering to them, forming clumps that can be easily burned. The remaining oil spill components are water soluble and destroyed by the natural bacteria in sea water. Chemist Adam Heller of the University of Texas at Austin claims that this process can clean up a spill the size of the *Exxon Valdez* in three days for about $75 million, a small fraction of the sum expended by Exxon.[25] However, in the final analysis, natural processes are likely to continue as the primary agent of oil spill cleanup in the coming decade.

CHAPTER 3

Organizational Perspectives

The greatest danger in times of turbulence is not turbulence; it is to act with yesterday's logic.

Peter Drucker

The principal organizations involved with the Alaska oil spill crisis are shown in figure 3.1. Each has its own view of the problem and its own agenda. Their perspectives differ not only from each other but also from the technical perspective discussed in chapter 2.

The stakes in the Alaska oil fields have been high indeed. From 1977 to 1987 it is estimated that the after-tax oil company profits were $40 billion, the revenues for the state of Alaska $24

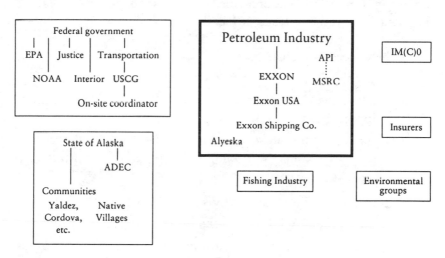

FIGURE 3.1
The Principal Organizational Actors

EPA	Environmental Protection Agency	ADEC	Alaska Dept. of Environmental Conservation
NOAA	National Oceanic and Atmospheric Administration	API	American Petroleum Institute
		MSRC	Marine Spill Response Corporation
USCG	U.S. Coast Guard	IM(C)O	UN International Maritime (Consultative) Organization

billion, and the revenues for the federal government $19 billion. The flow of oil constituted a financial gusher amounting to $400,000 per hour. From the outset, environmental concerns were answered by the industry with firm assurances that the pipeline would not endanger wildlife in Alaska, that operations in Prince William Sound would be the safest in the world, and that the industry could deal promptly and effectively with any oil spill, resulting in minimal effects on the environment.[1]

EXXON CORPORATION

As owner of the *Exxon Valdez* and a major partner in one of the most powerful industries in the world, Exxon Corporation played the central role in this crisis. To the profound relief of the governmental actors, both federal and state, Exxon at once took on responsibility for the cleanup.

It is safe to assume that the most important objective of this top *Fortune* 500 corporation is to maximize its profits from oil drilling and marketing operations. As a consequence of the Valdez incident and public outrage, it faced:

- heavy cleanup costs
- expensive litigation instituted by the affected parties
- constrictive changes in operating rules that add to the oil shipping costs
- denial of future oil exploration permits
- pressure to reduce favorable federal and state oil industry tax breaks
- boycott of Exxon gas stations by motorists
- disinvestment in Exxon stocks by individuals and funds
- impetus to accelerate development of energy sources other than fossil fuels, such as solar and nuclear energy as well as electric batteries for cars

Only the first three have proven significant. The perceived dangers appeared serious enough to justify a major effort at damage control, specifically a sizeable dollar expenditure. The addition of $1.1 billion to settle the federal and Alaska state claims to the cleanup cost of $2.2 billion means that the disaster so far has cost Exxon at least $3.3 billion.[2] On the other hand, tax benefits and insurance should ultimately reduce the figure significantly. For example, all but $100,000 (the criminal penalty) of the $1.1 billion federal and state settlement is tax-deductible. According to one economist, Exxon's write-offs may cost the taxpayer $300 million in lost revenue.[3]

Damage control clearly was the foremost near-term strategy. This strategy appeared to include five primary components. First, mobilizing vast resources to clean up the oil spill. Thirty-one hours after the grounding, Don Cornett, the top Exxon official in Alaska, said that it "doesn't matter if they are really picking up a hell of a lot of oil, at this point—it makes a real bad impression with the public, without any activity going on." An hour later, he was asked by an Alyeska official: "So it behooves us to have everything flapping in the breeze that we can, whether it's catching oil or not?" Cornett replied: "Yeah, absolutely."[4]

The resulting level of Exxon's response was summarized in its August 2, 1989, statement: a level of personnel in Alaska associated with the Valdez operations of about 11,000 people; a cleanup fleet including 50 landing craft and 25 maxi-barges; a water pumping capability of 140,000 gallons per minute (gpm) cold water and 12,000 gpm warm water.[5]

Second, focusing the blame on Captain Hazelwood of the *Exxon Valdez*. When asked "What have you learned from all this?" Lawrence Rawl, chairman of Exxon, responded: "Well, take the case of the captain of the ship. We can certainly minimize this type of thing from happening again."[6] Frank Iarossi, president of Exxon Shipping Company, blamed the disaster on "human error," presumably that of the Captain and Third Mate Cousins. Captain Hazelwood was publicly fired by Exxon within days of the accident for violating company policies, specifically, not being on the bridge and consuming alcohol within four hours of boarding the ship.[7] The Alaskan jury that tried Captain Hazelwood rejected the one-person cause argument, finding him guilty only on one misdemeanor count of "negligent discharge of oil."

Third, shifting the blame away from Exxon to other organizations. Asked "Why didn't you react immediately?" Rawl answered, "The basic problem we ran into was that we had environmentalists advising the Alaskan Department of Environmental Conservation that the dispersant could be toxic."[8] Likewise, when asked "Specifically, who stopped you from applying the dispersant immediately?" he said, "It was the state and the Coast Guard that really wouldn't give us the go-ahead to load those planes, fly those sorties, and get on with it . . . we could have kept up to 50% of the oil from ending up on the beach somewhere."[9] Transcripts of company conversations at the time cast doubt on this explanation, suggesting, instead, that "Exxon and Alyeska officials were confused about the location and type of dispersant stockpiles scattered around the state and were having trouble mustering aircraft to apply them."[10]

In Cordova, Dennis Kelso, commissioner of the Alaska Department of Environmental Conservation (ADEC), responded, "I think this is an arrogant disregard for the truth, and I think it's a systematic effort by Exxon to mislead Alaska and mislead America on Exxon's failure to deal with this spill."[11] Steve Cowper, governor of Alaska, observed: "Exxon is trying to give the state a black eye, probably to try to escape culpability on behalf of itself . . . Alaska is a long ways from the rest of the country and I'm sure that it's possible for a concerted public relations effort to put one over on the Lower 48."[12]

As late as August 3, 1989, Otto Harrison, Exxon's general manager in Valdez, insisted: "the state should share responsibility for fishing closures stemming from its zero-tolerance policy. That's a marketing decision, not an environmental decision."[13] John Sund, Alaska Oil Spill Commission member, and Dennis Kelso both disputed the validity of this assertion.

Fourth, communicating the impression that the cleanup operation is effective. Exxon data provided by its Valdez manager, Otto Harrison, indicate that, as of July 31, 1989, 197 miles of Prince William Sound and 508 miles of the Gulf of Alaska shore were treated. Exxon's May 24 estimate of impacted shoreline mileage was 209 miles in Prince William Sound and 521 miles in the Gulf of Alaska area. Thus it appeared that 94 percent of the impacted beach in Prince William Sound and 98 percent in the Gulf of Alaska shoreline were already taken care of, leaving every expectation that the job would be completed by Fall, 1989. It is natural for the public to assume "treated" means "cleaned up."

Fifth, announcing new operational procedures that will presumably prevent a recurrence of this crisis. On August 1, 1989, Alyeska presented a Tanker Spill Prevention and Response Plan for Prince William Sound. The cover letter states: "The plan reflects a commitment of personnel, equipment, and organization second to none in the world." Its innovative features are the creation of Community Response Centers and an Incident Command System organization for the Alyeska spill response team.

Public relations are clearly a key to the effective pursuit of such a strategy. However, despite its impressive show of action, Exxon stumbled badly. It failed to manage the public relations aspects of the crisis well. In 1986 a *Fortune* poll had ranked Exxon as one of America's ten "most admired" corporations. Only four years later, an article in *Time* appeared with the headline "Exxon's Attitude Problem."[14] The article noted that "the largest U.S. oil company is earning a reputation as a careless and callous despoiler of the environment." More recently, a poll of 200 money managers asked to choose their three or four favorite stocks for 1992 produced not a single vote for Exxon, a one-time favorite.[15]

When we look at low-likelihood/severe-consequence accidents such as Three Mile Island and Chernobyl, we find that each crisis was initially badly mishandled (by the General Public Utilities Corporation and the Soviet government, respectively). Until recently, the normal first impulse of management was denial of the crisis and cover-up.* Consider the comments of public relations expert John Scanlon:

> If the story is on the front page of the *Wall Street Journal*, there's a 30 to 40 percent chance that we'll get a call. We got a call from Exxon. And this is a great indication of how screwed up Exxon was, and how screwed up they are, in my judgment—they could never make a deal. We sat down to negotiate with one division; then they decided another division would handle it. Everybody was running around trying to protect their backs . . .
>
> Rawl should have been in Valdez . . . As a general rule, if there's an accident, the president should go. You get there as quickly as you can, and you embrace—you embrace the people . . . A human response.[16]

Indeed, Frank Iarossi, the president of Exxon Shipping Company, did arrive on the scene from Houston to take charge of the

---

*With a new emphasis on crisis management, the lesson has been learned in recent years by a number of corporations, such as Johnson & Johnson. They have established clear policies that mandate immediate recognition and public communication of disasters.

cleanup operations with minimal delay. But Exxon's corporate top management consistently displayed a tough, defensive attitude. "Exxon Strikes Back," the 1990 *Time* interview with Lawrence Rawl, was hardly designed to project a favorable image.[17] Even so, when the $1.1 billion settlement was announced, he said that he did not believe it would affect the company's image, which he described as being good. The attitude appears to persist. In August 1993 Interior Secretary Bruce Babbitt denounced Exxon's refusal to meet with fishermen protesting the slow pace of environmental restoration as "outrageous."[18]

Another contributing factor, for which Exxon management bears no responsibility, is the nature of the spill as a media subject. In contrast to a nuclear power accident or the national debt, it is easily understood and visually communicated. We cannot "see" nuclear radiation or a trillion dollars, but we can certainly see dead otters and birds, gooey rocks, and oil floating on the water surface. Close-up images of an otter or seal fighting for its life tears at the heart of a television viewer. An accident of this sort becomes a mega-media event, in other words, show business. Seeing a good story, national television began its coverage of the accident before Alyeska responded. Television newsman Dan Rather was able to show the spill to the East Coast on the national evening news before the response team had arrived on the scene at Bligh Reef, just twenty-eight miles from its Valdez base!

A further critical factor underlying Exxon's perspective is its corporate culture. Great entrepreneurial and technical skills have transformed the major oil companies from daring industrial pioneers to global giants. However, until recently they never were in the position where accidents in their operations could have impacts that would be widely perceived as catastrophic. In this regard the oil industry is in a position somewhat analogous to that of the utility industry.

Electricity generation became a commercial enterprise early in the twentieth century, as did oil production. Although the complexity of power generation and transmission steadily increased, the utility industry was not prepared for the level of complexity presented by nuclear energy. The Rogovin Report labeled the Three Mile Island accident a "management problem." The Kemeny Commission found that "[the utility] did not have sufficient knowledge, expertise, and personnel to operate the plant or maintain it adequately."[19] There was no adaptation in management to either the knowledge-intensive character of the new technology or the potential for catastrophe. Stringent minimization of error tolerance in internal operations and the criticality of external effects were not perceived as central. The utility companies' corporate culture contrasts sharply with the U.S. Navy's nuclear program as personified by Admiral Hyman Rickover. The differences between the Navy and utility industry in approach to construction and operations are startling.

The Three Mile Island crisis served to bring this situation into focus and resulted in major changes within the industry, for example, in operator training. It should be emphasized that the reason for these differences is not to be found in the fact that one is a governmental organization and the other is private. Catastrophes occur in government-run settings as well as in private ones. And so do examples of excellence in operating powerful systems that are subject to low likelihood/severe consequence incidents.

In the oil industry, corporate organization developed in an era of 18,000-ton deadweight (DW) tankers in World War II cannot be expected to be appropriate for today's supertankers, with a deadweight of upwards of 200,000 tons. A tenfold increase in size creates an entirely new presence—the possibility of low-likelihood/severe-consequence incidents. It is hardly surprising that a well-established, highly successful corporation fails to recognize for a long time that it is engaged in a new ball game that calls for a new style of management.

Profitable operation must be of foremost concern to any oil company and is aided significantly by rigorously minimizing costs. In 1986, when oil prices sagged, Exxon's flattened income and lagging stock price led Exxon to undertake a major internal restructuring. But it was designed to prop up profits and stock prices, not to create a management control system as sophisticated as its technology now demanded. The result was a drop of 28 percent in worldwide employment and implementation of many other cost-cutting measures. The work force was stretched thin, and morale declined as a result.[20] With the help of automation, sharp cuts were made in ship crews—from as many as forty in the late 1960s (for smaller ships) to the *Exxon Valdez*'s twenty-four in 1986 and nineteen in 1989.[21] This was in line with Exxon's aim for a sixteen-man crew on fully automated, Diesel-powered tankers by 1990.[22] The results are longer work hours, overworked seamen, and lower morale.*

Economic pressures have also significantly affected ship construction. Supertankers are preferred because one large ship is more economical than several small ships carrying the same total load. A single screw and shaft are more economical than twin screws and shafts. Small engines are more economical than large engines. A single hull is more economical than a double hull.

There is great pressure on ship captains to maintain tight time schedules. Shell has determined that, for its fleet of tankers, cutting one hour in port saves a total of $5 million annually. Tanker captains often take risky short cuts to make up time; this was the case when the *Torrey Canyon*, traversing the hazardous Scilly Islands, dumped 100,000 tons of oil on the British and French coastlines.[25] At other times, they avoid the use of expensive tugs in harbors and bays.

Walter Parker, chairman of the Alaska Oil Spill Commission, points to a common industry policy decision—not a captain's choice—to operate its tankers at sea speed in Alaskan waters for the same economic reason.** Not surprisingly, the traffic lanes in Prince William Sound have been increasingly ignored. The ice breaking off Columbia Glacier and entering the traffic lanes does not pose any danger to tankers provided they move slowly, say, five-knot speed; they can pose a danger at high speed. The *Exxon Valdez* was accelerating to sea speed and moving outside the traffic channel after dropping off the harbor pilot, thus seeking to avoid the ice and clear the Sound more rapidly.[27]

The overriding importance of profits to the oil company is perhaps best illustrated by the experience of Captain Hazelwood with the *Exxon Chester* in 1985. Traveling from New York to South Carolina, he encountered a freak storm off Atlantic City.

---

* The Exxon's officers and crew were very likely fatigued from doing double duty during loading operations the day before the accident.[23] Testimony before the National Transportation Safety Board suggests that Third Mate Cousins may have been awake and generally at work for up to 18 hours before the accident. However, Exxon insisted that "with respect to the grounding of the *Exxon Valdez*, there is no indication that manning levels or fatigue on the ship in any way contributed to the vessel running aground on Bligh Reef."[24]

**This policy follows a tradition dating back to the great sailing cargo vessels of the 19th century. The motto of the renowned German captain Robert Hilgendorf was "Im Interesse der Firma immer so schnell wie möglich!" (In the interest of the firm, always as fast as possible.)[26]

High winds snapped the ship's mast and knocked out the radar and electronics gear. Hazelwood calmed the crew, rigged a makeshift antenna, and guided his ship out of the storm. He returned with his damaged ship to New York and "to his surprise, ran into a brief storm of criticism from dollar-conscious superiors at Exxon who had wanted Hazelwood to continue the journey southward."[28] Nevertheless, Lawrence Rawl, chairman of Exxon, insists: "We're not rushing people when they're moving oil. We want them to slow down."[29]

Other instances have also raised eyebrows. In 1987 Exxon refused to comply with Montana's air quality standards. On December 24, 1989, an explosion at Exxon's huge Baton Rouge refinery suggested questions about the connection between tight budget constraints and failure to winterize pipes as well as between budgeting and manpower adequacy.[30] On January 2, 1990, an underwater pipeline leak caused a spill in Arthur Kill between New Jersey and Staten Island; it amounted to 13,500 barrels of refined oil (one-twentieth of the amount of the Valdez spill). The shoreline was blackened and marine life killed. A supervisor had overruled the alarm system three times as fuel gushed into the harbor. Although this action clearly contradicted the operating manual, "this doesn't mean he deviated from the norm," according to his colleagues.[31] On February 28, 1990, 30,000 gallons of oil leaked from a barge anchored at Exxon's Bayonne terminal. Of course, since the Alaskan spill, the mere occurrence of the name *Exxon* in any incident (accident, court-room news, etc.) immediately triggered media attention, thus compounding the public relations problem for the company.*

Lee Raymond, president of Exxon, finally ordered an internal inquiry to determine what was wrong with the organization.[32] He also created a new post, vice president-environment and safety, and filled it with a senior vice president. Even so, he still maintained that "our safety practices have been excellent, and we have drilled them and drilled them into our employees over the decades."[33]

Exxon's advantages as a large and tightly coupled organization came to the fore most effectively in mounting the very large cleanup operation. As soon as it took over from Alyeska, things

---

*On August 1, 1993, the name of Exxon Shipping Company was changed to Sea River Maritime Inc. and the word *Exxon* removed from all its ships.

began to happen. This was demonstrated by the rapid deployment of the *Exxon Baton Rouge* to transfer most of the oil from the stricken ship (transfer began at 7:36 A.M. on the second day), its rapid manpower mobilization, and effective installation of a communications center in the spill area. The local airstrip, which had handled about ten flights per day prior to March 24, was soon handling 750 to 1,000 flights a day. Exxon's capability was again demonstrated in the case of the Arthur Kill spill when Exxon's mop-up team quickly recovered about one quarter of the oil.*

Exxon has an annual revenue of $100 billion and in the period 1986–90 averaged a net annual income of $4.5 billion to $5.5 billion. The first *quarterly* net income reported for 1991 reached a record of $2.24 billion, the highest of any quarter since John D. Rockefeller organized the company in 1882.[35] Exxon has in excess of 7 billion barrels of proven oil reserves—more than twice the reserves of each of its six U.S. competitors.[36] Not surprisingly, the company believes that the costs arising from the oil spill will not affect its long-term financial situation.[37] As Lawrence Rawl, Exxon's chairman, confidently said after the proposed $1.1 billion settlement in March 1991, "it will not curtail any of our plans."[38] Indeed, two years after the spill (June 1991) Exxon ranked as the world's second largest public corporation in terms of market value—a most impressive achievement. Its sales in 1990 increased 22 percent over those of 1989, its profits 68 percent.[39]

To outsiders, the power of Exxon and, more generally, the oil industry, appears awesome. Historically, the companies have been linked together not only through industry associations and a large number of joint venture arrangements but also through interlocking directorates by way of common directorates in commercial banks. And it is reasonable to assume that, in recent years, former oil man George Bush has been a valuable, sympathetic contact in Washington. In the 1990–92 period, oil companies gave $1.3 million to the Republican National Committee and other Republican groups (and only $733,000 to the Democrats). In its final days,

---

*Other oil companies, such as Mobil, have initiated serious oil-spill prevention and cleanup activities of their own, including simulated spill-response drills. Mobil has created a special Spills/Disaster Preparedness Group at its Technical Center and its Mobil Shipping and Transportation Company conducts emergency-response exercises in various parts of the world two or three times annually.[34]

the Bush administration presented oil firms with hundreds of millions in royalty and interest payment breaks. American military intervention in the Persian Gulf and Somalia may also prove helpful to American oil companies that have a strong interest there.[40] Globally, the oil industry is able to dominate the United Nations International Maritime Organization (IMO). For example, the industry lobbied successfully against an IMO proposal to require double hulls.[41]

However, corporate insiders view the power of their opponents as onerous. They see themselves constantly under pressure to defend their operations from ever more strident energy conservation, environmental, and other anti-industry groups that exploit legal processes and the media to impede their operations. The insiders are bitter about the extraordinary amount of time they must now devote to compliance with increasingly burdensome rules. They see themselves being chased offshore by the climbing costs of operation foisted on them in the United States.

ALYESKA PIPELINE SERVICE COMPANY

Charged with the operation of the Trans Alaska pipeline and Valdez port complex, the Alyeska Pipeline Service Company is owned by a consortium of oil companies. The three largest shareholders are British Petroleum with 50.01 percent share, Arco 21.35 percent, and Exxon 20.34 percent. The State of Alaska insists that the oil companies control Alyeska's budget to a degree that makes them responsible for Alyeska's response to the spill. According to Robert LeResche, Alaska's oil spill coordinator, "We feel that Alyeska as a corporation was merely a sham, a corporate shell . . . behind which these partners have been hiding for the last 12 or 15 years."[42] And in a speech to the Alaska Chamber of Commerce, Bill Wade, president of ARCO, reminded the *Anchorage Daily News'* Michael Carey that "Bill Wade, Arco, and all the other Alyeska participants maintain the polite fiction that they aren't responsible for the consortium's follies. They treat Alyeska as a completely foreign operation for which they aren't accountable."[43]

The construction of the oil pipeline is considered to be one of the most expensive engineering efforts undertaken entirely by private industry. The original project cost estimate was $900 million;

the actual cost $9 billion! (The privately financed English Channel Tunnel is costing more than $13 billion.) At least $2 billion of the huge cost overrun was attributed by Alaska to waste and poor management on the part of Alyeska.[44]

In 1971 hearings before the Interior Department, L. R. Beynon of British Petroleum testified for Alyeska: "The contingency plan which will be drawn up will detail methods for dealing promptly and effectively with any oil spill which may occur, so that its effect on the environment will be minimal . . . operations at Port Valdez and in Prince William Sound [will be] the safest in the world."[45]

The initial 1976 Alyeska contingency plan was immediately considered inadequate. The ADEC regional supervisor for Prince William Sound opened his comments on the plan to the ADEC deputy commissioner on December 13, 1976, as follows:

> Alyeska's Valdez Terminal Oil Spill Contingency Plan, in almost every major facet, contains mistakes and inadequacies, demonstrates microscopic thinking, and worse, omits major functions that are necessary. In addition to the following general critique of major shortcomings, certain expletives are pencilled in the margin of the Plan. The initial Plan is so bad, the Department should consider prosecution for violation of Solid Waste regulations and anyone who reviews this Plan should get hour-for-hour Comp[ensation] Time as Sick Leave.[46]

In view of the huge cost overrun, subsequent cost cutting efforts in the operations of the consortium are hardly surprising.

> The pipeline operator's track record, as shown in internal documents, state records, talks with regulators, public testimony and interviews with current and former employees, paints a picture of a consortium that has long pursued a policy of cutting corners on the environment.[47]

Twelve years after the pipeline was completed, it was corroding. There have been fourteen spills along the line since 1977. None of the six that exceeded the 750 barrel threshold designed into the computer monitoring system triggered an alarm.[48]

The Office of Pipeline Safety initiated a probe of Alyeska in connection with allegations that large-scale corrosion was due to deferred maintenance. Specifically, there appeared to be (1) failure of protective coating and wraps, (2) an absence of regular corrosion checks, and (3) wearing through of storage tank bottoms resting on untreated salty sea sand land fill. The cost of pipeline corrosion repair alone may have totalled $1.8 billion over five years, and the state was concerned about the resulting reductions in state revenues from Alyeska.[49]

Charles Hamel, a former oil broker turned whistle blower, has provided much information about the pipeline safety and environmental problems to state and federal authorities. Alyeska hired a security firm, the Wackenhut Corporation, to mount an extensive undercover operation to find the sources of the leaks (in information, not in the pipeline). The elaborate sting set up a phony environmental organization to win Hamel's confidence. Former employees of the security firm admitted that Hamel's garbage was stolen and his telephone records were analyzed.[50]

A former senior quality-control inspector for Alyeska has insisted to congressional committees that Alyeska has faked X rays of pipeline welds and that inspectors have been intimidated and "blackballed" for pointing out problems. Alyeska claims the allegations reflect a difference of interpretation about pipeline procedures. A 1992 report prepared for a federally mandated watchdog group, the Prince William Sound Regional Citizens' Advisory Council, found air pollution due to benzene in Valdez to be as high as in many smoggy U.S. cities and potentially cancer-causing.[51]

An investigation by the *Wall Street Journal*, which can in no way be called "anti-business," paints a grim picture of the company. It reports that many oil-spill safeguards were quietly scrapped, promised ones were never implemented, and new regulatory controls were vigorously fought. According to this source, employees admitted they sometimes fabricated environmental records and doctored test results. Defenses against a major accident were allowed to fall into disrepair. A dedicated emergency twelve-man spill response team was disbanded in 1982. Effective air and water pollution controls were resisted, and continuous monitoring of harbor water was abandoned. A routine inventory of cleanup equipment in March 1988 found only half the emergency lights required and half of the required length of hose. Eight of the ten blinking barricades and 15 percent of the boom listed in the plan were missing. Drills for catastrophe responses were "a farce, a comic opera." George M. Nelson, Alyeska's president, termed these charges "largely discredited accusations." It should be noted that Nelson was replaced as president in October 1989.[52]

In 1988 Alyeska tried to control state inspections of the terminal by requiring advance notice of inspection visits and refusing to permit the ADEC to bring video cameras.[53] Alyeska also ignored state law in failing to notify Dan Lawn, as head of the Valdez office, that some cleanup equipment was not operational as specified in the contingency plan.[54]

Alyeska demonstrates the power of the oil industry:

> When individual regulators do lean on Alyeska, its response can be fierce. Dan Lawn, the state's top Alyeska inspector, was thrown off Alyeska's premises one day in 1986 . . . Alyeska . . . tried to get him fired and attempted to limit his access to the terminal . . . Says Mr. Lawn: "I would characterize their attitude toward regulators as utter contempt."[55]

On June 22, 1982, Alyeska told state regulators that the "estimated time of completion of spill cleanup of a 100,000 barrel spill would be less than 48 hours." But in recent testimony to a House Interior subcommittee, Alyeska insisted that it "never promised to pick up 100,000 barrels of oil in 48 hours." It had merely been talking about the manufacturer's rating for the equipment. Alyeska insists that the 1987 contingency plan, agreed to by the state, was designed to clean up only "the most likely" spill, calculated by a consultant team to be in the range of 1,000 to 2,000 barrels, in forty-eight hours.[56]

The General Accounting Office found in 1991 that

> Alyeska's ability to clean up a major oil spill is in doubt because the company has never conducted a full test drill. The oil companies are opposed to such a drill, the audit said, because it would require temporarily shutting down the pipeline. The drills have been limited to such practices as hiding a piece of black plastic, which is supposed to simulate oil, and then ending the drill once the plastic is found.[57]

After the *Exxon Valdez* oil spill, it took three times as long for Alyeska to respond as its contingency plan had postulated. The barge assigned to the response had been damaged by a windstorm several weeks earlier and was being repaired. George M. Nelson, Alyeska's president, insisted that the contingency plan did not require that the barge be loaded. He pointed out that the readiness factor was not relevant in this case, since the response team would only have been able to handle a 2,000-barrel leak even if it had been stationed at Bligh Reef.

No effort was made to boom off the tanker immediately after the spill. One ex-Alyeska employee explained that the Coast Guard prevented placement of the booms around the ship for fear oil fumes might create an incendiary gas bubble. Alyeska subsequently contended that key parts of the plan were mere "guidelines . . . that cannot really be extrapolated to the real world."* To this Dennis Kelso replied: "That's like saying the fire code is just a set of guidelines. It's just an incredible and appalling fabrication . . . Alyeska stands as a monument to a powerful and rich industry's fundamental failure to keep its commitments."[59] Even Lawrence Rawl, chairman of Exxon, agreed that "Alyeska was not equipped to handle an unfortunate incident like this one."[60] And the industry's own [API] task force report admitted: "The industry has neither the equipment nor the response personnel in place and ready to deal with catastrophic tanker spills . . . the industry is not prepared anywhere along the coastal U.S. to deal with a spill of [216,000 barrels—less than the Valdez spill]."[61]

LaPorte has found that organizations that deal effectively with knowledge-intensive technologies subject to low-likelihood/severe-consequence failures have very distinctive characteristics. Most importantly, "each organization has a strong, clear sense of its primary mission, operational goals and the technical means necessary to accomplish them."[62] Alyeska failed this criterion. But significant changes have since been initiated. In addition to changing its top leadership, $1 billion is being devoted to improving response capability. For example, Alyeska has converted three ships to emergency-response vessels (ERVs). They now escort tankers in the Valdez area and are outfitted with skimmers, boom, and cranes for immediate use in case of spill. A fully manned skimmer barge has also been placed on constant standby alert in Prince William Sound.[63]

Another major new step taken by Alyeska is its agreement to spend $2 million annually to fund a permanent fifteen-member watchdog group to monitor and investigate the company's oil-shipping operations. Alyeska has no say in the appointments, cannot cut funds, and must submit disputes to binding arbitration.

---

*The attitude persists three years later. Alyeska feels that it does not have a legal obligation to clean up oil once it is loaded onto tankers, contending that its emergency crews' responses are "voluntary." In 1992 a bill sharply limiting the liability of cleanup contractors was introduced in the Alaska House of Representatives and pushed by the oil industry.[58]

This Regional Citizens Advisory Committee has been headed by Ann Rothe, Alaska Director of the National Wildlife Federation and Scott Sterling of Cordova, one of the fishing villages.[64]

AMERICAN PETROLEUM INSTITUTE

On June 14, 1989, the API proposed a series of steps involving the spectrum of spill prevention, spill response, and research. Most significant was the concept of an industry-funded Petroleum Industry Response Organization (PIRO) consisting of a heaquarters in Washington, D.C., and five Regional Response Centers (none in Alaska). The headquarters was to coordinate a multi-region response and foster research and training. It would also keep a data bank of available response resources. Unlike Alyeska, each regional center was to be manned around the clock with skilled personnel and have a spill-response capability of 216,000 barrels. This proposal has since been further developed and implemented with the formation of the Marine Spill Response Corporation (MSRC). It is an $800-million, five-year effort. A quick-response network was set up in 1990 with five regional centers: (1) Seattle, Washington; (2) the New York-New Jersey area; (3) Port Everglades, Florida; (4) Lake Charles, Louisiana; and (5) Port Hueneme, California. Each center will presumably have the ability to handle a spill of the Alaska magnitude. The company will administer spill response and supplement existing cleanup capabilities, contracting out most of the work to other companies.[65]

U.S. COAST GUARD

The Coast Guard is a chronically overcommitted and under-funded military service organization. With 37,000 people, it is minuscule by comparison with the Department of Defense; the Navy has 600,000 and the Marine Corps 200,000 men and women. In fact, the Coast Guard is comparable in size to the combined staffs of the Congress and the White House (34,000). Since 1967 it has reported to the Department of Transportation (DOT), as does the Federal Aviation Administration. Traditionally, search and rescue (SAR) has been its foremost mission. With illegal immigration and drug traffic soaring in recent decades, enforcement of laws and treaties (ELT) has de facto become a far more important mission. The barrage of environmental legislation, including the National Environmental Policy Act, the Water Quality Improvement Act, the Coastal Zone Management Act, the Fisheries Conservation and Management Act, and the Marine Protection, Research, and Sanctuaries Act, add a further burden to the ELT mission.

Unfortunately, as one experienced Congressional aide notes, "the Coast Guard just doesn't have a constituency here [in Washington]."[66] Its fiscal year (FY) 1987 budget was close to $3 billion, and this was cut by $100 million in FY 1988. As a result, Admiral Yost had to cut back operations, including those at Valdez. Secretary of Transportation Samuel Skinner has proposed a $25 "user fee" (i.e., tax) on every boat in American waters. Although the expected $180 million could augment the operating budget of the Coast Guard, the receipts might well disappear into the general treasury.

The plain fact is that the Coast Guard does not have the resources to fulfill its missions, to monitor compliance with regulations, to enforce standards, and to apprehend violators. In the case of the Valdez incident, the Coast Guard failed to monitor the tanker after it veered outside the normal shipping lanes and did not communicate with the ship until after the grounding, about an hour after Captain Hazelwood's last radio transmission announcing his detour to avoid ice (11:25 P.M.). Surprisingly, the tanker was not then tracked by radar but only spotted after it ran aground. (The Coast Guard maintains that it was not required to track ships as far as Bligh Reef.)

In the trial of Captain Hazelwood, it was brought out that the *Exxon Valdez* third mate lacked a valid USCG license for Prince William Sound and that the two ships preceding the *Exxon Valdez* out of Valdez were also piloted by crew members without proper Prince William Sound licenses.

Organizationally, there is unhappiness in the Coast Guard with its location in DOT. One proposal to increase the Coast Guard's political clout is to establish a Federal Maritime Administration headed by a civilian undersecretary. Presumably an Under-Secretary could be more effective in pushing the Coast Guard's interests in DOT. Other proposals include shifting the Coast Guard to the Navy, shifting it back to the Treasury, and establishing it as an independent agency.

The Alaska Oil Spill Commission has offered the suggestion that the overworked Coast Guard be relieved of its oil-spill-response function and that this task be given to the U.S. Army Corps of Engineers.[67] For the Valdez oil spill, the role of the Coast Guard was a central one: it provided the on-scene coordinator (OSC) in the person of Vice Admiral Harold Robbins. In this role, he had to approve the cleanup plans of Exxon. In testimony to the Alaska Oil Spill Commission on August 3, 1989, he praised Exxon's efforts and expressed his conviction that the company would do whatever would be asked of it the following spring.[68] However, he provided USCG data on the cleanup that differed markedly from that of Exxon—1,081 miles of beach impacted and only one-third of the shoreline treated (cf. section on Exxon Corp.). The differences may be partly due to different modes of measurement. Indeed, a coastline maintains its roughness and irregularities no matter whether the observer is a satellite or a snail. Legitimate length estimates may thus vary widely.

The relation between the Coast Guard and Exxon is itself complex. In organizational terms, we have noted the enormous power of the oil industry and the weakness of the Coast Guard. An example of the relative power is seen in the alacrity with which the Coast Guard bowed to industry pressure to reopen Prince William Sound to oil shipping days after the spill. It is, of course, fairly common in our system to find regulating agencies bending to the will of those whom they are mandated to regulate. A case in point is the tie between the Federal Aviation Administration and the aircraft industry.[69] Furthermore, recalling the "revolving door" between the U.S. military and the defense industry, we should not be surprised that senior retired USCG personnel find second careers with the oil companies.

Thus the industry's new Marine Spill Response Corporation is headed by John D. Costello, a retired Coast Guard vice admiral. The Alaska Oil Spill Commission Report gives the Coast Guard's "unduly friendly relationship with industry" as one reason for its failure to provide proper oversight of the country's oil transportation system.[70] Recent actions by the Coast Guard to improve its operations include expanding radar coverage and establishing an around-the-clock watch-supervisor position in Valdez.[71]

THE FEDERAL GOVERNMENT

The U.S. Department of Justice is a key actor in the drama. It is central to an understanding of its involvement that the Reagan-Bush Attorneys-General Meese and Thornburgh transformed the department into arguably the most politicized one in administrations that were strongly probusiness.

On February 27, 1990, the Justice Department indicted Exxon on five criminal counts. This followed months of negotiation to effect a plea bargain under which Exxon would have paid $150 million into an environmental restoration fund, possibly adding $350 million later. The company would have retained veto power over how the money was spent. Under the proposed settlement, the federal government would also have agreed to drop any civil suits for damages against Exxon.[72] Under the criminal indictment the company would have had to pay a maximum of $640 million. The charges alleged that Exxon employed crewmen they knew to be "physically and mentally incapable of performing the duties assigned to them." Felony violations of the Dangerous Cargo Act and the Ports and Waterways Safety Act were alleged, as well as misdemeanor violations of the Clean Water Act, the Refuse Act, and the Migratory Bird Treaty Act. Attorney General Thornburgh told a press conference, "By pursuing criminal charges in this case, the federal government is sending a strong signal that environmental crimes will not be tolerated."[73] However, the draft minutes of the March 2, 1990, meeting between state and federal officials suggest that the Justice Department was less than wholehearted in its dedication to prosecute Exxon. In the minutes, Justice Department representatives appear to characterize the public's mood as unreasonably vindictive toward Exxon, saying "the public would not be satisfied unless Exxon's assets were essentially depleted and all board members put in jail."[74]

Environmentalists and Alaska officials publicly alleged that the indictment was a last resort and that the Justice Department had, in fact, been trying to work out a "sweetheart" deal with Exxon. In response, Justice Department officials said that the chances of winning $600 million in fines at trial were "very risky and uncertain." This provoked a new round of accusations from Alaska that the Department was playing public relations games to advance the "sweetheart" deal.[75]

Phase 2 began with the election of "independent" Walter J. Hickel as governor of Alaska in November 1990. In a private meeting in mid-January 1991, he proposed dropping all state and federal lawsuits against Exxon in exchange for $1.2 billion to restore Prince William Sound for the fishing and tourism industries. The federal government was also eager to settle. The civil case against Exxon could easily drag out over five years, and the criminal suit would involve new and untested environmental law aspects.[76] The legal costs would be enormous.

Most importantly, the president and Governor Hickel were anxious to open the Arctic National Wildlife Refuge to oil exploration. Although increasing the average fuel economy standard for new cars from 27.5 to 34 miles per gallon by 2001 would save more oil than could ever be pumped from the refuge, President Bush announced that he would veto any energy bill that did not lift the restrictions on Arctic oil development.[77] It was felt that the settlement would show the government was serious about penalizing Exxon and restoring Prince William Sound as well as would quell public anxiety about expanding Arctic oil exploration as ardently desired by the oil industry.[78]

Under the leadership of Samuel K. Skinner, Secretary of transportation, the key parties gathered on February 5, 1991, in Washington: William K. Reilly, administrator of the EPA; John Knauss, administrator of NOAA; Manual Lujan, secretary of the interior; Governor Hickel; Charles E. Cole, attorney general of Alaska; and Lawrence Rawl and Lee Raymond from Exxon. This meeting began a series of top-level negotiating sessions that led to the March 1991 proposal for a $1.1 billion settlement of the federal and state civil and criminal cases. The settlement would also immunize Alyeska from prosecution. Two issues almost caused a breakdown:

- the government's insistence that Exxon plead guilty to the criminal charge of violating the Clean Water Act, the Refuse Act, and the Migratory Bird Treaty, and pay a $100 million criminal penalty
- EPA's insistence that a reopener clause be included in the agreement; through Secretary Skinner's negotiating skill, a collapse on this issue was averted by the provision that Exxon might have pay to an additional $100 million after 2001 if more work in the Sound was needed.[79]

Once the agreement was announced, new obstacles material-
ized that combined to kill the settlement:

- The 5,000 unrepresented Alaskan native villagers felt that
  their rights to sue Exxon would not be protected. Fifteen
  native Alaskan villages sued the federal government in the
  Federal District Court in Washington, D.C., to prevent a
  settlement that would bar the natives from suing Exxon.
  The judge, Stanley Sporkin, ruled that "the court shall
  retain jurisdiction over this matter to ensure . . . that the
  plaintiffs' rights are protected."[80]

- On April 24, 1991, H. Russell Holland, Alaska Federal
  District Judge, rejected the agreement, focusing on the
  criminal penalty: "The fines that were proposed to me were
  simply not adequate. They do not adequately achieve deter-
  rence. I'm afraid these fines send the wrong message, suggesting that spills are a cost of business that can be
  absorbed."[81] In view of the boast by Lawrence Rawl,
  Exxon's chairman, that the settlement would "not have a
  significant effect on our earnings," the judge's action is
  understandable.[82] Using the latest quarterly earnings figure
  as a basis, the criminal penalty is equivalent to about four
  days' worth of profit! The judicial rejection of the fine is
  not subject to appeal, so the next phase will involve either a
  trial on the criminal charges or a new attempt at settlement
  encompassing stiffer criminal penalties.

- Five days later, Judge Sporkin announced that he was not
  satisfied that the rights of the natives were protected by the
  settlement.[83] Next, the Alaska House of Representatives
  rejected the settlement. The second phase ended when Governor Hickel and Exxon formally withdrew their approval
  of the settlement and Exxon withdrew its guilty plea.

On September 30, 1991, a renegotiated settlement was
announced by Governor Hickel. In it, Exxon was committed to
pay a slightly higher amount over a ten-year period—$1.125 bil-
lion, rather than $1.1 billion as called for in the original settle-
ment.

The forbearance of the federal government vis-à-vis the oil
industry is also reflected in the tendency of the responsible federal

agencies to rely on Alyeska to police itself. For example, these agencies did not know early that the pipeline was corroding. They relied on Alyeska's statements that there was no corrosion and did no independent checking. The Department of Transportation, the Environmental Protection Agency, and the Bureau of Land Management respond to this 1991 General Accounting Office charge by claiming they simply have not been given enough funding for regulatory personnel to do all the testing required.[84]

THE STATE OF ALASKA

The state receives 85 percent of its revenue from oil, and each of its permanent residents receives an annual check from the oil fund. In 1991 this amounted to $931.34, or $3,725.36 for a family of four. These facts inevitably weigh heavily in the relationship between the state and the oil industry. The oil companies also contribute substantially to the legislature. State Senator Drue Pearce, chairwoman of the Oil and Gas Committee, received oil money contributions of $36,145 directly and $13,703 indirectly (through oil contributions to the Republican Party) in 1987–88; the total amounted to 27 percent of all money she received. The state has a law giving large tax breaks to oil companies, and efforts to repeal it were defeated for several years by Republican-led coalitions that have run the Senate.[85] On the other side, an oil industry view is that the state taxes the industry too heavily and raises its taxes whenever it needs money.

The Alaska Department of Environmental Conservation (ADEC) is the agency chiefly responsible for ascertaining that Alyeska actually has the equipment, manpower, and know-how to clean up oil spills. At the time of the spill, its commissioner was Dennis Kelso. Legislation passed in 1988 gave ADEC clear authority over oil spill cleanup plans written by the company. But the state refused to fund the needed inspectors. Alyeska's marine terminal in Valdez is just one of ninety-three onshore oil terminals that must pass muster. And ADEC is responsible for more than 400 facilities, including tankers, barges, and drilling platforms. Yet there was no formal terminal and tanker inspection program. In 1988 ADEC asked for more than $500,000 to hire inspectors to review contingency plans and inspect facilities, but it was granted only $150,000.[86]

In 1990 a number of actions occurred. Steve Cowper, governor of Alaska, introduced four bills, calling for (1) state inspection of tankers, (2) state inspection of the Alyeska terminal without prior consent by Alyeska, (3) stricter spill response requirements, and (4) 500 percent increase in civil penalties for spills.[87] Alaska angrily rejected the Phase 1 U.S. Justice Department/Exxon plea bargain settlement proposal with the comment that "the state is being sold down the river" by the Justice Department.[88] The state was also concerned that Exxon was not being forced to spend enough to fund federal studies of the spill damage, a complaint echoed by the Interior Department. There was justified fear that available studies would provide insufficient data to determine long-term damage.

It will be interesting to see whether the renegotiated settlement paves the way for oil exploration of the Arctic National Wildlife Refuge (ANWR). The state of Alaska is pushing its case by advertisements in recession-hit California: "California Needs Jobs . . . ANWR's Got Them . . . By supporting [ANWR] you could help create 735,000 jobs in America, 80,000 jobs that stay right here in California." In asking readers to call their senators and congressmen to support the project, the advertisement carefully avoids explaining to conservation-conscious Californians that ANWR refers to the Arctic National Wildlife Refuge.[89]

LOCAL COMMUNITIES

Many communities were affected by the oil spill. Within a week of the spill, Valdez saw its population more than double. By summer 1989, it swelled to more than five times its normal size. The town's economy became totally disoriented, with some raking in windfall profits and others finding the sudden rise in living costs disastrous.

In Cordova there were bitter complaints by the Cordova District Fishermen United (CDFU) about the poor response of Exxon to early offers of concerned, knowledgable, and willing Cordova fishermen to help promptly to deploy booms: "By midafternoon on the day of the spill, over 50 boats from Cordova and Tatitlek were ready to go. We still never received a phone call back despite many more attempts to contact Alyeska."[90] The fishermen felt the urgent need to secure the five fish hatcheries in Prince William

Sound and went on a worldwide search for booms. Many foreign and domestic companies called back to say, to their surprise, that Exxon had told them they did not need any more boom at this time. Nevertheless the fishermen proceeded to obtain boom equipment to secure the hatcheries.

The Alyeska plan of August 1, 1989, suggested that the oil companies appeared to have learned a lesson from this experience. Further, Alyeska's new Regional Citizens Advisory Committee was designed to avoid a repetition of the noninvolvement on the part of the local population during the crisis, the basis for much anger against Alyeska. The proposed committee called for inclusion of a member from CDFU.

In the first year after the spill, Exxon paid more than $180 million to 13,000 fishermen. The oil spill may ultimately not be Cordova's most serious problem. Alaska's Native American firms are selling vast tracts of land around Prince William Sound to create tax losses. As many as 500,000 acres of spruce forest may fall to clear-cutting, further damaging the Prince William Sound fishing industry. "This logging is more devastating than the oil spill," says Tom Johnson, one of Cordova's commercial fishermen.[91]

Port Graham, 200 miles west by sea from the spill source, is typical of the isolated native villages that suffered, and are still suffering two years later, from the spill effects. The oil is mostly gone, but the effects remain: mussels are gone and a major source of employment, the cannery, remains closed. The natives feel that they were ignored in the 1991 negotiations by both U.S. and Alaska representatives. As the president of the local village corporation observed: "Governor Hickel said we're getting a settlement on behalf of all Alaska but all Alaska didn't have an impact like our people and our corporation did."[92]

In Tatitlek and other native villages, there was confusion as to what subsistence food sources were poisoned by the spill. In Chenega Bay there was panic as the oil-laden waves rolled in. In Seldovia, nearly all residents volunteered to design, build, and deploy their own log booms. When Exxon ignored their effort, the frustrated community became resentful. In Homer there was anger when Exxon sent a public relations man. In Kodiak the domestic violence rate tripled and the caseload for the Mental Health Department rose 700 percent during the first few months.[93] The natives were particularly incensed that they were hurting economically while Lawrence Rawl boasted that Exxon was not hurt at all by the settlement. They felt alone and abandoned—as usual.[94]

OTHER GROUPS AND ORGANIZATIONS

There were, of course, many other organizations involved with the oil spill crisis. Examples include environmental groups such as The Cousteau Society, the National Resources Defense Council, and the National Wildlife Federation. By 1993, Exxon and 100 insurance companies in more than thirty countries, led by Lloyd's of London, were suing each other over liability claims. Other indirect participants are major stockholders such as pension funds. New York City's Public Employees Pension Fund controlled 6 million Exxon shares, California's Public Employees Retirement System 8.24 million and its State Teachers Retirement System 7 million.[95] Their directors' attempts to pressure Exxon are exemplified by the statement of California's state controller Gray Davis:

> Sadly for Exxon investors, the company's stock price is lagging about 25 percent behind its competitors . . . The bill for these accidents [Alaska, Baton Rouge, Arthur Kill, Bayonne] is likely to far exceed $4 billion. That's $4 billion-plus of nonperforming assets. At some point, Exxon shareholders need to ask: ". . . When will Exxon recognize that environmental safety *must* be a key factor in its business planning and performance?"[96]

One of the most interesting consequences to date of the *Exxon Valdez* incident has been the impetus it gave to the drafting of guidelines and criteria for corporate conduct relating to the environment. The code of ethics is the joint product of various environmental groups, the Social Investment Forum, and the public-fund managers from New York and California. The code is named *the Valdez Principles.*[97] The hope is to use the principles to influence public opinion and potential investors to make informed judgments about corporations' ethical conduct. Subscribing corporations pledge to do their part to fulfill the following ten principles:

1. Protection of the Biosphere—minimize and eliminate emission of pollutants which cause environmental damage.
2. Sustainable Use of Natural Resources—insure sustainable use of land, water, and forest; conserve nonrenewable resources through efficient use; protect biodiversity.

3. Reduction and Disposal of Waste—recycle wherever possible; employ safe disposal methods.

4. Wise Use of Energy—employ safe and sustainable energy sources; conserve and maximize energy efficiency of products.

5. Risk Reduction—minimize environmental, health, and safety risks.

6. Marketing of Safe Products and Surfaces—sell products that minimize adverse environmental impacts; inform consumers of the impacts of products and services.

7. Damage Compensation—restore the environment from harm caused; provide compensation to persons adversely affected.

8. Disclosure—disclose accidents and hazards; protect employees who report them.

9. Environmental Directors and Managers—have at least one Board member qualified to represent environmental interests; appoint a senior executive to be responsible for environmental affairs.

10. Assessment and Annual Audit—conduct annual self-evaluation to determine progress in implementing principles; create independent environmental-audit procedures.

Two years after proclamation of the Valdez Principles, some small corporations have signed on. It is, of course, easy to see why Exxon would be unenthusiastic. Indeed, the statement of principle 4, "the wise use of energy," can be interpreted as a grave threat to most energy-marketing corporations. However, Sunoco is an oil company that did come on board in 1993.

There are also international organizations that are concerned with oil spills. The UN International Maritime Organization has promulgated worldwide tanker standards and safety regulations, but it has no power to enforce them. "Flags of convenience" are used by owners to register their ships in countries such as Panama, Malta, and Cyprus, which have low taxes and minimal regulations. In 1992 seven of these countries accounted for more than half the ships lost. The worldwide tanker situation is worse than that prevailing in U.S. waters. It is estimated that 20 percent of the world's tankers are unsafe. The Institute of London Underwriters refused to insure 85 percent of the 133 ships it inspected in 1992.[98]

SUMMARY

Many in the oil industry would argue that a spill in which there was not a single human casualty hardly constitutes a catastrophe. Exxon, the central and most powerful actor, responded immediately and massively. Its tight coupling and enormous resources gave it great leverage. However, the task of alleviating the disastrous effects in Prince William Sound was hampered by the less than satisfactory coordination among the various organizational actors, each one bent on protecting its own back and avoiding blame. Damage control was clearly the foremost concern of organizational thinking.

Although Exxon's cleanup costs ultimately will exceed $3 billion, this amount does not constitute a major financial burden for a company of Exxon's enormous size. The effectiveness of the cleanup is another matter entirely: it remains technologically far beyond the industry's current capability and an uncertain prospect for the future. Furthermore, the profoundly changed level of technological sophistication and risk accompanying the supertanker era has not been reflected in the industry's management style.

Alyeska's actions raise serious questions about its ethical standards as well as its ability to perform emergency response, one of its stipulated missions. Nor did its August 1, 1989, plan instill confidence that it recognized the need for fundamental organizational changes to become a truly high-capability response organization. More promising are the replacement of its leadership and the creation of the Regional Citizens Advisory Committee. As the company is controlled by the oil consortium, responsibility for changes in attitude ultimately continues to rest with the consortium members.

The establishment of the Marine Spill Response Corporation following the American Petroleum Institute's recommendation is the most constructive industry response to the disaster so far.

Constant overcommitment and underfunding reflect the absence of any constituency in Washington for the Coast Guard. Its weakness makes it difficult for the Coast Guard to stand up to industry pressures.

The judicial system's problem lies in the inherent nature of complex systems: it is impossible to pin responsibility on any one participant. The accident would not have occurred if any one of at least six minor failures had not happened. These failures involved the captain, the crew, Alyeska, and the Coast Guard, as well as Exxon.

The State of Alaska also has problems facing up to the industry in view of its almost total dependence on oil revenues. Its support for oversight and control of oil operations has been quite limited until recently. It has a difficult task standing up to both Exxon and the federal government with their enormous resources.

The fishing communities were badly hurt, but their voice, although articulate, is a small one among the organizational power players and easily drowned out.

A pervasive dilemma is the creeping routinization of organizational activities. In its early years, the Alyeska operation appears to have been well managed and there was a serious attitude toward spill response. Even minute spills were cleaned up promptly and completely. The Coast Guard also exercised firm control. Only a single tanker at a time was allowed in the Valdez Narrows, traffic lanes were observed, and three ship traffic control operators were on duty at all times—a supervisor, a radio operator, and a radar watch person. Later the Alyeska operation deteriorated, and the Coast Guard saw itself as being only a "provider of information."

How can an organization responsible for systems that may experience catastrophic failure, albeit with very low likelihood, maintain its initial high effectiveness over a long period of time? We shall address this question in chapter 5.

Creeping bureaucratization is also glaringly apparent in the disposition of the $1.125 billion that was to be provided by Exxon in the renegotiated settlement. For four years following the spill, the main beneficiaries have been the throngs of attorneys and consultants. Federal and state administrative costs and studies have been eating away at the funds marked for the restoration of Prince William Sound. With the trustees for the restoration fund mired in paralyzing debates, economic exploitation of the area through logging and recreational activities has proceeded apace. Of the first $202 million disbursed by the trust fund, only $45 million went to new environmental restoration and habitat protection projects. A major task for the protection of wild salmon runs and bald eagle nesting is to prevent wholesale clear-cutting of the rain forest. A beginning has finally been made in applying the trust fund to the purchase of coastal forest tracts. In May 1993, the trustees voted to spend $387 million to buy and preserve from logging 42,000 acres of native-owned coastal land on Afognak Island in the sound. In August 1993 negotiations were in progress to acquire scenic wildlife lands at Orca Narrows and Cordova.[99]

CHAPTER 4

Personal Perspectives

I had other things to do.

Lawrence Rawl, chairman of the Exxon Corporation

There are few things in this world that will tear your heart out quicker than the pained screams of sea otters as they try to get the oil out of their eyes.

Michelle Hahn O'Leary, Cordova, Alaska

HUMAN BEINGS AND PROBABILITIES

Human beings do not deal with probabilities as would perfect scientific calculating machines (see also chapt. 10). A consultant study done by Woodward-Clyde for Exxon calculated the "most likely" oil spill size to be in the range of 1,000 to 2,000 barrels, while an "unlikely" catastrophic spill of more than 200,000 barrels would occur only once in 241 years.[1] Since the Alaska pipeline life span is about thirty years, Alyeska felt comfortable in using a 2,000-barrel spill as a goal for cleanup capability in its 1987 contingency plan. This is like saying the most probable auto accident is a fender bender and car safety design should be based on this possibility. Alyeska emphasizes that this scenario was approved by the Coast Guard. It can also point to the spill January 5, 1989, by the tanker *Thompson Pass* during loading at Valdez. The 1,700 barrel seepage, caused by a twelve-foot crack in the ship's hull, fits right into the "most likely" range.[2]

The reality is that the very low probability of a catastrophic spill—once in 241 years—by no means negates the possibility that another such spill may occur in the next twelve months.

Human beings also have difficulty grasping very low probabilities. We are comfortable discussing probabilities of 50 percent, 20 percent, and 10 percent. We have trouble dealing with probabilities such as .0056, the probability of losing a ship in the year 1979 (400 ships lost in a worldwide fleet of 71,129), and .00004, the percent of load spilled by Valdez tankers since 1977 (based on 8,700 loaded tankers departing from Valdez).

Recalling that past events fade in our mind, it is inevitable that the *Exxon Valdez* disaster will be discounted as time moves on (also see chap. 9). It will disappear from front pages and television screens and most peoples' consciousness. In the same way, the environmental effects (e.g., disappearance of a species) are typically long-term, hence discounted relative to the near-term economic impacts (e.g., loss of jobs). *In cases of low likelihood events where the consequence of their occurrence is catastrophic, probabilities do not offer a basis for planning.*

EXXON'S TOP EXECUTIVES

When a Japan Airlines DC-8 crashed in Tokyo Bay as a result of pilot error, the president of the airline resigned. A heightened sense of personal responsibility is uncommon in American corporations. When the Union Carbide industrial accident occurred in Bhopal, India, Warren Anderson, CEO of Union Carbide, flew to the scene of the disaster. Lawrence Rawl, Exxon's chairman, stayed out of public view for nearly a week after the disaster and let others take the heat. He defended himself in a *Time* interview: "We had concluded that there was simply too much for me to coordinate from New York . . . I went on TV and said I was sorry. I said a dozen times that we're going to clean it up. But people keep saying that I don't commit. I don't know what the hell that means. What do you do when you commit? Do you hang yourself?"[3]

Rawl's remarks one year after the accident did not shift. Examples:

> On not going to Valdez: "It wouldn't have made any difference if I showed up and made a speech in the town forum. I wasn't going to spend the summer there."[4]

> On being charged with arrogance: "That bothers the hell out of me. Maybe 'big' is just arrogant. Or maybe I just get emotional and that's arrogant. Or maybe I say things people don't like to hear. Is that arrogance? You tell me."[5]

The photo of the chairman that accompanies the article vividly, if unintentionally, exudes an aura of defiance. A comparison with the bright matinee idol appearance of ADEC's Dennis Kelso inevitably yields the *perception* of the "good guy" underdog battling the tough "bad guy." A report of the critical February 5, 1991, top level meeting in Washington (see chap. 3) notes: "Mr. Rawl . . . was in a dour mood. After being asked to wait outside a conference room for 30 minutes while the government officials finished a meeting, Mr. Rawl became furious . . . Mr. Skinner said that when the Exxon chairman entered the room he lashed out at the negotiators, saying he was sick and tired of how the company had been treated by the government, the news media and the people of Alaska."[6]

Hours after the announcement of the $1.1 billion Phase 2 agreement engineered by Samuel K. Skinner, secretary of transportation, Rawl still expressed little regret over the spill, allowing in a news conference that, in the two years since the spill, his most painful moments were usually caused by bad publicity. In the words of the general counsel of the National Audubon Society, "Mr. Rawl's comments were incredibly arrogant."[7]

When Alyeska's top executive in Valdez, Chuck O'Donnell, was awakened at 12:30 A.M. by a call from the terminal informing him of the accident, he ordered a subordinate to head to the terminal and went back to sleep. A company spokesman insists that this was "in accordance with accepted consortium procedures for dealing with possible disasters."[8] The failure of the supervisor to follow the plant operating guide in the Arthur Kill oil spill was similarly explained as being "consistent with customary field practice."[9]

The Exxon executive who was, in a sense, in the eye of the storm was Frank Iarossi, president of Exxon Shipping Company. He strove intensively to bring all available resources to bear, only to find himself frustrated at every turn by his own corporate superiors as well as by the bureaucratic behavior of key Coast Guard and State of Alaska individuals.

CAPTAIN HAZELWOOD

As captain of the *Exxon Valdez*, Joseph Hazelwood was inevitably thrust overnight from obscurity into the limelight. It was never clearly established in the subsequent trial that the captain was drunk at the time of the accident. An alcohol test was administered ten hours later, and his actions immediately after the grounding were not characterized by those present as those usually associated with drunk behavior. However, Hazelwood's two earlier arrests for drunk driving, the repeated revocation of his driving license, his involvement in an alcohol rehabilitation program, and his admission of drinking before sailing on March 23, 1989, weighed heavily against him in the public eye. Polls showed that 70 percent of the populace believed Hazelwood "was guilty for sure" of some wrongdoing.[10] With the trial behind him, Hazelwood reflected that, like the tragic figure in *Les Miserables*, "It seemed like I was Jean Valjean with about a hundred inspectors chasing me around the sewers of Paris."[11]

Exactly one year after setting out on the fatal voyage, Hazelwood was convicted in an Alaska state court of the misdemeanor charge of negligent discharge of oil and sentenced by Karl Johnstone, the judge and a commercial fisherman in Prince William Sound himself until 1988, to spend 1,000 hours helping to clean up the beaches and pay the state a token restitution of $50,000. The judge lectured the defendant: "I think Captain Hazelwood knows the buck stops with him and he has to take responsibility."[12] However, the Clean Water Act of 1972 grants immunity to those who report oil spills to the authorities. Although this provision was meant to uncover minor spills, it was the basis for a reversal of the Hazelwood verdict by the Alaska Court of Appeals in July 1992. Captain Hazelwood had promptly notified the Coast Guard after the grounding and was thus not considered legally negligent.[13]

The inherent nature of complex industrial systems that are subject to accidents involving very low probability plus very severe consequences poses an unprecedented legal challenge to the judicial system. How is liability to be assigned in "normal accidents," where inherently many actors or none can be held responsible?[14]

NATIONAL LEADERSHIP

The possibility of serious problems was recognized before the Alaska pipeline was authorized. In the period 1969 to 1973 there were intensive debates between the oil companies and the environmentalists about the safety of the proposed oil transportation system. However, the 1973 oil crisis led to a quick resolution of the debate. Congress approved the bill authorizing construction of the pipeline and terminal, and President Nixon signed it. Even the White House signing ceremony gave inadvertent recognition to the situation. Nixon, after affixing his signature to the authorization bill, turned to the witnessing group of legislative and executive leaders and said: "The Alaska pipeline is on its way. The environment will be saved. Right? No problem. Right?" Getting no clear answer, Nixon repeated: "You have no problem on the pipeline. Right?" Someone answered: "No great problems." Nixon responded: "That's a way of saying there could be problems."[15]

Two weeks after the Alaska oil spill (April 7, 1989), President Bush held a press conference in which he, like Lawrence Rawl, focused on the "alleged human error of a pilot" as the cause of this "aberration." Although he campaigned in 1988 as "an environmentalist" and manifested righteous wrath at the pollution in Boston harbor, he took only mild interest in the largest spill in American history. He did not visit the scene of the disaster.

The key player at the national level proved to be Samuel K. Skinner, secretary of transportation. On his role in the Phase 2 settlement negotiations, he observed: "I viewed my job as a facilitator. You had a huge amount of egos and interests that had to be blended together. In my experience I've found that if the principals don't want to settle they look for an opportunity to get out. In this case everybody wanted a deal because they knew the alternatives didn't make sense."[16] He believed that a successful negotiation was possible only if it was conducted at the Cabinet level: "I said this case will not be settled by lawyers. First of all, they don't know how to settle it. Second, they have a built-in conflict of interest. This could go on for years."[17]

H. Russell Holland, the chief judge for the United States District of Alaska who rejected the Phase 2 agreement, is described by longtime acquaintances as an "ultimate professional," "a decent man," and "totally honorable." A Reagan appointee to the federal bench, he is neither strongly environmentalist nor prodevelopment. His decision was clearly influenced by the many letters he received criticizing the paltry size of a penalty that did not "adequately punish the defendant for the crime for which guilty pleas are offered."[18]

ALASKA'S DAN LAWN

Dan Lawn was hired by the Alaska Department of Environmental Conservation (ADEC) in 1979 and became an acknowledged expert on the Alyeska operations in Valdez. His outspoken criticism of Alyeska in the 1980s became a thorn in the company's side until Alyeska President Nelson finally tried to have Lawn's superiors remove this "troublemaker." He complained that Alyeska was unwilling to spend money and that it cut back manpower and let the system deteriorate. In a 1984 memo to ADEC he warned, "we can no longer ignore the routine monitoring of Alyeska unless we do not care if a major catastrophic event occurs."[19] With Alaska's oil revenues declining in consequence of the oil price collapse, monitoring was a badly underfunded activity and, according to Lawn, served as a signal to Alyeska that it could reduce its response capability with impunity.

A state test of the emergency-response procedures was conducted in the winter of 1986. It involved the spill of sixty barrels of oil (simulated by a spill of oranges). The company had failed a similar test in 1984. Lawn judged their performance in 1986 as less than passing.

Lawn was on the scene of the *Exxon Valdez* spill within three hours of the grounding, at 3:00 A.M. on March 24. Nine hours later he was still wondering, "where the hell's the response equipment?" However, his superiors saw Lawn as a "loose cannon" and subsequently demoted him for "lack of objectivity and professional manner toward those we regulate." He filed a grievance case and won, but was understandably bitter: his many warnings had gone unheeded until it was too late. To Lawn, neither the federal government nor the oil companies kept their early promises that any spill could be contained.

GOVERNOR WALTER J. HICKEL

Walter J. Hickel, Alaska's governor, played a significant role before and after the oil spill. Serving as second governor of the new State from 1966 to 1968, he was an early and ardent champion of the exploitation of Alaska's natural resources in general, and oil in particular. He promoted drilling in Prudhoe Bay, proudly recalling even now: "I told them to drill right there. I told them there was 40 billion barrels down there."[20]

After several unsuccessful political campaigns, this self-made millionaire was reelected governor in 1990 and once again championed exploitation of Alaska's natural wealth: oil exploration from wildlife refuge areas, timber from national forests, a natural gas pipeline, a water pipeline, and gold mining. He is intent on freeing Alaska from federal land use and environmental restraints. Hickel's effort to obtain a quick cash settlement of the oil spill was predictable for a man whose motto is "big projects define a civilization."[21]

CORDOVA FISHERMEN

The local fishery people showed the most emotional reaction, at times stunningly articulate. At the August 4, 1989, Alaska Oil Spill Commission hearing, a fisherman read a poem he had written on the subject. A particularly striking example was that of Michelle Hahn O'Leary of Cordova, representing the Cordova District Fishermen United (CDFU). Cordova is economically dependent on fishing. O'Leary's testimony reflected the strength of the bond between the sound and the people of Cordova:

> Our lifestyle is ruled by the tides and the fish that inhabit Prince William Sound, instead of by a clock ... 1989 is the 100th aniversary of commercial fishing in Prince William Sound. What a sad, destructive, and pathetic situation fishermen and all the members of the Sound face in this historic 100th year ...
>
> We cohabit daily with land otters, sea otters, seals, eagles, Canada geese, and many species of water fowl and shore birds that feed and haul out on the beach in front of the house. These are our companions and friends ... We find ourselves crying as we harvest oil coated dead sea otters and deck load our boats with birds doomed because they ingested toxic oil as they attempted to free their wings from the black goo.[22]

O'Leary subsequently served as secretary of the new Regional Citizens' Advisory Council of Prince William Sound.

Jeff and Claire Bailey fulfilled a lifelong dream in moving from Massachusetts to Cordova. Jeff (aged 32) crewed on fishing boats from April to September, while Claire (aged 30) ran the Killer Whale, a combination deli and café. In December she shut down the restaurant and filled in as X-ray technician in Cordova's Community Hospital. They lived comfortably on a $60,000 income. In their leisure time, he hunted moose and bear, and they went clam digging, hiking, and kayaking.

On March 24, 1989, their life changed. The State cancelled the herring season, although Exxon reimbursed the fishermen for their losses. Oil pollution cancelled his plans to harvest kelp from the bottom of Prince William Sound. Claire's café lost customers and help because two-thirds of the townspeople were off working on the cleanup. Exxon paid Claire $2,000, but the café continued to lose $500 a week. Conscience did not permit Jeff to go to work for Exxon—a costly decision. Jeff became politically active as he battled to obtain compensation from Exxon for the "indirect" losses the café suffered.[23]

There were long compensation delays as Exxon used complicated and debatable formulas to determine the fishermen's compensation for lost catches. The company wanted to base compensation on past catches; the fishermen preferred an average share of available catches. Despite the spill, fishermen netted 23.8 million salmon from Prince William Sound in 1989, more than the ten-year average of 20 million, but less than half of the 48 million fish anticipated before the spill.[24] In 1990 they took in 43 million and in 1991 the salmon runs were again expected to be of record proportions.[25]

Many Cordova fishermen felt Exxon ignored their concerns, but ADEC was willing to listen: "Dennis Kelso, the ADEC Commissioner, is a folk hero here. He walked into a party at the Cordova Telephone Cooperative late Thursday night and a cry went up, "Our hero!" The entire group applauded and cheered . . . Cordovans have cast him in the role of champion in a mythic struggle against a faceless villain, a part he plays well."[26]

The lavish cleanup effort poured large sums of money into the fishing villages, for example, $53 million worth of purchases and salaries into Valdez, a town with a population of 3,500. Local "spillionaires" were created, while the influx of transients exacerbated problems ranging from crime to sewage. The local Native Americans were seriously hurt, and the divorce rate in the general population rose. In Cordova alcohol and drug use increased by 28 percent.[27] According to Jean-Michel Cousteau, the social chaos created in the affected area by the spill are as significant as the ecological problems.[28]

SUMMARY

The human tendency to weigh the near-term more strongly than the more distant past or future means that there will be a steady lessening of interest in the oil spill as time moves on.

Exxon and consortium executives have intense loyalty to their companies, but a sense of personal responsibility for the oil spill is associated solely with Captain Hazelwood. The nature of complex systems confounds both pinpointing of responsibility and determination of losses, resulting in a legal quagmire. A "tough, get out of my way" attitude seems to radiate from the top of the corporation.[29] However, the company made a sizeable cleanup effort and compensated the fishermen and the towns for losses and problems resulting from the spill.

A self-styled "environmentalist," President Bush showed little personal interest in the crisis; he had Samuel K. Skinner, secretary of transportation, take on a key negotiating role.

Dan Lawn, ADEC's Valdez inspector, became an unpopular Cassandra with his outspoken criticisms of Alyeska and early warnings of a potential catastrophe. Dennis Kelso, ADEC commissioner, proved to be a most effective spokesman for Alaskans. Walter J. Hickel, governor of Alaska and a former presidential cabinet secretary, was motivated primarily by concerns for the Alaskan economy to offer the basic settlement proposal soon after his election.

Among the people whose lifestyle was sorely affected by the spill, the fishermen spoke out with surprisingly articulate voices, but their political power was modest.

Overall, two points stand out. First, the accident at Bligh Reef is seen in strikingly different ways through different eyes. To Lawrence Rawl it appeared to be an annoying diversion of attention from running Exxon's business. To President Bush, Governor Hickel, and other oil industry executives, it represented a threat to the opening up of the Arctic National Wildlife Refuge for oil exploration. To the Coast Guard's Admiral Yost, it was an overblown minor incident, as no one was killed. To Alaska's Dan Lawn, it was an outrage, because much of the damage was the result of inexcusably sloppy Alyeska operations. To environmentalists, it was an unprecedented ecological catastrophe; and to the native fishermen it was a horrendous threat to their way of life.

Second, the reflexive pattern of organizations to protect themselves by pinpointing individual scapegoats is perceived in both public and private sectors. Exxon focused its glare on Frank Iarossi as well as on Captain Hazelwood, while ADEC left Dan Lawn "twisting in the wind."

CHAPTER 5

Implications

I'm mad as hell, and I'm not going to take it any more.

Paddy Chayefsky, *Network*

IMPORTANT LESSONS

We shall now consider some implications of our analysis to date from T, O, and P perspectives. In effect we offer our integration of the various perspectives. Others, examining the same perspectives, may well use different weighting factors and arrive at different implications.

As a starting point, we must recognize that the system we have examined is open and dynamic, requiring continued monitoring and periodic reexamination. The long-term effects of a large oil spill, the achievement of significant improvements in cleanup technology, and the ability to adapt the organizational culture and style to the demands of potentially dangerous technologies will not be knowable for many years. Thus, we are facing a nonterminating task (guideline 8 in Appendix B).

Current knowledge about the effects of oil pollution, as well as the means to clean it up, is totally inadequate. It is indicative that we lack an unambiguous definition of the meaning of *cleanup*. We cannot expect that sudden discoveries will resolve the uncertainties. As with the discovery of cancer-causing materials, vital impacts of oil pollution on the exceedingly complex ecological system in Prince William Sound, as well as on the Alaskan social system, may require decades of careful field observation and laboratory research.

We noted in chapter 1 that it took 110 years to recognize an important impact of the construction of Canada's Welland Canal (1829), the destruction of Great Lakes fishing by the migration of the eel-like lamprey from the St. Lawrence River through the canal into Lake Huron and Lake Michigan.[1]

In the long term, natural processes prove far more effective than any human intervention attempted todate. But human concern is focused on the near term, so that a sense of frustration is shared by all parties. Exxon, which had responsibility for the cleanup, learned that throwing massive resources of money and manpower at the problem could not solve it. There is no reason to believe that a takeover of the cleanup operation by the federal government would have done any better.

In these circumstances, it is reasonable to assume that the acceptable level of "cleanup" will be determined by situation-specific realities, rather than a prior definition of what constitutes "cleanup." Let us step back and consider the basic options.

The Fail-safe Approach

Interest in this path is underscored if we recall that the *Exxon Valdez* ranked only thirty-fourth in worldwide accidental oil spill volume up to 1989, and, as the 1991 Kuwait oil fires remind us, much more severe disasters can occur. There are several alternatives:

1. Extend the Alaska pipeline through Canada to the U.S., thus obviating the need for shipping oil by sea. This possibility was seriously raised in the 1960s and dropped for debatable, if not spurious, national security reasons. In view of the required construction time and the oil reserves remaining in Alaska, such an expensive project is impractical at this point.

2. Have the actual cleanup capability determine acceptable tanker size. This would mean drastic reduction of tanker size and greatly increased tanker traffic. It would require a vastly improved and empowered marine traffic control system. It would also result in unacceptably inferior cost-effectiveness in oil shipping operations. It is therefore another unrealistic approach.

3. Stop Alaska oil shipping and shift to other fuels. The high level of carbon dioxide (CO_2) emissions produced by oil (and coal) play a major role in raising the atmospheric temperature (the greenhouse effect) and contribute heavily to stratospheric ozone depletion. A shift to less CO_2-producing natural gas would help to reduce the likelihood of a climatic crisis in the next century. Even more desirable would be a shift away from fossil fuels entirely. Nuclear and solar energy produce no greenhouse gases at

all, but current nuclear energy systems also have a potential for catastrophe. This leaves solar energy and nuclear fusion as attractive options at some time in the 21st century (see chap. 7). Finally, taking into account the huge investment of the oil and automobile industries in current fuel technology, it is obvious that this option is unrealistic until an environmental crisis is widely perceived to be imminent.

The Safe-fail Approach

The system can certainly be made less error-inducing and more safety-reinforcing. Research and development efforts may make the cleanup technology less primitive. Significant technical improvements in prevention and in response capability appear feasible. The proposal of the the American Petroleum Institute for a new industry response organization has been implemented with the creation of the Marine Spill Response Corporation (MSRC). Additional steps have been proposed and, in some cases, partially implemented by the private and public sector organizations. There are a number of examples:

- Alyeska: Escort vessels for all laden tankers in Prince William Sound, fully manned skimmer barge on constant standby in Prince William Sound, establishment of a Regional Citizens Advisory Council to draw the local population actively into Alyeska operations, and creation of Community Response Centers to help local residents with equipment to protect their shorelines. (Significantly, an escort tug was at hand and may have helped to avert the potentially disastrous grounding of the disabled tanker *Kenai* in Prince William Sound in October 1992 by pushing the bow of the ship away from Middle Rock with only 100 yards to spare. The tanker was carrying 35 million gallons of Alaskan crude oil.[2] The Regional Citizens' Advisory Council observes, and participates in, drills. However, its role remains unclear.[3])

- Coast Guard: Expanded marine traffic control (full, state-of-the-art radar coverage of Prince William Sound, manning for effective monitoring), strengthened licensing requirements and enforcement, alcohol testing of captains prior to departure, close examination of ship crew fatigue induced by manpower reductions, and promulgation of new regulations limiting work hours.

- Science and engineering community: Research on vessel configuration, (e.g., use of bow thrusters or twin screws for improved navigational capability), alarm systems for automatic pilots, bioremediation, and chemical dispersants.

- State of Alaska: Closer tanker and terminal inspection, stricter spill response requirements, increased civil penalties for spills.

- Department of Transportation: Improved National Contingency Plan.

- U.S. Congress: Requirement for double hull tanker construction, oil spill liability and compensation legislation.

Specifically, the spill impelled the Congress to end fourteen years of committee deliberations and enact the Oil Pollution Law of 1990, providing for the following:

- An eightfold increase in maximum federal liability of spillers for cleanup costs and damages, from $150 to $1,200 per gross ton, with unlimited liability if the spill is caused by gross negligence or willful misconduct (in the case of the *Exxon Valdez* this would have meant $112 million instead of $14 million)

- Operators to draw up contingency plans for worst case spills;

- A domestic fund to cover costs (up to $1 billion per spill) in excess of a spiller's limits of liability, financed by a five cents per barrel tax on crude oil

- Coast Guard required to maintain a national command center and several spill cleanup task forces

- Oversight of response required on the part of the president

- Slow phase-in of double hulls between 1995 and 2015

- $21 million expended annually for spill cleanup research

The law appears to have had a beneficial effect on reducing routine spills. In 1991, 55,000 gallons were reported to be spilled in American waters, the lowest documented total since 1978.[4]

As to the Exxon penalty, the disposition of the money was still being debated four years after the spill. More than $200 million has been used to reimburse government agencies and Exxon for additional cleanup, restoration, and studies. The trust fund committee is considering options, such as land purchase for wildlife habitat preservation, development projects such as fish hatcheries and port enlargement, and additional scientific studies.[5]

Many other steps can be taken to make the system less error-inducing and more safety-reinforcing. For example:

- Alter the marine insurance system to provide more incentives for shippers to operate safely
- Provide meaningful penalties for shippers who are found to have inadequate crews (undertrained, understaffed)
- Tighten up Coast Guard enforcement by providing specific budgets for inspectors and giving the service more incentives for effective inspection programs (patterned after the FAA)

These steps certainly move the system in the right direction, but in a real-world setting we must draw on the organizational and personal perspectives to penetrate more deeply.

ORGANIZATIONAL RETHINKING

We recognize that most organizations are motivated to make major changes in their structure only when they are in a state of crisis, and the oil industry is hardly in such a state. The federal government has the resources but not the motivation; the state of Alaska has the motivation but not the resources. Thus organizational rethinking presents a major challenge.

Recognizing the weakness in cleanup technology, such rethinking must concern itself with both spill prevention and cleanup. There must be a match between technical and organizational capabilities in both areas. Contending with an accident of very low likelihood and very severe consequence requires unique standards of human technical and management abilities as well as a strong sense of individual responsibility on the part of those involved in shipping-related operations. Consider two examples:

- Prevention—Normal operations must be imbued with a sense of intolerance for noncatastrophic errors. The individual at all levels must see detection of, and alerting to, a problem and potential failure as a personal duty. Cover-up must be punished.

- Response—The response teams must maintain a continual high state of readiness. They must be able to resist the normal tendency of creeping complacency and routinization after public interest and concern diminishes as the *Exxon Valdez* accident recedes in the mind.

However, the error-inducing quality (see chap. 2) may still bedevil the system after technical improvements are in place:

- The improvements may create such an aura of improved security that control of the ship is more often left to lower rank crew members—"with these improvements, anybody can run the ship"

- The added costs to the oil companies of paying for the improvements may lead them to cost-cutting measures elsewhere, for example, further crew reduction and design economies effected in new ship orders that increase risk of ship failure

- Deterioration of local and shipboard response-related knowledge and capability may occur because of reliance on the newly established MSRC as a panacea

Also, accelerated research is likely to move in a vacuum unless there is a close linkage established with a knowledge-intensive operational organization.

The bits-and-pieces approach offered by the various proposed improvements does not per se leave us with an organization that insures effective integration of improvements and long-term maintenance of a high state of readiness and capability. Nor does it give us a high degree of knowledge about all aspects of the system's operating characteristics and environment. There are many ways sources of failure can escape recognition. We can never eliminate them all, but we can do better than the present compartmentalized approach to safety in complex systems. With the present approach, is anyone authorized to examine the system couplings and interactions for weaknesses and take action? For example, should productivity pressures on ship captains be loosened to reduce their need to take imprudent risks? Who can order the requisite changes?

For *both* prevention *and* cleanup, the concept of the 'high-reliability organization' with superbly motivated and trained personnel deserves serious consideration. In the area of prevention, this means it has the authority to stop tanker operations or override captains and terminal managers in pursuing safety concerns. It has the autonomy to practice readiness and to order drills. It has a very high level of system knowledge, and it is able to monitor research and propose and test new technological developments. We refer to an organization that can shift from routine to crisis management easily, that is, one whose organizational hierarchy is flattened instantaneously. In the area of response such an organization has a frequent drills that are held to very strict standards and evaluations. In both prevention and response, the organizational culture reflects an uncommon sense of personal responsibility.[6]

We find examples of high-reliability organizations in both public and private domains. Some utilities (electric, telephone), some airlines, and some medical organizations offer examples of such vigilance in the private sector. In the public sector this can be observed in the nuclear submarines, nuclear aircraft carriers, the national air traffic control system, the Los Angeles County Fire Department. One particularly interesting example is the Strategic Air Command, which was kept on an effective high state of alert continuously for many years to be ready in minutes for the unlikely event of a Soviet nuclear attack. A command post was kept airborne at all times.

An interesting project on high-reliability organizations[7] deals with some of them and finds they share distinctive characteristics (table 5.1).*

---

*LaPorte recounts an interesting observation on an aircraft carrier during practice takeoffs and landings. A sailor on the deck suddenly waved off incoming aircraft. Upon being summoned to the bridge, he explained to the captain his reason: he had misplaced some tools on the deck and was afraid an aircraft could suffer damage upon landing. Instead of scolding him, the captain praised him for recognizing his personal responsibility. For an instant the organizational pyramid was flattened. Note the striking difference between this approach and that of Alyeska described earlier.

TABLE 5.1
Characteristics Exhibited by High-Reliability Organizations

1. Complex and demanding tasks are performed under considerable time pressure with near-zero error rate and almost total absence of catastrophic failure.

2. There is a strong, clear sense of the primary mission, operational goals, and technical means to achieve them.

3. Well-grounded understanding of powerful, knowledge-intensive systems.

4. System mechanisms and processes are nearly completely specified.

5. Production units are complex, linked into large operating networks.

6. Networks can adjust to surprise.

7. Consequences of failure are perceived to be great.

8. Near-failure is considered to be almost as difficult to tolerate as actual failure.

9. Error regimes are specified as deviation from norms and basis for identification and alert.

10. There is wide agreement in society on events to be avoided.

11. Monitoring capability detects external effects of operations.

12. Organization possesses error-absorption and damage-containment capabilities.

Source: T. LaPorte, "High Reliability Organizations Project" (unpublished memorandum, Department of Political Science, University of California, Berkeley, 1989).

In our context, such an organization will exhibit a balance of tight coupling and loose coupling (chap. 2) that transforms oil shipping from an error-inducing to a safety-reinforcing system. Examples include tighter coupling in forced adherence to strict rules and looser coupling in the organizational flexibility to shift rapidly from hierarchical (vertical) to nonhierarchical (horizontal) type. With the organizational perspective; there is a practical question always close to the surface: Where are the appropriate points of leverage for implementing recommendations for change? In view of the stature of the oil industry, it would appear to have very strong leverage in comparison to other organizational actors. Therefore, a private organization emanating from the American Petroleum Institute, such as MSRC, could be a driving force in creating high-reliability organizations.*

The previous joint industry operation, Alyeska, must be kept clearly in mind as a nightmarish example of what not to do. We recall that Alyeska

- Experienced a deterioration of operational capability over time
- Appears to have had little autonomy
- Practiced coverups
- Has not been managed and staffed by a breed of personnel characteristic of high-reliability organizations

Possibly, talent from known high-reliability organizations can be brought in to address the challenge of developing such new organizations. This is where the personal perspective is crucial. Leadership of a new kind is the first requirement.

Let us now consider the organizational aspect of research and development. A crisis often initiates a burst of research and experimentation. The *Torrey Canyon* disaster prompted work on detergent toxicity. It is expected that bioremediation will command more attention now because of the Alaskan oil spill. There is agreement on the part of industry and the government that research in cleanup technology is needed. The API report and the report to the president both stress this point.

API originally proposed to let the new response organization manage a modest $30–35 million five-year program (about 1 percent of the expected costs of the Alaska spill). Their plan gave top priority to prevention and mitigation of shoreline impact (about one-third of the total), but prevention of oil loss from, or oil retention by, the ship merited only 2 percent of the R&D budget. This raises the same question just posed. If the consortium is creating another Alyeska-style organization reflecting the industry's past culture, the prospects for effective R&D and interaction leading to implementation appear cloudy.

---

*The Alaska Oil Spill Commission insists that the government must take overall charge of the response to any major spill: "never again should the spiller be in charge of a major oil spill." It states that even the American Petroleum Institute agreed with this conclusion.[8]

What is missing here is any sense of an industry-institute/university working partnership. If a high-reliability organization is to function, the level of technical knowledge throughout the organization must be on a higher plane than has been acceptable in the industry in the past. Close, sustained interaction between operations and the research/engineering effort is highly desirable. Current personnel rotation and economic constraints must, and can, be lifted to make it possible. This interaction would specifically facilitate the transfer of technology, usually the weak link in the technological innovation process.

There is, of course, the option of a federal program. The federal government has the experience and funds to undertake a strong R&D program. But the motivation is lacking:

- There is no Sputnik orbiting the earth, (mis)interpreted by the public as an ominous sign of a Soviet military threat, loosening the federal purse strings.

- The military-industrial establishment has overwhelming political leverage, exemplified by its ability to increase military R&D from 51 percent of total federal R&D to over 70 percent in the 1980s. The medical-industrial complex also has political leverage in R&D. Strangely, the oil industry's undoubted political power has not been effectively asserted in a similar way to strengthen federal R&D in oil transportation and spill related technology.

- Oil transportation and spill related technology does not have the glamor in the science/engineering community as does information or space technology.

Alaska's small population does not translate into high leverage in Congress. But an advanced technology center in this field does not appear a realistic option for Alaska. A coalition of states including Alaska also does not appear feasible; the only other West Coast state for whom R&D in this field should be of major concern is California, site of the Santa Barbara and Huntington Beach spills. The two states are probably too dissimilar in resources needed for such R&D to permit consideration of such a union.

INDIVIDUAL ATTITUDES

Profound organizational changes of the kind indicated here are likely to succeed only if there is also a corresponding change in individual attitudes. Each individual has his own perception of risk (see chap. 6). The CEO of Exxon, the tanker crew, and the Cordova fisherman clearly have very different perceptions of the Alaska oil spill. Similarly, the traditional oil shipping organization and the high-reliability organization discussed in the preceding section have disparate perceptions of desirable operating styles.

Improvements in technology and organization can reduce the frequency of catastrophic spills and mitigate their impact. But the effectiveness of an organization depends on its individual members. A high-reliability organization cannot be expected to succeed if the participants do not share its perspective. Ideally, a corporate leader with a vision of an ethical shipping system can communicate that vision to his associates and sweep them along, that is, have their individual perspectives coalesce with his organizational one. The reality is that the myopia determining the short-planning horizon of most corporate managers, stockholders, politicians, and citizens generally threatens to make the far-sighted leader appear a Don Quixote fighting windmills. The benign neglect of Presidents Reagan and Bush in matters of environmental protection and corporate ethics has also had its effect.

The importance of individual ethics is seen when we realize that the world's oceans are being constantly damaged by oil (and other substances) dumped surreptitiously. For example, vessel tanks are cleaned and residue is quietly slipped into the water. *The publicized oil spills pale in comparison with the quantity of oil slowly and insidiously converting the ocean into a convenient garbage dump through feckless human action.* Away from the coast, it is physically impossible for any organization, governmental or private, to effectively police such behavior on the part of ship crews. Only individual ethics can ultimately prevent marine environmental deterioration.

A NOTE ON OTHER ACCIDENTS

We began part 2 with some general observations about the proliferation of physical hazards in a world of expanding population and technology. Before concluding the Alaska oil spill case study, we now pause to ask: Is this case typical of industrial catastrophes in its integral blending of technical, organizational, and personal factors? All indications point to a positive answer. For example, organizational weaknesses and human quirks were of primary significance in the case of Bhopal, Three Mile Island, and Chernobyl:

- With regard to Three Mile Island (1979), the President's Commission concluded: "the fundamental problems are people-related problems and not equipment problems . . . wherever we looked, we found problems with the human beings who operate the plant, with the management that runs the key organization, and with the agency that is charged with assuring the safety of nuclear power plants."[9]

- In the chemical (methyl isocyanate) accident which killed more than 4,000 people at Bhopal (1984), both Union Carbide of India Ltd. (UCIL) and the local government were afflicted with remarkable inertia. UCIL ignored early warning signals and in-house safety audits. The state government failed to monitor the chemical in question; indeed it had only fifteen factory inspectors for 8,000 industrial plants. On the personal side, the general manager of an adjacent plant was able to safely evacuate about 1,400 workers as soon as the leak was detected. The reason: this exceptional individual had been a brigadier in the Indian army and his military training provided the needed crisis management capability.[10]

- Investigation of the Chernobyl nuclear plant accident (1986), which exposed millions to undesirable levels of radiation and caused thousands of cancer-related deaths, showed both technical flaws in the reactor design and a series of human errors: "If at least one violation of the six would be removed, the accident would not have happened. The engineers psychologically did not believe that such a sequence of improper actions would be committed. Such a sequence of human actions was so unlikely that the engineer did not include [it] in the project. Is that human or technical?"[11]

Thus it appears that the lessons of the Alaska oil spill extend far beyond one locale, one company, one industry, and one country. Ausubel points out that the Chernobyl experience also teaches some unique organizational lessons that carry over to other accidents. First, existing government and industry have neither the competence nor the credibility to do what is needed. A new organization, the Pripet Research Industrial Association, was established to manage decontamination and research on the site; it has 6,000 employees. Second, the scope of the accident was such that it is difficult to imagine how one can meaningfully prepare for it—a million cubic meters of soil have been moved so far and the scope of the water supply problem is staggering. Third, such an organization must have longevity to deal with both acute and chronic problems with high effectiveness for decades. For example, it must monitor the state of the sarcophagus around the damaged reactor for decades. How will the Soviet structural breakdown affect this ability? There are fifteen Chernobyl-type reactors still operating in the former Soviet Union. Another accident, fortunately minor, occurred in March 1992.[12]

Another recent examination of the Three Mile Island, Bhopal, and Chernobyl accidents focuses on deficiencies in the four generic system control structures:

- Inherent controls: original design to insure a stable system that is easy to control (T type)
- Engineered controls: procedures designed in to maintain stability and automatically carry out certain transitions and protective functions (T type)
- Societal controls: laws, regulations, norms, practices (O type)
- Personnel controls: operational instructions, work practices (O/P types)

Each of the three industrial systems exhibited deficiencies in each of the four kinds of control structures. For example, personnel control failures are exemplified by operators at the plant who were unaware of problems with a failed open pressure relief valve (Three Mile Island), lax management attitudes regarding obvious deficiencies in safety (Bhopal), and major plant tests initiated without reactor safety clearance (Chernobyl). Engineered control

failures include misleading indication of pressure-relief valve position (Three Mile Island), unavailable safety systems at the plant
(Bhopal), and the possibility of disconnecting crucial safety systems (Chernobyl).

The control structures are, of course, interrelated: tightening
one often leads to loosening another. The study notes that "It is a
relatively recent recognition that organization and management
can have an influence on the safety of technical systems . . . a
complex system will require many different views, depending on
the control task under consideration . . . The complexity of the
technologies can be managed only with a consideration of multiple perspectives (technical, organizational, personal)."[13]

A NOTE ON OTHER
ENVIRONMENTAL MANAGEMENT PROBLEMS

Conflict in environmental management is almost invariably a consequence of differing perspectives. Thus the use of technical, organizational, and personal perspectives has proven helpful in
addressing issues such as the hard versus soft energy controversy
and industrial pollution control measures.[14] Specifically, such
analysis leads to identification of strategies to resolve the conflict.

In the case of the pollution control issue, for example, we
have the perspectives of industry (cost-effectiveness), government
(social desirability), and environmentalists (moral imperative). By
viewing pairs of perspectives and developing interventions linking
them, the sources of conflict are reduced. For example, providing
incentives for pollution-free technologies involves the government-industry link, while facilitating public participation in waste
reduction and reuse programs involves the government-environ-
mentalist link. Only by confronting the conflicting perspectives
can we explore the full range of constructive interventions.

It should not be surprising to find that natural catastrophes,
such as Hurricane Andrew, pose a similar challenge. *Technology*
provides far earlier warning of the hurricane, but *organizational*
inadequacies are reflected in the readiness of the Federal Emergency Management Agency (FEMA) as well as in state and local
governments that sanction construction, enact weak building
standards, and tolerate poor code enforcement in the most
exposed locations along a coast from Florida to Maine that is

subject to frequent severe storms. The consequences are unnecessary confusion in the initial response and excessive property loss.
Individual attitudes focus only on the "here and now" and ignore
prudent long-term planning. They justify organizational complacency and exploitative greed to maximize near-term profits. This
focus will be explored further in chapter 9.

FINAL WORDS

The aim of our analysis of the *Exxon Valdez* oil spill has been to
show the value multiple perspectives in dealing with the complex,
low-likelihood/severe-consequence system of shipping oil from
Alaska. Each perspective draws forth significant insights not
obtained by the others. We observed, for example, the primitive
nature of cleanup technology (T), the need for high-reliability
organizations (O), and the influence of key actors like Lawrence
Rawl and Samuel Skinner (P).

   *It is a daunting challenge to bring the oil transportation sys-
tem close to a safe-fail status. Technological changes without
organizational changes are not likely to be very effective. Neither
will organizational changes without changes in individual attitudes—at both senior management and ship crew levels. Ethics
must be not a mere afterthought but an integral, inseparable
aspect of the whole.*

PART 3

About Multiple Perspectives

CHAPTER 6

T, O, and P

The man who neglects the real to study the ideal will learn how to accomplish his ruin.

Niccolo Machiavelli

While the hundreds of systems studies performed during the past decades were regarded as "the way to go," they rarely went anywhere!

Ida Hoos

We have met the enemy, and he is us.

Walt Kelly, *Pogo*

In part 2, we found multiple perspectives useful in examining the Alaska oil spill. In this chapter we focus on the concept and discuss the technical/analytic (T), organizational/institutional (O), and personal/individual (P) perspectives.

THE TECHNICAL PERSPECTIVES (T)

Basic Characteristics

Science and technology represent the most successful "religion" of modern times. From Galileo to the Apollo manned lunar landing, from Darwin to recombinant DNA, its methods have yielded dazzling triumphs. They form the paradigm for the technical perspective. The T worldview is typified by the following characteristics:

- Problems are simplified by abstraction, idealization, and isolation from the real world around us. There is the implicit assumption that the processes of reduction and simplification permit "solution" of problems.

89

- Data and models compose the basic building blocks of inquiry. Logic and rationality as well as objectivity are likewise presupposed. Order, structure, and quantification are sought wherever possible. Observation and model building, experimentation and analysis are usually aimed at improving predictive capability. Validation of hypotheses and replicability of observations and experiments are expected. The attainment of elegant models and best or optimal solutions is particularly prized.

The power and success of the "technical" worldview and its value in yielding remarkable insights and excellent predictions in science and engineering remains unchallenged. Its extension beyond their borders is therefore understandable. Economics and the social sciences have striven to adopt the same paradigms. The impressive tools developed, particularly since World War II, under such labels as "operations research," "systems analysis," "decision analysis," "management science," and "econometrics" illustrate the technical perspective in action.

Limitations

The approach works well for those problems beyond science and engineering that are tame, docile, or well structured. Examples are factory or blood-bank inventory management, optimal urban fire station site location, airline scheduling and seat pricing, and economic input-output analysis.

The strong urge to organize, to structure, and to model larger systems is exemplified by the work of James G. Miller and Jay W. Forrester. Miller perceives all living entities—from a simple biological cell to a supranational system—as systems comprising twenty subsystems, eight processing matter and energy (e.g., ingestion), ten processing information (e.g., memory), and two processing all three (e.g., reproduction).[1] These generalizing efforts are remarkable tours de force. But categorization or imposition of a structure also acts as a constraint, even a straitjacket. It fixates our thinking along one groove or one way of seeing. The danger is compounded by the fact that the structure of a complex system proves to be far more important than its state at any time. In other words, the linkage among the elements of the system are of overriding importance[2].

In applications of the technical perpective to fields such as urban planning, transportation, criminal justice, health care, and technological impact assessment "one . . . assumption was that the "hard" part of a problem—which could be expressed in mathematical terms—could usefully be isolated from the human and organizational elements which could thus be eliminated from the analysis."[3]

Jay Forrester, an electrical engineering professor at the Massachusetts Institute of Technology, borrowed a tool from electrical engineering, *system dynamics*, to develop a computer model of the corporation (*Industrial Dynamics*), the city (*Urban Dynamics*), and the world (*World Dynamics*). His work was the basis for the popular 1972 Club of Rome book *The Limits to Growth*. In this model, five quantities—global population, industrial production, natural resources, agricultural production, and pollution—were considered. Data from 1900 to 1970 were collected and mathematical relations among these five quantities were devised. They were then adjusted until the computer, beginning with the 1900 data, could use the relations to reflect the real-life behavior of the five quantities to 1970. The same relationships were then used to run the model and forecast the behavior of the same five quantities 130 years ahead—to the year 2100. This world model gave rise to a whole cottage industry of *global modeling,* generating copious academic debates among modelers offering competing versions.

There are some startling assumptions built into such models. First, in Forrester's own words: "All systems that change through time can be represented by using only levels and rates. The two kinds of variables are necessary, but at the same time sufficient, for representing any system."[4] In other words, Forrester is convinced that any corporation, any city, even the world, is a system that can be described by a set—indeed a small set—of quantities and their rates of change over time. It is a stunning statement.

Second, the structure of the model, defined by the entire set of relationships, remains unchanged over its run from 1900 to 2100. The world model in *The Limits to Growth* computes the occurrence of global catastrophes in the 21st century reminiscent of the time of the Black Plague in Europe. A good portion of the world population disappears as a result of the combination of worsening overpopulation, pollution, and resource depletion. Will human beings really march like lemmings to self-annihilation? The model implies a societal rigidity or lack of confidence in human creativity and adaptability that is most disturbing.

Such models are put to better use for judicious probing of "what-if" scenarios. Indeed, the runs do provide some valuable insights and constitute a useful learning technique. Unfortunately they tend to be taken too seriously; they are oversold and used to generate questionable forecasts. As Nobel laureate Herbert Simon, recently observed,

> The fundamental conclusion drawn from the model—that exponential growth cannot be sustained indefinitely—is entirely true, has enormous impact for public policy, and could have been inferred from textbook treatments of linear dynamic systems without any computation . . . (the lesson the report aimed to teach could have been explained with a three-equation model computed on a pocket calculator) . . . There may even be actual harm in carrying out such a modeling exercise when the data and equations of the model are, at best, only very approximately accurate. It may give skeptics entirely too much ammunition for questioning even those conclusions that can be validly drawn from the model.[5]

Similar conclusions had been formed earlier about Forrester's urban model.

> The model claims to show that public programs that benefit the poor only make the situation worse . . . many people—Milton Friedman being a notable example—have been saying this for a long time . . . Rather than illuminating [the structure of the problem], Forrester's model buries what is a simplistic conception of the housing market in a somewhat obtuse model, along with some other irrelevant components. What is most disturbing about this model is the gulf between what it is and what Forrester claims it is.[6]

System dynamics models were soon seen as little more than cargo cult models, primitive belief systems based on the products of modern technology.* Even one of the great success stories of systems analysis, nuclear strategy models, which had gained fame for the RAND Corporation and seemed to lend themselves to endless computer modeling and games, are now seen in a different light:

---

*Cargo cult models originated in Melanesia and propagated the belief that the spirits of the dead would return and bring them cargoes of modern goods. They gained strength in World War II when Western military aircraft deposited and left much equipment on the islands.

> The best minds of the defense community have been drawn toward nuclear analysis, but so were the best minds to be found in the monastery, arguing the Albigensian heresy in the fourteenth century . . . [But because no one has experience fighting a nuclear war, our theoretical scenarios have] no more foundation in fact than other theologians' fiery visions of hell.[7]

Today we recognize the failure of T models to consider higher order complexities reflected in nuclear proliferation, the phenomenon of nuclear winter (nuclear blasts causing massive fire and smoke, reducing sunlight and temperatures), the fog of battle, and the impossibility of adequately testing systems such as the Strategic Defense Initiative (SDI or Star Wars).

It seems that the analyst is in danger of perishing between two options: addressing open, unsolvable, exceedingly complex problems or reducing them to closed, solvable, but irrelevant ones. The dilemma is reminiscent of Buridan's ass, which starved to death because, being equally far from two piles of hay, it was unable to decide which of the two piles to go to.

Being human, modelers succumb to the Pygmalion phenomenon. The sculptor king of Greek mythology fashioned a beautiful statue of a woman and then fell in love with it. Responding to his plea, the goddess Aphrodite brought the statue to life and Pygmalion married his model. Today's modelers, mesmerized by the ability of modern computers to bring their models to life, also become enamoured of their creations. The models have become real to them.

One obvious way to avoid taking any single model too seriously is to use several models instead of one. This also helps to overcome any one model's ever-present limitations, such as artificial boundaries, unwarranted assumptions, and oversimplifications. Multiple models are commonly used within the realm of the T perspective. In classical physics, there is great value in using both wave and particle models of light, both Newtonian and Einsteinian universes for mechanics. In aircraft development, the project engineer, the aeronautical engineer, the electronic engineer, the engine builder, the interior designer, and the market analyst all look at the same aircraft using distinct T perspectives. Representing different disciplines, they use different models and data. Nevertheless, they all operate with the same T paradigms.

Prediction versus Insight

We must clearly distinguish two functions of models: *(a)* prediction; the ability to draw predictions from a mathematical model, and *(b)* explanation or understanding, an abstract thinking aid revealing or illuminating some aspect of system behavior in a simple way or unlocking an insight.

The abilities of science in modeling systems can be illustrated as follows:

- excellent explanation and excellent prediction: celestial mechanics
- excellent explanation and poor prediction: evolutionary biology
- poor explanation and excellent prediction: quantum mechanics
- poor explanation and poor prediction: economics[8]

In the practical world, we are frequently more interested in a good forecast than in a good explanation. But, as Herbert Simon explains, "The rapid rise in the last decade of chaos theory . . . has shown the fundamental reasons why such prediction may be impossible, now and forever. These are linked to the complexity of many systems of interest. Nature is capable of building, on a scale of microcosms or macrocosms or any scale in between, systems whose complexity lies far beyond the reach of our computers and supercomputers—present or prospective."[9] Or, as John Casti observes, prediction requires computability and mathematically only a small subset of all possible functions is computable. It is therefore plausible that mathematical descriptions of many natural or human phenomena are inherently uncomputable. The more a system is susceptible to human influence, the lower is its predictability.[10] We shall return to the important subject of system complexity in chapter 12.

Conclusion

The classic dream of the scientist and technologist, mastery and control of our complex systems, our physical and social environment, is inherently unrealizable—our power will remain quite limited. We are rediscovering the wisdom of Goethe's Dr. Faust:

I have, alas, studied philosophy,
Jurisprudence, and medicine, too,
And, worst of all, theology
With keen endeavor, through and through—
And here I am, for all my lore,
The wretched fool I was before.
Called Master of Arts, and Doctor to boot,
For ten years almost I confute
And up and down, wherever it goes,
I drag my students by the nose—
And see that for all our science and art
We can know nothing.[11]

The technical or T perspective has proven highly useful in developing complex technological systems. But we already saw in the case of the Alaska oil spill the significance of O and P perspectives. It is not an isolated case. Indeed, no complex real-life systems are purely technical. The T perspective has serious limitations when one is dealing with systems that involve technology *and* human beings, systems that are of foremost concern as we face both technology and population explosions. We can and must *simultaneously* sweep in perspectives to augment T, perspectives that use other modes of inquiry and other ground rules.

THE ORGANIZATIONAL PERSPECTIVES (O)

Human beings are supreme social animals. Since the dawn of their existence, they have organized themselves into social groups and societies. The individual surrenders some of his rights and accepts responsibilities in exchange for the benefits offered by membership in a group or *organization*. In its more generalized form, we have the *institution*. It is shaped by a common culture. It may foster a set of distinctive values, beliefs and myths, sanctions and taboos, functions and standard operating procedures (SOPs), ceremonies and rituals, flags and uniforms, even a language for insiders (e.g., bureaucratic, legal, black English, Madison Avenue). All these bestow pride and prestige on its members ("the chosen people," exclusivity), promote bonding and loyalty. An institution may be issue-specific and informal, for example, an agreement or set of rules that defines the practices of its participants. The law of the sea may be considered a social institution.[12]

The life of the organization is in many cases expected to endure beyond the membership of the individual, thus providing a sense of stability and an anchor of security. As civilization has advanced, more and more groupings have evolved. The modern individual is a member of many overlapping groups, such as family, gang, company, bureaucracy, town, state, church, school, union, social club, addiction support group, lobby, and professional society. The individual constantly faces tension between his freedom and the constraints associated with his group attachments—parental and school discipline, religious sanctions, company and club rules, and state laws. The group constantly faces tension in its drive to survive and grow, protecting its identity and fulfilling its functions. It must have the ability to maintain a unique collective perspective despite the diverse individual perspectives of its members.

The organizational or institutional perspective sees the world through a unique filter. To illustrate how it differs from the technical perspective, consider three examples. First, recall our discussion of Forrester's use of the electrical engineering technique of system dynamics to model a corporation—"industrial dynamics"—and provide important insights about the enterprise, particularly its production, distribution, employment, and advertising activities. The brilliant Florentine civil servant Niccolo Machiavelli in the *Discourses* (1531) and *The Prince* (1532) provided very different insights about organizations and how to control them:

> Of no little importance to a prince is his choice of ministers, who are good or bad according to the prince's intelligence. In forming an opinion about a ruler's brains, the first thing is to look at the men he has around him, for when they are adequate and loyal he can be considered prudent, because he recognizes those who are competent and keeps them loyal. When they are otherwise, the prince is always to be estimated low, because the first error he makes, he makes in choosing advisers.[13]

> Men go from low to high fortune more often through fraud than through force.[14]

> Public affairs are easily managed in a city where the populace is not corrupt . . . Where this goodness does not exist, nothing good can be expected.[15]

Anthony Jay observes how apt such insights are for modern management:

> Machiavelli . . . is bursting with urgent advice and acute obser-
> vations for top management of the great private and public cor-
> porations all over the world . . . It means looking at the corpo-
> ration in a new way: looking not through the eyes of the
> accountant and systems analyst and economist and mathemati-
> cian, but through those of the historian and political scientist.

Referring to a passage in *The Prince*, he writes:

> The guiding principle is that senior men in take-over firms
> should either be warmly welcomed and encouraged, or sacked;
> because if they are sacked they are powerless, whereas if they
> are simply downgraded they will remain united and resentful
> and determined to get their own back . . . Since reading that
> passage I have tried out Machiavelli's principle on several man-
> agers who have had to cope with takeovers; they are with him
> to a man.[16]

Machiavelli and Forrester are both looking at organizations, but
from very different angles and with very different paradigms.

Second, we also noted in chapter 1 that Graham Allison
examined the Cuban missile crisis using multiple perspectives. In
his rational actor perspective, the analyst considers "the United
States" and "the Soviet Union" as unitary decision makers, each
having national goals, alternatives for action, and desirous of a
rational, value-maximizing choice.

Allison's organizational process perspective recognizes that a
government is not monolithic, but composed of organizations,
each with its own parochial priorities and perceptions. For example, in the case of the Cuban missile crisis, it was at first puzzling
why the movement of missiles into Cuba was cloaked in secrecy—
lumber ships were used, and the Cuban harbor where they were
unloaded was cleared of all Cubans—while the uncamouflaged
preparation and construction of the missile sites was easily identi-
fiable in aerial photographs obtained by U.S. U-2 surveillance
flights over Cuba. The Soviet Union certainly knew we could
carry on such flights near our shore; after all, they had shot down
a U-2 over the Soviet Union years earlier. The mystery was solved
when it became clear that responsibility for the two tasks was
assigned to different Soviet organizations. The responsibility for
transport security arrangements was given to two agencies that
practice secrecy as a standard operating procedure: shipment to
Soviet military intelligence (the GRU) and port security clearance

to the KGB. However, site preparation was the responsibility of the Soviet Air Defense Command's surface-to-air missile construction teams and they followed their own procedures. Missile sites had never been camouflaged in the Soviet Union, so there was no thought of changing the procedure in this case.[17]

Third, in engineering, we find that technological risk cannot be understood purely in technical terms such as equipment mean time to failure and probability analysis. The Kemeny Commission on the Three Mile Island nuclear accident recognized the central role of human problems in operation, management, and governmental oversight in the anatomy of the accident (chap. 5). The Alaska oil spill case (chaps. 2–5) offers overwhelming evidence of the need to go beyond the T perspective in the management of technological risk.

These three examples suggest the distinctive character of the O perspective. The term *institution* usually refers to a type, the term *organization* to a specific entity. Thus, the American corporation is an institution, Ford Motor Company an organization. Both terms encompass all kinds of collective human groupings— informal to formal, small to large, temporary to permanent, egalitarian to hierarchical, private to public sector, intimate to transnational. A given decision problem may involve a diverse group of organizational actors, both affected and affecting parties. Each perceives and defines the system in a unique way and makes its own assumptions about the other actors. Contradictions thus must be expected. Chapter 3 illustrates the variety of organizational perspectives that must typically be included.

The organizational perspective focuses on *process* rather than *product*, on *action* rather than *problem-solving.** The critical questions are, Does something need to be done, and if so, what? and, Who needs to do it and how? rather than, What is the optimal solution? There must be a recognition that top-down imposition of solutions may well fail if there is no bottom-up support. Thompson and Warburton, who were sponsored by the United

---

*Systems philosopher West Churchman points out that the language of logic and mathematics, the foundation of modern science, leaves out teleological thinking, that is, it does not focus on design or purpose or management. For example, the logic statement "*a* belongs to *b*" as in "all cats are animals" does not refer at all to purposes. The same statement "*a* belongs to *b*" in teleological thinking might mean that the management of *b* must consider *a* in making its plans—clearly an action-oriented statement.[18]

Nations to study environmental degradation in the Himalayas, conclude that the classic development approach has been to sound the alarm and then tell the country what the solution is, in other words, take the T approach.

> It has not worked . . . because it has ignored (as if it were merely a detail of implementation) the deep political, economic and cultural structure . . . What is needed is a more sensitive approach, an approach that places "mere details"—the institutions that constitute the deep structure—at the very center of the stage . . . There is, we concede, a fair-sized break between the traditional single problem/single solution approach and the one we have developed here. There are many ways to characterize this break but perhaps the best is in terms of the shift it makes from *product thinking* to *process thinking*. The systems frame is no longer a model of the problem but simply an evaluative mechanism . . . We need more than one perspective. The approach by way of plural institutions and divergent perceptions meets this need.[19]

An important aspect of all organizations is the internal flow of information. For example, one sometimes finds a dramatic difference in vertical and horizontal communications (see the discussion of China in chap. 8). In the U.S. there is a dramatic difference between low-tech and high-tech operations. This accounts in part for the difference between the nuclear Navy and the nuclear power industry (chap. 3). The former was organized by Admiral Hyman Rickover as a high-tech operation, that is, as a high-reliability organization; the latter was an extension of an old, relatively low-tech industry (see chap. 5 for a more detailed discussion of high-reliability organizations).

In the O perspective we deal with power. How can conflicts among subunits be turned to constructive use? There is no intensive search for analytic tools; in fact, there often is a mistrust of "academic" techniques. They are viewed either as unrealistic or as unpredictable and uncontrollable. It may come as a surprise to the T-oriented analyst that the typical organization chart is a poor guide regarding the locus of power in organizations: "Real power does not lie in documents and memos outlining your terms of reference and area of jurisdiction: it lies in what you can achieve in practice. The boss's secretary can wield great power, like the king's mistress, without any authority at all—or at least not the sort you can show anybody."[20] In organizations that operate with potentially hazardous technologies, a crisis may instantaneously change the structure from a hierarchical to a flat organization in which the power of previously "low" levels is enhanced and equalized with that of previously "high" levels (see chap. 8).

The world seen from the O perspective is ideally an orderly progression from state to state with occasional minor perturbations along the way, for which experience and procedural manuals have the answers. Rules and procedures are there to be followed, thus assuring the integrity of the organization. These characteristics are in varying degrees reflected in organizations ranging from the Vatican's Curia to the American corporation, from the nation-state to the Sierra Club.

A system involving diverse institutions may depend for its very survival on contention among them. Each institution tries to muster credibility for its perspective, its definition of the problem. Often institutions and the total system can continue to exist only if enough uncertainty is generated by these institutions to swamp the contradictions inherent in their pluralized positions. In other words, uncertainty is not merely the absence of certainty but a state socially generated by institutions to protect their legitimacy.[21]

Unlike the technical perspective, the O perspective is politically sensitive. High-tech companies have been known to undertake a contract to keep their most important asset, the engineering force, motivated even if it does not result in a profit. The true reason is unlikely to appear in the corporate annual report. Politically sensitive aspects are not likely to be documented in print. Drawing forth these perspectives may well require one-on-one interviews.

The dialectic approach characteristic of the organizational perspective is reflected in the history of energy resource forecasts in the United States.[22] The deep division between industrial interests and conservationists on oil and gas resources was already apparent in the early 1900s. In 1908 the U.S. Geological Survey (USGS) forecast total U.S. oil resources between 10 and 24.5 billion barrels and indicated we would run out of oil between 1935 and 1943. In 1974 USGS estimated oil reserves between 200 and 400 billion barrels. Each side seized on these estimates to confirm its policy stand. Many forecasts have been made since then, and, except for the World War I and II periods, each faction has habitually accused the other of manipulating the forecasts for its own purposes. Table 6.1 suggests the different organizational views on resource forecasts. It becomes clear that the forecasts are the ser-

TABLE 6.1
O Perspectives on Oil Reserve Forecasts

	When prices are high	*When prices are low*
Industrialists favor	High forecasts "major new supplies can be found if prices are high"	Low forecasts "higher prices are needed to bring on more supplies"
Consumers favor	Low forecasts "oil is no longer the solution"	High forecasts "no need to raise prices"
Conservationists favor	Low forecasts "high prices encourage overproduction"	Low forecasts "low prices encourage overconsumption"

Source: A. Wildavsky and E. Tenenbaum, *The Politics of Mistrust* (Beverly Hills: Sage Publications, 1981) © 1981. Reprinted by permission of Sage Publications, Inc.

vants of policies already determined or preferred rather than being prerequisites for policy formulation.

Many organizations find it difficult to accept failure. An aura of infallibility is seen as necessary to maintain its authority, its mythology, or the members' unswerving loyalty. This applies particularly to governments, ideologies, and religions. Therefore, there may be great resistance to experimentation, risk taking, innovation, and tolerance of other organizations' perspectives (see chap. 11).

THE PERSONAL PERSPECTIVES (P)

The third type of filter is the personal perspective. It views the world through a unique individual. This perspective sweeps in aspects that relate individuals to the system and are not captured by technical and organizational perspectives.* The individual's

---

*In the preceding footnote, we illustrated the distinction between T and O in the interpretation of the statement "*a* belongs to *b*." We can extend the illustration also to P. Here "*a* belongs to *b*" is often understood as a personal bonding.

perspective is thus determined in part biologically and in part socially.* It is a dynamic amalgam drawing on influences such as biological parents, social peers, education, religion, profession, and marketing pressures. The company man, the obedient soldier, the religious disciple, and the unquestioning party follower have all largely subordinated their personal perspective to that of the organization. At the other extreme, the creative artist and charismatic leader, the entrepreneur and maverick are galvanized primarily by their own unique perspective.

The individual can make a crucial difference. An effective leader can impose his perspective on that of his followers and organization, changing a corporation or a society. An independent individual can march to his or her own beat, defy the collective, or invoke creativity.

From Pericles to Churchill, from Lincoln to Martin Luther King, from IBM's Thomas Watson to Microsoft's Bill Gates, individuals have provided leadership. From Socrates and his love of wisdom to Rachel Carson and her focus on the environment, individuals have set examples. Leaders' perspectives have unique qualities: they are farsighted and have a vision of the future; they are able to communicate that vision effectively to others and thus gain their support; they are willing to take "fuzzy gambles" and considerable risks. Today, many American public and private organizations are "overmanaged and underled."

For much of Western cultural history, a personal perspective was a luxury of the elite. In medieval times, the vast masses were peasants with no societally recognized individuality. The Renaissance initiated a basic change in this attitude. In modern times, increasing concern has been expressed over the danger of ignoring the individual and losing him/her in the aggregate view.

Two hundred years ago, Adam Smith made this point:

> The man of system . . . seems to imagine that he can arrange the different members of a great society with as much ease as the hand arranges the different pieces upon the chessboard; he does not consider that the pieces upon the chessboard have no other principle of motion besides that which the hand impresses upon

---

*Recent research, involving identical twins separated at birth and growing up in totally different environments, suggests that genetics accounts for more than half of their life patterns, that is, commonalities prevail after many years, often a quarter century, apart.

them; but that, in the great chessboard of human society, every single piece has a principle of motion of its own altogether different from that which the legislature might choose to impress upon it.[23]

In the epilogue to *War and Peace*, Tolstoy addressed himself to the eternal question, Were Napoleon and Tsar Alexander the cause of the effects they produced, or was the movement of nations produced by the activity of all the people who participated in the events? His answer: "Morally, the wielder of power appears to cause the event; physically it is those who submit to the power . . . The cause of the event is neither in the one nor in the other but in the union of the two."[24]

Tolstoy recognized the importance of the unique individual and felt he had to present the invasion through the eyes of individuals, not organizations or abstract forces as historians traditionally do. His unresolved dilemma was the difficulty of integrating a vast number of P perspectives. He could not do what, say, the physicist does in dealing with a vast number of particles in a gas, that is, integrate the individual effects into a very meaningful gas law. Focusing on Napoleon is clearly not the answer, although he was certainly an important individual.

In our time a commanding figure is Mikhail Gorbachev:

It took a generation for the forces favoring reform to gather and for a broad base of support to develop . . . Still, these forces were dormant, inchoate, until Gorbachev set them free, galvanized them, and gave them direction and form. They were the tinder, he the match. The kind of astonishing political explosion that took place in the mid-eighties required both ingredients; until the match was struck, that undergrowth of social disenchantment lacked fire and force. But without a vast reserve of incendiary disaffection, the match could have been struck and the flame of reform would have flickered out and died.[25]

Churchman makes the following relevant observations:

Economic models have to aggregate a number of things, and one of the things they aggregate is you! In great globs you are aggregated into statistical classes.

Jung says that, until you have gone through the process of individuation . . . you will not be able to face the social problems. You will not be able to build your models and tell the world what to do.

From the perspective of the unique individual, it is not counting up how many people on this side and how many on that side. All the global systems things go out: there are no trade-offs in this world, in this immense world of the inner self . . . All our concepts that work so well in the global world do not work in the inner world . . . We have great trouble describing it very well in scientific language, but it is there and is important.[26]

My thesis is that we are often driven to act on the basis of moral feelings, which basis has little to do with objectives or maximization of some measure of performance. Such a thesis attacks the foundations of economics and operations research; one cannot appropriately express moral feelings by writing a constraint equation in a mathematical program. One cannot do so because, once the constraint is written, we can infer a price for morality, which is inappropriate. Another way to say the same thing is that there is no trade-off for committing evil: once evil is done, it cannot be reduced by good acts. Only forgiveness diminishes evil.

Given a choice of explanations [of human moral feelings, I prefer the explanation of men like] Kant . . . For Kant, the self is not the ego alone; there is also the moral will which operates not to attain goals, but rather by its own free legislation. The Good Will is beyond cost-benefit motivation and similar "utilitarian" motives; it is our most precious inner gift. At the end of his *Critique of Practical Reason*, Kant is ruminating about man's place in the immensity of the universe, and summarizes his feelings thus: "two things fill my heart with never ending awe: the starry heavens above and the moral law within."[27]

His comment on the question of objectivity is also telling:

One of the most absurd myths of the social sciences is the "objectivity" that is alleged to occur in the relation between the scientist-as-observer and the people he observes. He really thinks he can stand apart and objectively observe how people behave, what their attitudes are, how they think, how they decide . . . [it is a] silly and empty claim that an observation is objective if it resides in the brain of an unbiased observer.[28]

Subjectivity creeps in with the preference for quantification and computer modeling. Variables that are difficult or impossible to quantify are cast aside: quantitative analysis drives out qualitative analysis. The computer, a prosthesis of the human brain, reflects the subjectivity of its programmer. The Pygmalion phenomenon, that is, the modeler's love affair with the model, itself manifests subjectivity.

Equally incisive is the amusing, but penetrating, description of a big-name scientists' international symposium by novelist Arthur Koestler:

> Scientists pose as dispassionate servants of Truth, free from all emotional bias, while ambition and jealousy steadily gnaw away at their entrails . . . each of them possesses a small fragment of the Truth which he believes to be the Whole Truth, which he carries around in his pocket like a tarnished bubble gum, and blows up on solemn occasions to prove that it contains the ultimate mystery of the universe. Discussion? Interdisciplinary dialogue? There is no such thing, except on the printed program. When the dialogue is supposed to start, each gets his own bubble gum out and blows it into the others' faces. Then they repair, satisfied, to the cocktail room.[29]

Cause and effect is a fundamental explanatory paradigm of the T perspective. As cyberneticist Heinz Von Foerster tells us, it is inoperative in explaining the behavior of social systems. The law that transforms the past cause into the present effect is itself changed by the very effect it produces. Thus it is not very predictable. Indeed, we must learn to see things we cannot explain. Furthermore, we have a blind spot: we do not see that we do not see.[30]

Historian Arnold Toynbee illuminates the situation in another way:

> I have been thinking in deterministic terms of cause and effect . . . Have I not erred in applying to historical thought, which is a study of living creatures, a scientific method of thought which has been devised for thinking about inanimate nature? And have I not also erred further in treating the outcomes of encounters between persons as cases of the operation of cause and effect? The effect of a cause is inevitable, invariable, and predictable. But the initiative that is taken by one or the other of the live parties to an encounter is not a cause; it is a challenge. Its consequence is not an effect; it is a response. Challenge and response resembles cause and effect only in standing for a sequence of events. The character of the sequence is not the same. Unlike the effect of a cause, the response to a challenge is not predetermined . . . and is therefore intrinsically unpredictable.[31]

One uniquely personal trait is intuition. In discussing inventions in mathematics, Jacques Hadamard writes: "That those sudden enlightenments which can be called inspirations cannot be produced by chance alone is already evident . . . there can be no doubt of the necessary intervention of some previous mental process unknown to the inventor, in other words, of an unconscious one."[32]

More recently, Nobel laureate Herbert Simon and associates explored the differences between experts and novices in solving physics problems. They found that the expert is mentally guided by large numbers of patterns serving as an index to relevant parts of the knowledge store. These patterns are "rich schemata that can guide a problem's interpretation and solution and add crucial pieces of information. This capacity to use pattern-indexed schemata is probably a large part of what we call physical intuition."[33]

Each individual has a unique set of patterns that inform his or her intuition. In calling on the P perspective, we are thus augmenting the conscious, logical T process by opening ourselves to the deeper mental levels that store patterns of great potential value. Salk specifically stresses the need to cultivate both intuitive and reasoning realms—separately and together. Indeed the evolution of the human mind depends on this binary relationship.[34] Of course, business leaders have always appreciated the value of intuition:

> Walk through an office, and intuition tells you if things are going well.

> On each decision, the mathematical analysis only got me to the point where my intuition had to take over.[35]

The individual is a product of his or her own unique genetic material and environment. The latter clearly subjects him or her through education and social contacts to T and O input. Thus the individual's P perspective is itself a unique, and usually unbalanced, mix of T, O, and P.* In some persons, the analytic T domi-

---

*The mix may well be determined by the particular development and interaction in the individual mind among three groups, or seven specific forms, of human intelligence posited by Gardner: object-free (linguistic and musical), object-related (logical-mathematical, spatial, and bodily-kinesthetic), and personal (by which we perceive the psychological dimensions of ourselves and others).[36]

nates, in others, the organizational O, and in still others, the personal P. We all know "absent-minded professors," "organization men," and "strong personalities." Robert McNamara, the "whiz kid" systems analyst who became president of Ford Motor Company and U.S. secretary of defense, is an example of an individual dominated by the T perspective. He was a brilliant manager of numbers (rather than people), in Senator Goldwater's words, "an IBM machine with legs." In typical T fashion, he was convinced that "running any large company is the same."[37]

Among U.S. presidents, we sense T strength in Jimmy Carter, O strength in Lyndon Johnson, and P strength in Ronald Reagan. Carter did not understand the importance of working with the Congress, an ability Johnson had par excellence; Reagan exuded charm and communicating skill. It is our surmise, albeit unprovable, that an effective leader exhibits a good balance of T, O, and P characteristics informing his or her own perspective. Intellectual ability must be matched by "street smarts" and good intuition.[38]

It is interesting to note that, in 1992, The Futures Group, a think tank specializing in strategic planning and policy analysis for both public and private sectors, developed Personality Intelligence Profiling for client companies. It focuses competitor assessment on what future strategic decisions are likely to be made by key individuals within the competitive company. The individuals' education, career history, personality type, and personal motivations provide vital input in determining factors such as risk-taking propensity, team-building ability, and vision.[39]

Finally, it should be noted that recent T-type thinking acknowledges that uniqueness or individuality of components is a basic property of complex adaptive systems. The deep sub-substructures of such systems shape individual differences while generating similar top behavior. The combination of individual variation and hierarchical dynamics permits the system to act both as a standard system and as a unique system.[40]

ORGANIZATIONAL VERSUS PERSONAL PERSPECTIVES

It is by no means always easy to distinguish O and P. Frequently one cannot be sure whether the person explaining the organization's perspective is not, in fact, giving his or her own. Indeed, many organizations strive for congruence of their organizational perspective with their members' personal perspectives. The assumption is that this maximizes the motivation of the members and thereby the effectiveness of the organization. In other words, what is good for the organization is seen by its members as being good for them, and vice versa.

P perspectives can be very helpful in clarifying O perspectives. By probing individual beliefs of a group, we can separate those that are widely shared from those that are not. We can also determine whether the O perspective being offered misrepresents the members' P perspectives.

Another aspect of the organizational/personal relationship is brought out by Ascher and Overholt in their concern with political risk forecasting for business. They draw a clear distinction among the policy maker's "rational information needs," "political needs," and "psychological needs." Rational (T) needs refer to the meaning and content of information, its degree of certainty, and the policy recommendations embedded in, or implied by, the information. Political (O) needs focus on the ability to be a convincing advocate of preferred policies, to be correct whenever possible, and when wrong, not to be disastrously so, that is, to avoid adverse political repercussions for the policy maker. Psychological (P) needs include the yearning for simplicity and certainty, reducing the anxiety of making decisions based on inadequate information, as well as the desire to reinforce and confirm existing views. The authors are convinced that long-range strategic thinking is qualitatively different from short-range tactical thinking, frequently to the extent of requiring different personalities, alerting us that P is as important as T and O.[41]

The two types of perspective are often in philosophical conflict. Jean-Jacques Rousseau's vision that "man is born free but is everywhere enchained" by institutions contrasts with Thomas Hobbes' vision that institutions are all that save man from a life that is "solitary, poor, nasty, brutish, and short."[42]

Garrett Hardin's "Tragedy of the Commons" suggests that there is an inevitable conflict between the individual and the organizational perspective. He refers to the old British commons or pasture which was open to all. Each herdsman seeks to maximize his gain by grazing as many animals as he can on the commons and selling them. But after some period the carrying capacity of the commons is reached and each additional animal then contributes to overgrazing, the negative effects being shared by all herdsmen. Each herdsman is locked into a system that forces him to increase his herd without limit in a world that is limited—and the result is ruin for all.[43] Today, the collective need for energy conservation conflicts with the individual's love of the automobile. Why should an individual act responsibly and drive a small car when surrounded by profligate energy wasters who enjoy gas guzzlers? His or her energy saving only serves to give fellow citizens more to waste.

The perspectives thus echo the tension between the individual and the society—between individual and collective responsibilities, or between individual obligations and entitlements from the collective, or between managing oneself and being managed. The dominance of individualism encourages creativity and entrepreneurialism but also egocentrism and anarchy. Autonomy and self-management mean the individual is responsible for his or her own actions. The dominance of the society favors collective undertakings and welfare but also leads to conformity and control. The biblical commandment "thou shalt not kill" prescribes a moral precept for the individual, whereas the society honors those who kill to protect it or to further its political goals. In the latter case, the system stipulates the purpose of the individual and he or she give the all-too-familiar explanation "I am not responsible for my deeds, I am only following orders." Here we touch on what may well prove to be the most significant problem of evolving societal systems in the coming decades. We will return to it in chapter 9.*

COMPARISON OF T, O, AND P

We have described three types or classes of perspective, T, O, and P. In table 6.2 we compare them and show that they are clearly distinguished by their paradigms. It is a corollary that O and P differ from T not only in their way of focusing on problems but also in the way they obtain their input and communicate their output, as well as in their concern with "facts" and "perceptions."

Within each category there may be many different perspectives, corresponding to the various actors, parties, and models that view a system, problem, or situation. Indeed, *we may define*

---

*For an interesting cultural anthropologist's categorization of multiple O and P perspectives, see the work of Magoroh Maruyama on "polyocular vision" and "mindscapes."[44]

TABLE 6.2
Paradigms Associated with the Three Perspective Types

	Technical (T)	Organizational (O)	Personal (P)
Worldview	Science-technology	View unique to group or institution (formal, informal)	Individuation, the self (hardest to access)
Objective	Problem solving, product	Action, process, stability	Power, influence, prestige
System focus	Artificial construct	Social	Biological, psychological
Mode of inquiry	Sense-data, analysis, idealization, abstract models	Consensual and adversary	Intuition, thinking skills, adaptive learning, experience
Ethical basis	Logic, rationality	Justice, fairness	Morality
Planning horizon	Far	Intermediate	Short, with exceptions
Other	Cause and effect	Agenda (problem of the moment)	Challenge and response
	Problem simplified and bounded	problem delegated and factored	Hierarchy of individual needs (survival to self-fulfillment)
	Conceptualizations and theories (order, chaos, feedback, systems evolution, self-organization)	Political sensitivity, loyalties	Compassion, altruism
		Reasonableness	Greed, hate

	Technical (T)	Organizational (O)	Personal (P)
Other (continued)	Need validation, replicability, assumption of objectivity	Standard operating procedures, routinization of techniques	Filter out inconsistent images
	Optimization (seek best solution), choice among alternatives	Satisficing (first acceptable solution)	Can cope with only a few alternatives
	Quantification	Incremental change	Fear of change
	Trade-offs, cost-benefit analysis	Compromise and bargaining	Leaders and followers
	Use of averages, probabilities	Reliance on experts, internal training of practitioners	Creativity and vision by the few, improvisation
	Uncertainties noted (on one hand, on the other)	Uncertainty used for organizational self-preservation	Need for certainty, beliefs, and illusions
Communication	Technical report, briefing	Insider language; outsiders' assumptions often misperceived	Personality and charisma desirable

the complexity level of a system, clearly a property of its observers (see chap. 12), *as proportional to the number of different perspectives on it.* By this we mean perspectives that cannot be made equivalent.[45]

In view of our Alaska oil spill case study (see part 2), it is appropriate that we also show how the perspective types compare in their application to risk analysis. The fundamental distinctions in the way T, O, and P conceptualize risk are illustrated in table 6.3.

Another brief illustration shows how the three types of perspective yield quite different perceptions on the significance of trade deficits.

T: The academic economist's perspective suggests that *trade deficits are a "nonproblem."* Suppose group A prefers to consume more now than it produces, while group B prefers to consume less than it produces. Then at a future time, A has must produce more than it consumes, while B consumes more than it produces then. Presently, A has a trade deficit, while B has a trade surplus. B thus acquires ownership of some of A's future consumption; that is, A is now a debtor, B a creditor. Put differently, A borrows now from future generations who are unrepresented in this decision process. The underlying assumption is that economic growth will be strong, that A's future generations will be so wealthy that the burden of paying back to B what is owed will be light. In this perspective there is no right or wrong, simply a rational free choice of producing now and consuming later, or vice versa.

O: The political perspective identifies a trade surplus with national power, a trade deficit with national weakness. Power has precedence over profit. Trade is an alternative source of power to military strength in pursuing national goals. Political intervention to exercise this power may include industrial policy and trade barriers. In chapter 10 we will see that economic competition among superpower blocs may replace the cold war of military competition. In this case, therefore, *trade surpluses are desirable, trade deficits undesirable.*

P: The Bible tells us "save for a rainy day." The Protestant work ethic warns: "don't live beyond your means." Churchman, a systems philosopher, feels that it is immoral to reduce the options of a future generation—which we do when we borrow from it and burden it with a large debt. In this perspective, *trade balances are desirable.*

It is evident that the historic economic perspective fails to capture the complexity of the trade deficit issue. All three perspectives are likely to interact in the decision-making process.[46]

INTERACTIONS AMONG THE PERSPECTIVES

Recognizing the significance of the individual perspectives that bear on a given system, we also must consider their interaction and integrate or synthesize them to derive deeper insight for decision and implementation.

Historically, multiple perspectives must be as old as human societies. The coexistence of individuals has always involved at least two and usually more perspectives. The individual has usually been involved simultaneously not only with the family but at the very least with the civil community (the clan, the tribe, the town, the state) and the religious community (priesthood, temple, church). Only the T perspectives are a relatively new phenomenon.

Therefore, the interaction of perspectives and their integration are hardly novel activities. It is, of course, true that many organizations strongly discourage multiple O perspectives, viewing them as competition for the individual's loyalty and consequently a threat to their strength. This explains inquisitions, excommunications, wars between church and state, union busting, witch hunts, loyalty oaths, show trials, and brainwashing of sect members. Today, complex systems involve many interacting elements and generally imply many interacting perspectives. At the same the limitations of the human brain are being challenged by the artifacts of information technology: richer interactions are possible and desirable.

One of the most familiar examples of the interplay and integration of perspectives is the American courtroom. Each witness in a trial provides a perspective. The prosecutor selects and "cross-cues" certain perspectives, then integrates them in his summation into a single one that suggests that the defendant is guilty. The defense attorney selects others and goes through the same process to show that the defendant is not guilty. The jury as decision maker may accept either integration or go back to the original testimonies and do its own interplay and integration.

Figure 6.1 schematically suggests the basic relationships

TABLE 6.3
Three Perspectives on Risk

Technical (T)	Organizational (O)	Personal (P)
One definition of risk for all	Definition customized to organization or group	Individualized definition
Data and model focus	Reliance on experts and precedent	Importance of personal experience
"Acceptable risk" criteria: logical soundness, open to evaluation	"Acceptable risk" criteria: practicality, political feasibility	Strong influence of media coverage of accidents
Probabilistic analysis "expected value"	Problem denial or "stonewalling"	Perceived horrors (AIDS, cancer, Hiroshima)
Actuarial analysis; statistical inference	Inertia; warnings ignored	Time for consequences to materialize (discounting long-term effects)
Fault trees; "mean time to failure"	Legal concerns: avoid blame, limit costs, litigation	Job loss
Quantitative life valuations; cost-benefit calculus	Financial consequences; threat to product line	If gains involved: avoid risk; if losses involved: take risk
Margin-of-safety design; fail-safe principle	Compatibility with standard operating procedures	Misperceptions of probability concepts

Technical (T)	Organizational (O)	Personal (P)
Failure to grasp "normal accidents"*	Crisis management capability needed	Peer esteem (drugs with teenagers)
Intolerance of "nonscientific" risk views	'High-reliability organization' concept†	Freedom to take voluntary risk (private flying, skydiving)

*C. E. Perrow, *Normal Accidents* (New York: Basic Books, 1984).

†T. LaPorte, "High Reliability Organizations Project" (Unpublished memorandum, Department of Political Science, University of California, Berkeley, 1989).

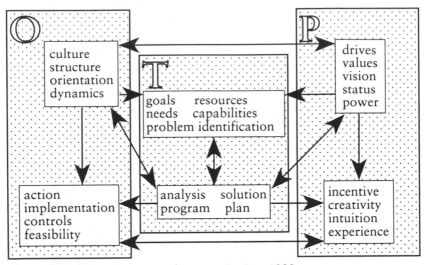

Source: E.R. Alexander, personal communication, 1988.

FIGURE 6.1
Perspective Interactions

among the perspective types we have considered. Figure 6.2 is a simplified schematic version we will use in parts 4 and 5.*

The interaction process itself can be concretized and formalized only to a very modest extent. There is a potential Catch-22. A technique to analyze the interaction process is likely to be drawn from a single perspective type. It can be very helpful in contributing and uncovering further insights, but it must never be viewed as definitive or comprehensive.

Such formalized procedures are sampled in Appendix A. In their concern with formal technique and structuring, the first two of the three procedures outlined remind us strongly of the charac-

---

An intriguing possibility of developing this schematic in conjunction with the discussion earlier in this chapter of the individual as "a unique and usually unbalanced mix of T, O, and P" is suggested by an analogy in physics. Consider a force or weight located at each of the three vertices T, O, and P in figure 6.2 and acting jointly on a hypothetical "body" to produce static equilibrium. If the three are equal, the "body" would rest in the center of the triangle, a condition posited in our discussion as a quality associated with effective leadership. The normal imbalance among the forces results in the "body's" displacement from the center position in the triangle, in accordance with the rule that the vector sum of the three "forces" must be zero. The equilibrium "body" position would thus indicate the individual's T-O-P balance or imbalance. For a typical academic, for example, the equilibrium position would be much closer to T than to O or P.

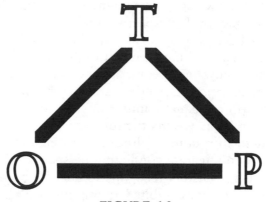

FIGURE 6.2
A Schematic for T–O–P

teristics of the T perspective. The third has at least a hint of T in its mapping representation. More informal approaches such as conferences to develop consensus or dialectic confrontations, as in legal proceedings, are clearly O-oriented means to handle inter-action and integration of perspectives. There is, of course, a large literature on organizational processes, such as negotiation, and we will not try to encompass it here.

Only uninhibited cross-cuing and feedback assure that impor-tant information is not overlooked. Bringing differences among the perspectives to the surface facilitates constructive resolution and integration. The perspectives are dynamic, that is, they can change over time, so that the cross-cuing and integration process may require iteration. The entire procedure overcomes the atti-tude captured in Will Rogers' dictum: "It ain't what you don't know that hurts you, it's what you know that ain't so."

CONCLUDING REMARKS

Evolution of life from the "primeval soup" toward the knowledge society proceeded for millenia before human beings evolved. Replicating molecules strove for negentropy (higher organiza-tion), range (expansion in space), and control through collabora-tion and Darwinian competition. Much of this heritage (accumu-lated genetic knowledge?) resides in human beings, individually and collectively, even though not yet encapsulated in formal sci-ence and technology. Thus O and P perspectives may automatically sweep in "knowledge" that augments the knowledge developed in recent times by T perspective analysis.[47]

Experienced decision makers know that planning and action, or solution concept and implementation, draw on different perspectives. They know that people are not moved by reason (T) alone, that organization (O) and personal communicative or persuasive power (P) are equally important for success. They value diverse perspectives as well as the interactions among them, and they integrate them routinely without any weighting formula. In fact, the ability to integrate conflicting perspectives is the key to effective leadership. History recognizes leaders in both public and private sectors who have exhibited an impressive ability to absorb conflicting perspectives, learn from them, and integrate them. Improvisation rather than a formal process characterizes the integration. Franklin Roosevelt, for example, exhibited a talent for learning—pragmatically trying out policies, abandoning them when they did not work, and playing them up when they were successful. Such leaders unabashedly adapt their own perspective, or persuade others to do so, in order to achieve consensus for action.[48]

To underscore this point, we conclude the discussion of multiple perspective interaction and integration with a comment by President John F. Kennedy: "The essence of ultimate decision remains impenetrable to the observer—often, indeed, to the decider himself . . . There will always be the dark and tangled stretches in the decision making process—mysterious even to those who may be most intimately involved."[49]

For those interested in applying the multiple perspective approach, ten general guidelines are provided in Appendix B.

In part 2 (chaps. 2–5) we applied multiple perspectives to the Alaska oil spill. In this chapter, we discussed the concept itself. In part 4 (chaps. 7–11) we turn to a much broader application. Rather than being local and retrospective, it is global and prospective. We shall focus particularly on the interactions among T, O, and P to draw forth insights that are likely to be of paramount significance as we approach the year 2000.

PART 4

The U.S. at the Edge of the 21st Century

No one really knows enough to be a pessimist.

Norman Cousins

More than any other time in history, mankind faces a crossroads. One path leads to despair and utter hopelessness. The other, to total extinction. Let us pray we have the wisdom to choose correctly.

Woody Allen, "My Speech to the Graduates"

INTRODUCTION

OPTIMISTS AND PESSIMISTS

Human beings are often divided into pessimists and optimists. Pessimists see plenty of problems and few solutions; optimists see plenty of solutions but few problems. Pessimists see every new human being as an additional load on an already strained biosphere; optimists see every new human being as a potential boost in brainpower.

If we were to travel back in time, we can imagine, if it is not a contradiction in terms or an oxymoron, a "reflective caveman" pondering his fate and that of the clan. As a reflective analyst, he—or she—is a natural pessimist. Prospects for long-term survival are not good. The constant threats to survival and the limited number of animals within hunting range do not bode well. Only very few fellow human beings can be supported by the available resources. No matter how reflective he or she might be, it is impossible to envision the possibility of feeding the clan through one of humankind's greatest inventions, farming. This remarkable innovation dramatically increases the carrying capacity of the land, providing a larger, more controllable energy supply.

In a similar fashion, the "reflective analysts" of the newer agricultural societies must have been pessimists as well. They must have felt overwhelmed and depressed by the small amounts of arable land that was available to support large numbers of hungry mouths. They did not, and perhaps could not, imagine the ability to support huge populations through industrialization.

In looking forward, we again have a choice between optimists and pessimists. The approach of the millenial year 2000 is giving rise to a veritable cottage industry of book production. They range from the feel-good forecasts of Cetron's *American Renaissance* and the 21st-century ecstasy of Naisbitt's *Megatrends 2000* to grim warnings about "the endangered planet" and *The Rise and Fall of the Great Powers*. We are told by the happiness purveyors (à la Émile Coué) that every day in every way things are getting better and better. On the other hand, the doomsday prophets look ahead to global economic and ecological collapse.

Many analysts, in particular, are pessimistic as they perceive limits to the carrying capacity of spaceship Earth. They question its ability to sustain the demands made upon it by the staggering growth of its human population. The worldwide depletion of natural resources and increasingly serious pollution problems provide plenty of evidence to argue for potential ecological catastrophes. The pollution problems are exacerbated by long delays—first in their impact, then in the recognition of the impact as a serious danger, and finally in the collective response. If we look at the record of the last 130 years, it becomes evident that the industrial age has raised the earth's average surface temperature significantly.[1] Soil disinfectant chemicals released in the Netherlands in the 1970s and banned in 1990 will reach their peak concentration in the groundwater only in 2020.[2]

The systems analysts' computer models show that, at the present pace, the global system of population, resource use, and waste production will "overshoot" the carrying capacity of the biosphere some time in the 21st century (fig. 9.1). Unless the system damage can be quickly repaired, for example, by resource renewal and regeneration, the models forecast a collapse with devastating global impacts on the standard of living and life expectancy.[3]

Finally, the pessimists point out that historical shifts from hunting to agricultural to industrial societies in every case had the luxury of a long conversion time, at least hundreds of years. Such changes today are compressed into decades and strain social learning and adaptation capability beyond reasonable limits.

At the other end of the spectrum, the optimists have an abiding faith in human creativity and the ability to overcome limitations. They see human beings as "the ultimate resource," poised at the dawn of the knowledge society*, an era promising astounding advances in information technology and a better quality of life. They would note that we are also moving into "an age of substitutability." We will have access to unlimited, clean energy (solar and/or fusion) in the 21st century and that will permit us to dig deeper into the earth for many resources. For virtually all others in limited supply, we can develop materials in near-inexhaustible supply as substitutes. We can even custom-design materials to have specified properties.[4]

The optimists further point out that democracy has been steadily gaining ground since 1800. The fraction of the world's population living in democratic societies has increased from less than 1 percent in 1800 to 40 percent in 1986. Extrapolation of this 190-year trend yields an estimate of 60 percent by 2025 (for details see fig. 8.1).

In hindsight, both pessimists and optimists can claim to be right. Many tribes and societies did outstrip their areas' carrying capacity and disappeared. There is ample evidence of a pattern of the rise and sudden collapse of civilizations. But in the aggregate, it would obviously have been better to bet on the optimists: vibrant new societies have sprung up where aging and unadaptable ones have faltered. Human evolution has moved forward in a kind of societal relay race. Both optimists and pessimists are useful in one important way. They suggest the breadth of the spectrum of possibilities for the path into the 21st century.

Human societies have been evolving rather successfully by self-organization and selection. We are only now beginning to gain some understanding of this remarkable process, a combination of randomization and organization, chaos and antichaos.** We are not *fail-safe*, but *safe-fail* (see chap. 2). That is, we are not designed, as it were, never to fail, but rather to overcome our failures and learn from them (by "feedback"). We also experience

---

*Futurist Alvin Toffler sees agricultural society as "the first wave," industrial society as "the second wave," and knowledge society as "the third wave." The first is founded on land and water, the second on energy and raw materials, the third on information and communication.

**See, e.g., the current development of "complexity theory."[5]

random events, such as the unplanned encounter that leads to a lifelong partner in marriage or business and the inspiring teacher who makes a lasting impact. Both randomness and our ability to learn from our failures play a vital part in our ability to survive and to adapt to a changing environment.

The authors share a *long-term* optimism. However, in the *near-term*, we find ourselves in the position of the forecaster who, during an interview by a reporter, was asked: "Are you optimistic or pessimistic about the future?" The wise man thought briefly and said: "Basically I am optimistic." The reporter countered: "Then why do you look so worried?" After a thoughtful pause, the wise man replied: "I really don't think my optimism is justified."

We not only see serious problems ahead in the near-term but also are plagued by disturbing questions of a very basic nature. An analogy may be helpful: We are embarking on a long journey by boat on a swift, turbulent, and unexplored river. Should we drift with the current or try to steer the boat to avoid shoals and rocks? It is the prospect of partial control of our destiny that motivates our look at the future.

What is a good strategy for our society to propel us beyond the dangerous shoals and rocks? The global system is becoming interlinked and interdependent as no system in history. Consequently, we need a depth of insight into complex systems that is totally unprecedented. We cannot expect to address 21st-century issues with a 19th- or 20th-century mindset. We must move beyond the modes of thinking that served us well—one is tempted to say "all too well"—in the past. We call on multiple perspectives and their interactions in our search for some sorely needed insights.

CHAPTER 7

The Technical Perspectives

The nation-state is becoming too small for the big problems of life
and too big for the small problems of life.

D. Bell, "The World and the U.S. in 2013"

Life is full of golden opportunities disguised as irresolvable
problems.

John Gardner

We begin with a brief look back at the United States since the end
of World War II and at the issues generated for the 1990s. Using
the technical perspective, we next turn to the forecasting process,
noting invariances, S-shaped curves, and cycles. Looking ahead,
the critical significance of population and technology trends
becomes apparent. We focus particularly on the overarching role
of information technology for the coming decades.

FROM WORLD WAR II TO THE PRESENT

World War II established the United States as the unquestioned
leader among nations. Twice in this century it had come successfully to the aid of England and France, suffering little damage in
the process. By 1945 it was the world's only nuclear superpower,
the richest and most influential country. In Winston Churchill's
words, it stood "at the pinnacle of world power." It produced
nearly half of the world's gross output, set the world's monetary
standard, and partly underwrote the recovery of Germany and
Japan. The country could take pride in the finest infrastructure,
including the best transportation and communication network as
well as the best staffed universities. America had always had a
love affair with technology.[6] By 1945 it led the world in the new
fields of computers and nuclear energy.

Conversion from a war economy to a peace one was effectively accomplished. Defense spending dropped from 40 percent of the gross national product to 6 percent, but the G.I. Bill, the Marshall Plan, and pent-up consumer demand greatly facilitated a smooth transition. The United States enjoyed huge domestic markets and was blessed with an incredible resource base. It had the best trained and most productive work force. Of the total, nearly 40 percent were industrial workers, while the rapidly increasing "information" workers already approached the 30 percent mark. Agriculture, dominant until 1907, was now reduced to less than 15 percent.[7] America's factories produced the most advanced and highest quality goods. Its workers enjoyed the highest wages and standard of living anywhere on the planet. This translated into a strong foundation of standardized, predictable consumer needs. The ready mass market for new products such as automobiles, refrigerators, and television sets justified huge manufacturing plants and equipment. Great economies of scale could be achieved. The United States became the most successful economic machine and the greatest consumption society the world had ever seen.

Internationally, the technology and economics of transportation and communication were such that, in effect, the two oceans served as buffers against foreign competition. The United States was more nearly self-sufficient than any other industrial country, and it could easily import any needed raw materials. Since other countries lacked the manufacturing facilities of the United States, they had few alternatives but to ship their raw materials to this country in return for high-value finished products. It was committed to Keynesianism, that is, to government intervention in the economy where necessary to prevent the excesses of laissez-faire. It played a leading role in the 1944 Bretton Woods Conference, which created the World Bank and the International Monetary Fund. Domestic and foreign policy was steered by the predominantly WASP (white Anglo-Saxon Protestant) establishment in forums such as the Council on Foreign Relations, headed by John J. McCloy, a Wall Street lawyer.

The one developing external threat, the Soviet Union, was a distant second. Nevertheless, its aggressive stance under Stalin soon fueled a near-hysterical fear that primed the reestablishment of a very large American military-industrial complex. The main objective of U.S. foreign policy became the global containment of

Communism, a goal realized with remarkable success (except for isolated, militarily weak pockets such as Cuba). American troops were spread around the world as guarantor of stability in a kind of Pax Americana. By 1985, there were still over half a million U.S. troops stationed abroad. Under the Reagan and Bush administrations, the U.S. mounted a series of punitive expeditions against Third World dictators—the Grenada invasion in 1982, the Libya bombing in 1986, the Panama invasion in 1989, and the Kuwait invasion to defeat Saddam Hussein in 1991. They are reminiscent of Rome's punitive expeditions against the barbarians during the latter days of the Roman Empire.

It is interesting to note as an aside that since the American Revolution, the United States has experienced a regular pattern of wars: about once every generation (fig. 7.1).* Not shown is the most important 20th-century confrontation, the forty-year Cold

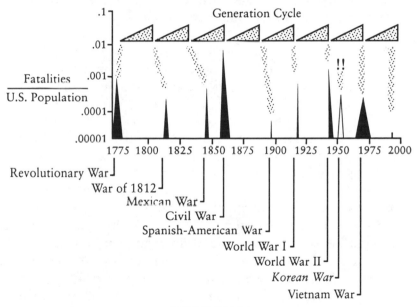

FIGURE 7.1
America's Wars
One per generation – plus Korea

---

*The Korean War, sandwiched between World War II and Vietnam, represents an exception. One explanation offered involves the discounting phenomenon, to be discussed in chapter 9. It appears to take one generation to dim the memory of the last war. Another suggests that the strong tend to display their power in a "hot war" at least once every generation to ensure its continued recognition.

War against the Soviet Union. This success exacted a very heavy price: the cost is now estimated at *$6 trillion*. As early as the mid-sixties, President Johnson discovered that America could not afford his Great Society Program and pursue the Cold War and the Vietnam War at the same time.

If progress was the ultimate shining light of American society, then technology was its guiding hand-servant. And if the tools and techniques, the computers and systems analyses, worked well for the defense sector, why not apply them to the civilian sector? What better conception of a stable, smoothly running world, all of whose inputs—raw materials, labor, infrastructure, consumer tastes, and demand—were under control, than a "machine"? And if machines in the small ran smoothly, what was there to prevent the machines from running even better if they were all hooked together into one large machine? Given the crushing advantages and overwhelming success the United States enjoyed, what could possibly derail the American dream machine?

A mixture of hubris and complacency, overconfidence in the technical perspective when dealing with human beings, individually and collectively, and the stresses produced by the new globe-shrinking technology and the global population explosion, makes fertile soil for the seeds of failure (see chap. 1). By 1992, the United States was still number one, but its lead had been reduced considerably. As Satchel Paige, the famous black baseball player used to say, "don't look back, someone may be gaining on you." Technology has now transformed the world into a global village. A ten-minute daytime telephone call from Los Angeles to New York cost (in inflation-adjusted dollars) $38.20 in 1950 and $2.45 in 1991. The high-speed trains (TGV) have shrunk France, and jet airliners have shrunk the Atlantic and Pacific Oceans.

While the U.S. undertook the burden of world policeman, the two totally devastated World War II enemies rose like a phoenix from the ashes. Americans saw them as potential bulwarks against the Soviet threat; they saw the opportunity to build a new economy. With America's help (e.g., the Marshall Plan) the effort would ultimately accomplish by peaceful means what they had failed to achieve by military force—dominant regional power.

Japan, in particular, brilliantly—and ruthlessly—turned problems into opportunities. The lack of natural resources became the opportunity to develop energy and material conserving processes. The destroyed infrastructure became the opportunity to build new plants that were smaller, more efficient, and flexible.* Thus they could efficiently accommodate (1) rapidly changing tastes and (2) the trend to product customization. They could take advantage of the globe-shrinking technology (transportation and information) to enter distant markets, quickly establish dealer networks, and create efficient supplier linkages. They learned that "less can be more" (see chap. 8).

Meanwhile, in the United States, manufacturing and marketing management saw no need to tinker with success. The motto seemed to be "if it ain't broke, don't fix it". America's decaying infrastructure shows the result of such beliefs. In the words of Joseph Coates, "We are faced with a steadily expanding wave of institutional incompetence, organized incapacity, and a genteel decline into seediness."[8] The New York subway system, collapsing highway bridges, and shrinking public library services provide ample evidence of the deterioration. Of the 1.2 million miles of interstate and other major highways, 52 percent are in bad condition, according to the Federal Highway Administration.[9] While motor vehicle travel increased 27 percent between 1977 and 1987, road mileage has increased only 8 percent from 1965 to 1990. While airline passenger travel has vastly increased, few major new airports have been built.

While benign neglect characterizes America's concern with the upkeep of its infrastructure, other affluent countries have been invigorating theirs. Germany has invested much more heavily in highway technology, with the result that its roads are in superior condition and built to last twice as long as ours. It is planning to spend more than $325 billion in the 1990s on transportation improvements alone. France has undertaken a massive rebuilding of its road and rail network, its telecommunications system, and its power-generating structure—at an expected cost of $250 billion. The TGV trains have been averaging 168 MPH and have carried 140 million people without an accident. The entire French telephone network has been rebuilt at a cost of $80 billion and is now the world's most digitized switching system. More than 5.5 million people have free videotex terminals that provide access to home banking, do-it-yourself travel reservations, and other services. At a cost of more than $14 billion, the joint British-French

---

*In the period 1965–85 Japan invested 5 percent of its gross national product on its infrastructure, while the U.S. invested less than 1 percent.

Channel Tunnel is the largest privately financed civil engineering project of modern times. In 1992, the twelve EEC countries approved a plan including $37 billion in continental infrastructure spending beyond that of individual governments and private enterprises.[10]

In this country, 1,245 toxic waste sites have been identified, and a superfund was created by the Congress to clean them up. The fund was financed by a tax on the chemical and petroleum industries, as well as by assessments on corporate polluters. After twelve years and $11 billion expended, only 84 of the sites have actually been cleaned up. Meanwhile, from 1986 to 1989, the insurers whose polluting clients want them to pay for cleanups have spent $1.2 billion just for lawyers.[11]

President Bush proclaimed that good health care is every American's right, but the fact is that the number of Americans without any health insurance rose from 29.9 million in 1980 to 37.4 million in 1987. This country may have the most advanced health-care technology in the world, but the care available to the poor leaves this society far behind other affluent countries, such as Germany and Sweden.

Education is almost a national disgrace in the United States today. American children are simply not challenged academically. They go to school only half of the days in a year; by contrast, Japanese children attend three days out of four. American children spend less than half an hour a day on homework, while their Japanese counterparts spend more than two hours.*

American children are said to spend over twenty hours per week watching television and less than five hours doing homework. The nation's public schools have spent over $2 billion on some 1.5 million computers; the total spent on computers, software, training, and staffing, amounts to 20 percent of the combined amount spent on all books and other instructional materials. Still, more than 20 percent of America's teenagers are functionally illiterate and in high school mathematics even the top 5 percent of students do no better than the average of all Japanese students. Nearly half of the nation's 191 million adult citizens are not proficient enough in English to write a letter about a billing error or calculate the length of a bus trip from a published schedule.[13] The recent revela-

---

*Other significant comparisons: On an average day 124 Japanese and 2,064 U.S. teenagers become pregnant; 559 Japanese and 4,477 U.S. youths are arrested.[12]

tion that most Americans are unable to program a simple VCR, given full instructions, is, to say the least, embarrassing.

A recent comparison of achievement in science subjects ranked students in seventeen countries (table 7.1).* The U.S. still has the highest percentage of high school graduates, 82 percent, among the 25 to 64 age group. But among younger adults, ages 25 to 34, its 86.6 percent is now surpassed by Germany (91.5 percent), Japan (90.6 percent), and Switzerland (88.4 percent).[15]

TABLE 7.1
The Foundation for Future Scientists—Two Tests

			Ranking		
	Japan	Hong Kong	Singapore	Great Britain	U.S.
10-year-olds (grades 4–5)	1	13	13	12	8
14-year-olds (grades 8–9)	2	16	13	11	13
science students (grades 12–13)					
biology	10	5	1	2	13
chemistry	4	1	3	2	11
physics	4	1	5	2	9

			Test Score*		
	S. Korea	Taiwan	Israel	England	U.S.
9-year-olds mathematics	75	68	64	59	58
13-year-olds mathematics	73	73	63	61	55
9-year-olds science	68	67	61	63	65
13-year-olds science	78	76	70	69	67

*Test scores are given as percentages.
Source of ranking data: *New York Times,* Mar. 24, 1991.
Source of test scores: *Los Angeles Times,* Feb. 6, 1992.

---

*Even within this country, schools mirror the ethnic factor. Since 1969, the makeup of the nation's most prestigious high school for science, Stuyvesant High School in New York, has changed drastically: by 1991, 51 percent of the students were Asian.[14]

A 1993 United Nations "Human Development Index," designed to measure human well-being in terms of health (life expectancy), education, and income (purchasing power), places Japan first and the U.S. sixth in the world.[16]

Dilution of school entrance and graduation requirements to capture the least common denominator of the student population ill serves the high-skill demands that will be made on tomorrow's work force. Indeed, the combination of infrastructure and work-force skill neglect places the nation in a weak global competitive position. Why should businesses locate in the U.S. when they can find a better infrastructure and workforce elsewhere?

Not only are the performance standards in most schools far too low, their compartmentalized structures are incompatible with the needs of today's world. We are entering an era dominated by information technology and unprecedented competition in a global village. America's archaic educational systems consume more than $400 billion annually and seem to drown in administrative hierarchies, while learning is largely stuck in a 19th-century industrial-age, assembly-line mode.

While the new economic contenders welded a strong bond between government, banking, industry, and labor, the U.S. officially resisted any inkling of a cohesive "industrial policy." In actuality, of course, the anti-trust laws constituted part of a de facto industrial policy, as did the bits-and-pieces legislation, such as tobacco farmers' subsidies and timber industry exploitation of national forests, effected by powerful business lobbies.

In nondefense research and development, there has been a remarkable phenomenon in the last twenty years. Japan and Germany have increased their spending from 1.9 percent and 2 percent of GNP, respectively, in 1971, to 3.0 percent and 2.8 percent in 1989. By contrast, the United States figures are 1.7 percent in 1971 and 1.9 percent in 1989.

While the U.S. maintains its lead in basic research, it is rapidly losing its dominance in technology. Using an index based on the number of influential patents, the U.S. had a 2:1 edge over Japan in 1983; by 1989 this lead was reduced to 6:5. In one sector after another, the United States has lost market share. While Americans are keenly aware of the situation in automobiles and consumer electronics, the impressive Japanese technological climb is an across-the-board phenomenon.[17]

This country is no longer in the dominant economic position it was after World War II. In 1989, it only ranked eighth in per capita GNP with $21,100; Switzerland was first with $30,270; Japan, third with $23,730.[18] Of the world's thirty largest public companies in terms of market value in June 1991, thirteen were Japanese and thirteen were American. Of the world banks with the twenty highest asset levels in 1990, thirteen were Japanese, and none were American.[19] In 1988 Japan attained 38 percent of the world's banking assets, with Europe maintaining 35 percent and the U.S. slipping to 15 percent.[20] Japan now invests 20 percent of its GNP in new plants and equipment, $5,320 per capita in 1991, compared to $2,177 for the U.S.[21] By 1997, Japan may become the world's leading manufacturing power, and, shortly after 2000, the leading economic power. But the success of Japanese corporations also carries with it the seeds of failure, as their growing size has some ominous implications—bigger does not mean better (see chap. 8). Indeed, by 1993 the Japanese economic engine flashed unmistakable warning signals—increased unemployment and reduced capital outlays. A drive for "leaner and meaner" organizations to help regain the momentum seemed likely.

Even though it now faces fierce foreign competition in its own domestic markets, the U.S. is still producing more than 20 percent of the world's gross output. The entertainment and aerospace sectors remain major export producers. In 1991, motion picture company revenues from foreign sources constituted 46 percent of the total; by 2000, that share is projected to rise to 70 percent.[22] With the aid of English as the *lingua franca*, American movies, *Playboy*, MacDonalds, jeans, and Disneyland continue to spread influence globally. The trillions invested to maintain world leadership in military systems wrought a historically unprecedented, high-tech military-industrial complex. Only in the field of weapons technology did one hear time and time again of the need to be a generation ahead of anyone else. Elsewhere, short-term thinking has dominated long-term considerations. Efficiency with its short-term payoff has been pursued at the expense of innovation with its longer-term benefits.[23] But technological superiority in the military field is clearly not broad enough to yield the desired economic benefits for the nation as a whole.

The United States is now running a trade deficit—even in high-tech goods—and it has become the world's leading debtor nation. The annual interest on the national debt alone constitutes 14 percent of the federal budget and is projected to exceed $200 billion beginning in 1993. Entitlements, interest payments, and defense together consume a staggering five-sixths of the federal budget.

The Soviet Union tried for decades to compete with America for global influence, squandering its resources in the cold war and Afghanistan. Innovation in the domestic sector languished despite a fivefold increase in research and development manpower from 1960 to 1988. As awareness of the failure of its system deepened, it set its European satellites free and attempted the harrowing task of fundamental internal restructuring. While global economic growth slowed to 1 percent in 1990, that of the Soviet Union and Eastern Europe plummeted to -6 percent. So far, the transition has produced Russian versions of Michael Milken but not of Bill Gates.

The losers of World War II, Japan and Germany, now form the core of burgeoning European and Far Eastern power blocs. The world of 1992 thus consists of one military and three economic superpowers.

Global population growth since 1950 has been awesome. The world population growth rate is 1.8 percent annually, with the figure for the developed countries and the Soviet Union 0.6 percent, for the developing countries 2.1 percent. Today 92 million additional people crowd the earth each year. The high birth rates of the poor and low birth rates of the more affluent are leading to a rapidly growing population of poor, young people. The widening disparity between birth and death rates is simultaneously increasing the world's elderly population, with more than 70 percent of the 1 billion who are who are over age 60 living in developing countries by 2020. Even now, over 80 percent of the world's population is living in poverty. According to the World Bank, in the 1980s the average income declined in more than forty Third World nations with a combined population of 800 million. Among the consequences in large regions of Asia and Africa are (a) increased illiteracy, malnutrition-induced mental deficiencies, starvation, disease;* (b) collapse of national governments, civil war, anarchy; and (c) intensified pressure for massive population movements from poor into wealthy areas (Europe, U.S./Canada), sharpening societal confrontations.

---

*Sub-Saharan Africa has two-thirds of the world's adult HIV and AIDS cases, estimated to rise to 11.4 million and 3.3 million, respectively, by 1995, according to a 1992 Harvard University study. A common attitude is reflected in the comments "you just live with it—it's your fate" and (on using condoms) "life with precaution is no life."[24]

Such population movements may lead to either *(a)* an overall economic decline of the wealthy regions, more accurately, the affluent experiencing "downward mobility" as the poor are modestly climbing or *(b)* creation of high protective walls by the enclaves of affluence to maintain themselves. If present policies continue, the U.S. population will vault from 250 million in 1990 to a staggering 400 million in 2050.

CALIFORNIA: A HARBINGER OF THE FUTURE?

Greater Los Angeles has often been labelled the city of the future. Its population was 102,000 in 1900 and is expected to exceed 16 million by 2000. It gives us a foretaste of an important effect of technology that we shall elaborate on throughout part 4: fostering simultaneous separation and integration. The first city designed specifically for the automobile age, it early became a sprawling collection of balmy, laid-back suburbs in search of a center. Its low density population distribution made mass transit a priori a money-losing proposition. The 740 miles of freeways have now integrated the huge area, but maintained the growth of distinct suburbs, each with its own shopping malls and other amenities. Today Angelenos remain total strangers to most parts of the metropolis, traversing it daily on the freeways without ever setting foot in most neighborhoods. Freeways have also encouraged both job and residential mobility, and neighborhoods change their ethnic complexion remarkably quickly. But the freeways have not kept up with traffic growth; in 1988, on "good" mornings, commuters moving not more than 15 MPH clogged 300 of the 740 miles. By comparison, only thirty miles of freeways were similarly overburdened twenty-five years earlier.[25]

This harbinger of the future is not only a hub of high-tech entrepreneurs, consultants, technologists, entertainers, and traders with the burgeoning Pacific Rim. It is a transnational community with the nation's largest port area, San Pedro–Long Beach. In 1988, 75 percent of the large downtown office buildings were owned or controlled by foreigners. The city is a node in the networks of five energetic global ethnic groups, giving it a strategic position of cultural, technological, and economic influence (see chap. 8 on "global tribes"). Even so, business is finding the area becoming less and less attractive as operating costs steadily escalate along with the community's social problems.

In view of the vital role of the aerospace industry in Los Angeles County, the end of the Cold War is creating a severe impact. It is estimated that, from 1989 to 1995, over half a million jobs are being lost. Besides the high unemployment rate, Los Angeles currently has about 68,000 homeless people. The city also has the second largest populations of Filipinos, Koreans, Mexicans, and Salvadorans of any city in the world. By 2000, Los Angeles will be less than 50 percent white, mirroring the world as a few islands of First World prosperity (the "la-la land" of Beverly Hills, Bel Air, and Brentwood) float in a roiling sea of more than one hundred struggling, fluid Third World enclaves or ghettos speaking eighty-two languages. The schools will then have a student mix of 51.4 percent Hispanic, 26.6 percent White, 12.4 percent Black, 6.9 percent Asian, and 2.7 percent other. About 38 percent of the class of 2000 will begin school with only partial English proficiency.[26]

According to the 1990 U.S. census, the decade of the 1980s saw the foreign-born fraction of the Los Angeles population increase from 18.5 percent to 27 percent; by 1993, it was estimated to include one of every three metropolitan area residents. During the 1980s the Hispanic part of the total population increased from 24 percent to 33 percent. As they crowd into an established area, non-Hispanics flee to other suburbs. Consequently, Hispanic students are now more segregated than blacks are in Alabama or Mississippi.

Los Angeles has an illegal immigrant community which, with its 250,000 locally born children, comprises a million people and generates costs of over $1.75 billion annually in education and emergency health care. About 5,000 illegal immigrants stream daily into the area, mostly bearing false documentation.

The combination of the widening chasm between rich and poor, the economically struggling middle class, the tensions among the ethnic groups, and the uncontrolled growth, constitutes a recipe for urban frustration and violence. Twice, in 1965 and 1992, this seething mix has ignited in firestorms of destruction. Gangs offer a desperately needed, distinctive identity to more than 100,000 ethnic youths (starting at age nine). They have nothing to live for but to "belong," feud with other gangs, and kill. When there is looting, they steal ghetto blasters and television sets that dispense a homogenizing culture. The more successful teenage drug entrepreneurs drive top-of-the-line BMW and Mercedes cars.

Just as Los Angeles is often described as the city of the future, the rest of California is observed for clues. The state already has the eighth largest economy in the world, ranking between Italy and Canada. It also reflects the widening gap between rich and poor. In the 1980s the median income grew by 17 percent (in constant dollars) while the number of children in poverty rose 40 percent. During the same decade, there were 6 million newcomers, 40 percent of them foreign immigrants, raising the state's population to 31 million. By 1990, one out of three Californians was speaking a foreign language at home. Of California's under-18 population, 45 percent are now Latin or Asian. What else distinguishes this trend-setting state? *Everyone seems to be a member of a minority:*

> Society in California is less a society than a congregation of subcultures, many of them with a membership of one.[27]

Even a very densely populated community, such as New York, has de facto walls separating ethnic groups. In the Crown Heights section of Brooklyn, black and Hasidic families live side by side, but not together. Anne Deavere Smith writes:

> They live on the same street, but it seems two different worlds . . . We are a country of strangers. And we are having a great deal of difficulty with our differences. Because ultimately, we lack the ability to look at specific human beings.[28]

These two quotations read like a prescription for civic disintegration.

Honolulu, another strongly multicultural city, offers a more positive image: ethnic diversity with significant civic unity.

POPULATION, TECHNOLOGY, AND SUSTAINABILITY

Human beings and their behavior dominate the shaping of the world of the 21st century. A doubling of the global population to more than 10 billion during the new century is highly likely, while technology presents us with an unpredictable mix of new opportunities and new threats.

Globally, rapidly intensifying farming, housing, transportation, and industrial activities are consuming the earth's resources much faster than they are being created. We are destroying or damaging the physical environment from the Amazon to the Arctic and from the sea floor to the stratosphere. Overconsumption by the rich societies and overpopulation by the poor ones combine to produce ever more waste and pollution. By 2025, 80 percent of the world's population (or 6 billion people) will crowd the tropical areas of the world, accelerating the deforestation process.[29] Forests are disappearing at a rate of 17 million hectares per year, an area of about half the size of Finland.[30] A recent report by the U.N.'s Food and Agriculture Organization shows that the rate of deforestation in the tropical world accelerated 80 percent in the 1980s with tropical forests shrinking 9 percent (despite the ambitious 1985 Tropical Forestry Action Plan of the U.N. and World Bank).[31] During the same period the per capita arable land declined 1.9 percent annually.[32]

Providing food for the burgeoning population will require either massive advances in biotechnology or ever greater reliance on fertilizers and pesticides, which will, in turn, add to environmental deterioration. The only presently feasible way to increase agricultural yield per hectare is the massive use of fertilizers, such as ammonium nitrate. They are making agriculture a prime source of stream pollution globally.

Industry is another major source of pollution. Mercury from industrial plants is emitted into the air, then by rain into distant waters and into the bodies of fish, where it accumulates. The Mediterranean and North Sea coasts have become dangerously polluted. Every day more than 40 million gallons of partially untreated sewage is dumped into New York's Hudson and East rivers. In China only five of fifteen rivers tested near large cities can sustain fish. The deterioration of the oceans is observed in the sharp rise of diseases in its inhabitants (cancer-afflicted beluga whales and turtles, harbor seals with viral infections, and poisoned fish), as well as the dying coral reefs.[33]

One-third of the world's population has inadequate sanitation, 1 billion people have unsafe drinking water, and 1.3 billion people live in urban areas with high levels of air pollution. Lead exposure causes 20 percent of the hypertension cases in Mexico City and a four-point drop in the IQ of Bangkok children by age seven.[34]

The developed countries account for 80 percent of the world's energy consumption, and, of this amount, 80 percent is derived from hydrocarbons. Let us therefore look in more detail at the main waste product, carbon dioxide. From 1800 to 1987 47 percent of all carbon dioxide came from the use of fossil fuels, and 52 percent came from forest clearing and soil emissions. Of the 47 percent, 60 percent came from coal, 30 percent from oil, and 10 percent from gas; 85 percent of the fossil fuel emissions were produced in the industrialized countries of the northern hemisphere. More than 50 percent of the carbon dioxide now comes from three areas: China, the U.S., and the former Soviet Union.[35]

In 1988, the U.S. spewed 4,800 metric tons of carbon dioxide into the air, the most of any nation, and was surpassed on a per capita basis only by East Germany.[36] Carbon dioxide has clearly become the dominant contributor to the atmospheric greenhouse effect that is associated with global warming. Levels of carbon dioxide in 1992 are 26 percent higher than the preindustrial concentration and continue to climb.[37]

In the stratosphere, the protective ozone layer is being destroyed by industrially produced nonbiodegradable chlorofluorocarbons (CFCs). This layer has effectively shielded the earth from ultraviolet (UV-B) radiation. The recently observed increase in this radiation damages living cells and impairs the food chain beginning with phytoplankton.[38] The Environmental Protection Agency estimates that there will be at least 200,000 additional U.S. deaths from skin cancer over the next fifty years due to the ozone problem. The amount of CFCs in the atmosphere is expected to rise until at least 2000, with ozone destruction continuing for several decades in the 21st century. First alarms about the ozone layer were sounded in 1974, and ozone losses have exceeded the predicted values since the mid-1980s. Even so, uncertainties were cited to justify years of inaction by the Reagan administration. DuPont, the inventor and dominant global producer of CFCs, delayed the development of substitutes until 1986. There is a lesson here for technology management: waiting for incontrovertible evidence may create grave problems of an ethical nature.[39]

The rapidly expanding scale of energy and chemical production is leading with increasing frequency to industrial catastrophes (e.g., Three Mile Island, Chernobyl, Bhopal, and the Alaska oil spill). In Eastern Europe, for example, more than forty nuclear reactors are so unsafe that they are deemed catastrophes waiting to happen. A corollary is the deepening dilemma of disposal of dangerous wastes. The safe disposal of unneeded chemical and nuclear weapons and waste is now proving much more difficult than ever anticipated. A 1990 Nuclear Regulatory Commission report on radioactive waste disposal concludes: "Studies done over the past two decades have led to the realization that the phenomena are more complicated than had been thought. Rather than decreasing our uncertainty, this line of research has increased the number of ways in which we know that we are uncertain."[40]

Our earth is clearly becoming less healthful for us. Social adaptation to such changes may successfully mitigate their effect, but it may require a time scale far greater than that of a few generations. Natural systems always find ways to correct major imbalances among their subsystems, typically starvation, adaptability, and survival of the fittest.* Consider natural calamities such as forest fires. Their devastating effect ultimately dissipates as nature reseeds itself and new forests arise. The same is likely to happen with human-induced environmental destruction, although the recovery time may require centuries. Thus, it is presumptuous to claim that "the earth is an endangered planet." However, for generations yet unborn, life will be very different. Our perspective today on its quality may well seem as quaint as that of our rural ancestors on life in a 20th-century city.

In economic terms, we cannot dismiss the environment any longer as an "externality" that can be assumed to be in infinite supply. A consequence of the inundation of the planet with waste and pollution is that air, water, and energy can no longer be considered free goods. All natural resources are assets and constitute wealth, just as dollars do. As nonrenewable or slowly renewable resource reserves are being diminished, their cost will rise. More and more money will be required to regenerate the wealth lost. How can economic growth be sustained in the 21st century when the population and technology explosions strain the environment ever more severely?

"Sustainability" may be defined as the management of our resources to meet the human demands of the present generation without decreasing the opportunities for future generations. In

---

*This may be true even of planetary systems, as suggested by the ability to correct imbalances, as claimed for the earth and its atmosphere by the Gaia Hypothesis. The theory notes the remarkable constancy of the amount of oxygen in the atmosphere over millions of years. It posits that a significant rise in atmospheric carbon dioxide will stimulate more natural oxygen production to redress any resulting imbalance.[41]

other words, it should leave future generations at least as well off as we are, not in every detail, but in an overall sense. It does not at all mean that we must leave the earth as we found it.[42]

Much uncertainty pervades this subject. For example, we know neither the values and definition of wealth of future generations (see chap. 8), the technological breakthroughs, or the currently unrecognized harmful long-term impacts of our technologies. We are not even sure of the long-term impacts that are currently recognized as detrimental. We should remind ourselves that global cooling with extensive Northern Hemisphere glaciation was forecast in the 1970s.[43] Now we see global warming as the threat. Even so, we do have considerable knowledge that permits us to move in the direction of sustainable development. A "least-regrets" strategy should consider the following possibilities: dematerialization, decarbonization, energy conservation, and other steps.

Dematerialization

The concept of 'dematerialization' may be defined in a number of ways, such as the reduction of "embedded energy" in industrial products, the decline of the weight of materials in industrial end products, or the reduction in the amount of waste generated per unit of industrial products. The subject of dematerialization mirrors the intricate system relationships: frequently those efforts that would seem to move us toward dematerialization may subsequently produce more material, that is, materialization. Factors that influence the process include size, quality, waste generation, ease of manufacturing, production cost, ease of repair and replacement. Consider a few examples:

1. Weight reduction is feasible. Between the years 1978 and 1988, American cars became about 400 pounds lighter. But they are less safe: a 2,000-lb car crashing into a 2,000-lb car is twice as likely to cause serious or fatal injury to the driver than a similar crash involving two 4,000-lb cars.

2. Lighter and smaller appliances may wear out faster, leading to more frequent replacement and thus more materialization.

3. Higher quality products are expected to facilitate dematerialization. The use of 100,000-mile tires should reduce solid waste from tires by 60 percent to 75 percent.

4. Decentralization (to be discussed in chapter 8), including population and industrial plant dispersal, leads to more roads, more households with more appliances, hence materialization. The popular personal computer is also adding to materialization.

5. Affluence generally suggests more materialization, at least in its early phases. The acquisition of microwave ovens has not meant replacement of conventional ovens, thus adding to materialization.

6. Much of the new technology should aid dematerialization. Consider the introduction of fiberglass cable to replace copper cable. One ton of copper cable is replaced by 25 kg of fiberglass cable, and its production requires only 5 percent of the energy needed for the production of copper wire. High-strength, lightweight alloys are replacing heavy steel and cast iron in construction. Materials engineering on a molecular scale and biotechnology also move us in the direction of dematerialization. One estimate concludes that the amount of industrial raw materials needed for one unit of industrial production is now no more than 40 percent of what it was in 1900, and the trend is accelerating. Information technology increases the "value added" to materials in manufacturing. Not surprisingly, the ratio of nonrenewable resource consumption to output has been declining steadily in the developed countries, and an acceleration in the trend is anticipated as we become more immersed in the information technology society.

The development of new technology cannot be ipso facto assumed to aid dematerialization; past forecasts have proven notoriously unreliable. Communication technology has not cut back transportation, and the telephone has not reduced letter writing. In 1986 the U.S. Post Office handled a record number of 45 billion pieces of bulk mail. Radio and television also have not cut down the use of paper. The paperless office, once forecast as a concomitant of the age of information technology, is now considered about as likely as the paperless toilet. In fact, the U.S. consumption of writing and printing paper increased in the period 1959 to 1986 from 7 million to 22 million tons. In a single year, 1986, businesses in the U.S. bought 200,000 photocopiers. From

1960 to 1982 this country increased the total amount of its wastepaper by 75 percent. Today, wastepaper constitutes 30–35 percent of all solid waste. In the future it would be appropriate to sweep the cost of waste disposal or recycling into product costs.[44]

Decarbonization

Carbon dioxide emissions and their effect on the atmosphere were discussed earlier in this chapter. The new century will see a shift in the primary source of carbon dioxide production, from the developed countries to the developing ones; for example, China will surpass the U.S. Three-quarters of China's energy is derived from coal, and, by 2020, that country will account for half of the world's coal production.[45] The base scenario shows that by 2020, the developing countries will emit 2.8 times their 1990 amount, while the OECD countries' emissions should remain at about the 1990 level.[46]

Significant climate changes in the 21st century are a distinct possibility. By 2030, global warming may raise temperatures in central North America 2.0°C or more, with rainfall up to 15 percent higher in winter and 5 percent to 10 percent lower in summer.[47] By 2100, U.N. estimates project a rise of about 2.5°C.[48] However, a countervailing climatic cooling trend, human-induced or natural, may also occur, upsetting such forecast. For example, the burning of tropical biomass (currently billions of tons annually) may be creating enough smoke to mask and delay the global warming trend.[49]

Stabilization of the air pollution level requires serious reduction of emissions. If we cut carbon dioxide emissions 2 percent per year beginning in 1992, the further temperature rise may be confined to less than 0.5°C; if we wait until 2010 for such reduction, stabilization may not be reached before a 1°–2°C rise or an even larger one occurs.[50] The rise of natural gas as the preferred fossil fuel source in the early part of the 21st century (see fig. 8.1) would also help, as it emits less carbon dioxide than coal. Using gas to generate 100 kwh of electricity produces only 63 percent as much CO_2 as does the use of coal.

At the same time, technological innovation is making human beings less vulnerable to climate changes, fashioning a society that is essentially "climate-proof." There is a long history substantiating this trend. In ancient days, cisterns and dams were constructed to lessen the vulnerability to water shortages. Greece and Rome imported food across the Mediterranean, breaking the reliance on local food production limitations imposed in part by climatic circumstances. Technological innovations in the last hundred years have given us light bulbs (1879) to expand activity indoors and at night, refrigeration (1873) and air conditioning (1902, 1906) to facilitate life in hot climates. Radar and satellites provide warning of weather disturbances. Technology has effectively made transportation systems insensitive to climate. Underground malls and telecommunications have had the same effect. Information technology facilitates climate control as well as automation (thus activities in inhospitable environments).[51]

Lessons learned in space operations provide valuable new insights on human adaptability to adverse climatic conditions. Adaptability of the society's capital stock to external changes is seen in the turnover time needed. For example, Japan's business corporations renew their capital stock over periods ranging from 22.5 years in textile manufacturing to 9.8 years in electrical machinery. Infrastructures present more of an adaptation problem: it took 65 years to build 90 percent of the U.S. railroad network (from 1855 to 1920).[52] It is reasonable to expect further significant technological innovations that will decouple human beings from climatic changes and constraints in the 21st century.

Energy Conservation

In general, per capita energy demand in industrialized countries rises linearly with per capita GNP. Major exceptions are found in the U.S. and Canada, which are noticeably more profligate. For example, the per capita demand for these two countries in 1989 was nearly twice that of West Germany. Another major exception is Japan, which departed from the trend in 1970 and by 1989 managed with a per capita demand less than half that of the U.S. While Japan's GNP doubled from 1973 to 1990, its energy demand increased only 26 percent by reason of energy conservation and total quality control. The aluminum, coal, steel, and basic chemical industries were the leaders. Such decoupling of energy from the GNP is virtually essential for sustainable development over the next several decades. We should also note that two countries, Canada and France, derive most of their energy from non-fossil fuel sources, France 87 percent (75 percent nuclear) and Canada 74 percent (58 percent hydropower).[53] Worldwide, 17 percent of electricity is now produced by nuclear fission.[54]

It has been widely suggested that market price mechanisms be used to achieve this goal of reducing energy consumption. Environmental costs and benefits should be accounted for in economic transactions. For example, a carbon tax would effectively lower carbon emissions and promote energy conservation. Considering that the price of energy in no way reflects its role as foremost producer of carbon dioxide emissions and the greenhouse effect, it is clearly underpriced. For industry, energy costs constitute only about 3.5 percent of the total costs. An annual increase of 5 percent in fossil fuel costs would thus add only 0.175 percent annually to industry costs.[55]

The problem with such pricing approach is that it flies in the face of the perceived abundance of energy, which has kept coal, oil, and gas prices remarkably low. The short-term advantage of low prices decisively outweighs nebulous long-term costs of environmental fouling and resource restoration (see the "discounting dilemma" discussion in chap. 9).

Other Steps

Obvious changes that would enhance sustainable development are (1) stabilization of the global population and, if possible, negative population growth; and (2) increased efficiency in industrial production and restraint in personal consumption. The objections to any such long-term, sustainable development strategy on the part of the religious and business sectors are very formidable and cannot be addressed by the technical perspective.

The most intriguing wild card is a breakthrough in energy technology. Suppose, at some point in the future, one near-inexhaustible, low-cost energy source were developed. It might be based on nuclear fusion or direct solar energy conversion. This would affect all other resources. Only 0.3 percent of the quantity and 14 percent of the value (excluding fossil fuels) of all nonrenewable resources are derived from resources in limited supply. Furthermore, most of the uses of these limited materials are substitutable by materials in near-inexhaustible supply, such as sand, stone, and iron. With cheap and plentiful energy, the additional energy needed to process less accessible or lower-grade substitutes would be readily affordable. We would, in effect, be entering an "age of infinite substitutability."[56] But then, the resource process-ing and resulting levels of industrial activity might create unten-able pollution problems.

The foremost insight that we can draw from this discussion is an appreciation of the great complexity of the global population-technology-sustainability dynamics and the need for far more eth-ical risk management. The inevitability of decision making under uncertainty must be accepted and reflected in developing alterna-tive resources and economic diversity.

FORECASTING LONG-TERM TRENDS

One way to look ahead is to examine data from the past to see if they indicate regularities or patterns that, one might expect, will continue in the coming decades. There are at least three interest-ing types to consider: invariance, the s-shaped curve, and cycles.

Invariance

The expression "the more things change, the more they are the same" effectively describes situations we can regard as invariant. Examples of constancy include[57]

- the U.S. mortality rate: nine deaths per thousand annually, an apparent asymptote or limit (based on a steady trend since 1900)
- the age at which mammals die, if measured by heartbeats: one billion for mice, dogs, sheep, elephants, and, until dra-matic improvements in medical care and living conditions occurred in the last few hundred years, human beings as well (life expectancy prior to modern times was twenty-five to thirty years; it is just above forty today, if measured from conception rather than birth)
- annual motor vehicle deaths: twenty-four per 100,000 (dur-ing the years 1930–1990)
- the average number of miles driven annually by American cars: 9500 (1940–1987)
- acceptable time for daily personal travel: about one hour, be it by foot, car, or aircraft*

---

*In Los Angeles, many commuters are now forced to travel four hours daily, but few would consider twenty hours per week "acceptable" commuting time.

The last item is of particular interest because Homo sapiens is a territorial animal and the personal territory is thus defined by the mode of transport.* Walking at three miles per hour, our ancestors felt comfortable making a daily round trip to a point 1.5 miles away. This corresponds to the territory of a typical village, about seven square miles. It is, in fact, close to the size of old villages and the area encompassed by the walls of ancient towns such as Persepolis and Rome, or, up to 1800, Berlin. For the car culture, 25 MPH travel speed corresponds to territory that is about seventy times larger or 490 square miles (the size of Berlin in 1950). Aircraft and magnetic levitation (Maglev) trains can enlarge the city area by another factor of 100. We already see people use shuttle flights to commute from Boston to Washington or from Los Angeles to San Francisco. With Maglev trains the future woman might easily live in Zürich and commute to work in Copenhagen or live in Toronto and spend a day shopping in New York.[58] Note that all the invariances listed are maintained despite a remarkable variation in contributing parameters. Thus the annual motor vehicle deaths have been constant despite speed limit changes, safety technology changes, car performance changes, and car number changes.

The S-shaped or Logistic Curve

The most familiar growth behaviors are linear and exponential. Linear growth means a constant increment per unit of time. Exponential growth denotes a constant percentage growth per unit of time, that is, growth is proportional to the value already attained, as with continuous compounding of money (fig. 7.2a). Both of these growths have no limit and are therefore realistic in a finite world only for short periods.

A more interesting example of a regularity is the growth and saturation pattern often seen in biology, for example, the growth of bacteria in a petri dish. Such growth initially looks like exponential growth, but it becomes slower and slower. Unlike exponential growth, it has a limit (fig. 7.2a). The same curve describes

---

*Control of territory has always been limited by the ability to move people and information. Ancient empires never became larger than an area allowing fifteen-day mail service from the capital. Infrastructure development (roads, bridges, etc.) was motivated by this territorial imperative, be it the Roman or the Inca Empire.

the pattern of substitution of one technology for another. J. C. Fisher and R. H. Pry, of the General Electric Company, showed in a famous paper in 1971 that the behavior is the same for such diverse substitutions as synthetic for natural rubber, detergents for laundry soap, and margarine for butter.[59] A logistic or S-shaped curve describes the growth of the fraction of the potential market captured by the new product (fig. 7.2a). It begins at zero and becomes asymptotic to a limit as the substitution is completed or the "niche" for the product is filled. It is remarkable how many phenomena of growth over time follow this curve. Examples are the cumulative number of explorations of the Western hemisphere, the Gothic cathedrals built, and the compositions by Mozart over his lifetime. In each case the cumulative curve has phases corresponding to birth (start of growth), accelerating growth, maturity (boundary between accelerating and decelerating growth), decelerating growth, and death (end of growth, niche filled).* In its remarkable generality, this ubiquitous behavior pattern, or natural growth law, underscores the fact that *the earth is a closed system of finite size, in which nothing of a matter or energy nature can grow without ultimately bumping up against some bound.* This means that the seemingly explosive global population growth (chap. 1) must slow and cease some time in the future. *However, information and knowledge are not of a matter-energy nature and no bound is apparent for them.*

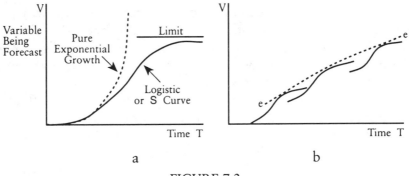

FIGURE 7.2

Logistic or S-Shaped Growth Curves

---

*Interestingly, AIDS growth also follows an S-curve. The data suggests that the maximum or niche for this disease in the U.S. is, fortunately, a low 0.95 percent of all deaths. In 1988 the AIDS deaths comprised 0.85 percent, suggesting that the S-curve is nearing its limit.[60]

With regard to limits, we find that in the transportation and energy sectors

- The curve for world air traffic growth indicates a likely limit of 360 billion annual ton-kilometers by 2020, a figure that implies a maximum increase of 64 percent over the 1990 level of air traffic
- Cumulative U.S. oil discoveries reached about 180 billion barrels in 1991 and will reach a limit of 220 billion barrels by 2040; in other words, the historical data suggest that 80 percent of all oil has already been discovered
- Annual U.S. energy consumption will increase about 66 percent from its 1980 value of $75 * 10^{15}$ BTUs to $125 * 10^{15}$ BTUs by about 2013, with an ultimate limit of about $187 * 10^{15}$ BTUs reached only in 2150[61]

A limit in a technological capability, possibly a physical constraint, is often overcome by shifting to a new technology. The limits of propeller engines in aircraft was penetrated by jet engines. Thus the long-term trend consists of a series of S-shaped curves (fig. 7.2b), with the envelope (ee) a better basis for forecasting than any one S-shaped curve.

In many cases, there are more than two technologies involved at one time. Using share of the total market as the measure, we find that B takes over from A, and C begins its march while both A and B are still being used. The operating principle is, first in (A), first out. In energy use, we find that coal is replaced by oil, which then is replaced by natural gas as leader in world market share. Coal and oil are both still in use as natural gas proceeds on its climb to market share dominance.[62] Similarly, for the transportation infrastructure, canals are replaced by railways, which give way to roads, which yield to airways. We thus learn that, as per cent of total length, roads passed railways in 1915 and airways will exceed roads by 2030.

The demand in air travel keeps increasing and the ever-more-crowded skies are likely to lead to a new technological substitution. The limit in the ability to handle air traffic volume and the limit in aircraft size are likely to spur the introduction of magnetic levitation (Maglev) trains to replace aircraft for short and medium distances. Today's Shinkansen train in Japan and TGV in France are the precursors. Aircraft will then focus their service on the long-haul traffic. The role of cars must also change in the crowded global village of the 21st century. Their unsuitability to densely populated areas has already become glaringly obvious to many commuters in cities such as New York, Los Angeles, Tokyo, and Mexico City. The automobile industry will be increasingly confronted with market saturation.[63]

Cycles

The sequence of substitutions hint at a regular cyclic pattern. In the 1920s, the "bourgeois" Russian economist N. D. Kondratieff found that Western societies experience a pattern of long waves, each consisting of recession-depression-revival-prosperity that repeats every fifty to sixty years (see fig. 7.3 [top]).[64]

We can go further and divide the past century and a half into techno-economic periods. Each has had associated with it an overarching technology: railroads in the 1800–56 cycle, steel in the 1856-1916 cycle, oil in the 1916–69 cycle. Now we are in the information technology cycle (fig. 7.3).[65] Similarly we can relate the prime energy source used in the world with these cycles. Coal reached its peak in fractional market penetration in 1920, oil in 1980—again nearly the same cycle time. On this basis one would predict natural gas to reach its peak about 2030 and nuclear energy to do so about 2090. The use of a nuclear fusion process for energy conversion and hydrogen for its transportation is conceivable one hundred years from now. If population growth moderates significantly by then, this would suggest sustainable development as feasible in the 22nd century, if we can just survive the 21st. In other words, in the very long run, energy will not be the limiting factor in societal development.

Innovations can also be tied to the long wave pattern. They cluster at the end of the economic decline period and lead to new products and industries replacing old ones during the next growth phase. For example, the 1930's depression saw the appearance of the computer, jet engines, radar, and television, each creating an entirely new industry in the growth period following the depression. Saturation clustering of these technologies then initiates the next economic decline.[66] *Continuation of this pattern suggests the current cluster of innovations that will create major new industries during the 1995–2024 growth period.* Candidates include

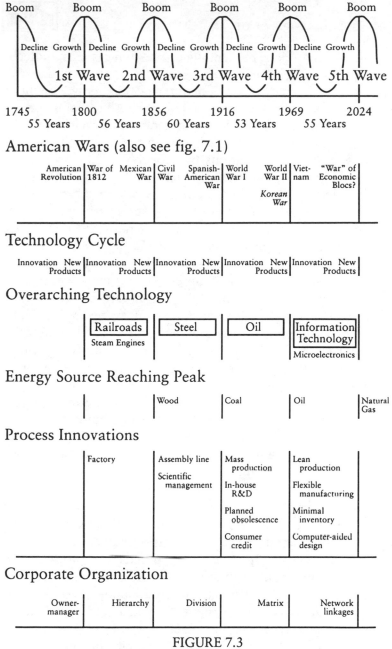

Boom   Boom   Boom   Boom   Boom   Boom

Decline Growth | Decline Growth | Decline Growth | Decline Growth | Decline Growth

1st Wave   2nd Wave   3rd Wave   4th Wave   5th Wave

1745	1800	1856	1916	1969	2024
	55 Years	56 Years	60 Years	53 Years	55 Years

American Wars (also see fig. 7.1)

American Revolution	War of 1812	Mexican War	Civil War	Spanish-American War	World War I	World War II / Korean War	Vietnam	"War" of Economic Blocs?

Technology Cycle

Innovation New Products	Innovation New Products	Innovation New Products	Innovation New Products	Innovation New Products

Overarching Technology

Railroads / Steam Engines	Steel	Oil	Information Technology / Microelectronics

Energy Source Reaching Peak

	Wood	Coal	Oil	Natural Gas

Process Innovations

Factory	Assembly line / Scientific management	Mass production / In-house R&D / Planned obsolescence / Consumer credit	Lean production / Flexible manufacturing / Minimal inventory / Computer-aided design

Corporate Organization

Owner-manager	Hierarchy	Division	Matrix	Network linkages

FIGURE 7.3
56–Year Cycles: A Basis for T-Type Forecasts

Sources: T. Modis, *Predictions* (New York: Simon & Schuster, 1992); C. Marchetti, "Primary Energy Substitution Models: On the Interaction between Energy and Society," *Technological Forecasting and Social Change,* vol. 10 (1977), pp. 345–56; C. Perez, "Structural Change and Assimilation of New Technologies in the Economic and Social Systems," *Futures,* Oct. 1983, pp. 357–75; J. J. van Duijn, *The Long Wave of Economic Life* (London: George Allen & Unwin, 1983).

made-to-order materials, new energy conversion systems, biotechnology, optoelectronics, and microelectronics.

It is intriguing that societal phenomena such as wars exhibit similar cyclical patterns. In figure 7.1, we saw that major American conflicts have tended to follow a generation cycle. We shall take two generations to span about fifty-six years. The Civil War was followed in two generations by World War I; World War I was followed in two generations by Vietnam (fig. 7.3). If the same pattern continues, *World War II could be followed two generations later (before 2024) by another war, perhaps in the guise of an economic conflict among the three economic superpower blocs—U.S./Canada, Europe, and the Far East* (see chap. 8). The societal environment and its cyclical variation also appears to correlate with political differences observed in successive generations. Dominant "idealist" and "civic" generations are born in the wake of the trough, raised during the upswing, reach adulthood in the downswing, and assume political leadership after the next secular crisis. Recessive "reactive" and "adaptive" generations are born in the downswing, reach adulthood in the upswing, and assume majority leadership during the next decline. Strauss and Howe have identified eighteen generations of Americans by this regular pattern, beginning with the idealist Puritans born in the period 1584–1614.[67] Finally, we observe in figure 7.3 that in industry both process innovations and organizational innovations experience the same rhythm of change, as new concepts take root in each cycle.

All of this discussion of long-term trends implies a certain inexorability and engenders considerable humility concerning the ability of governmental and private decision makers to influence the course of events. Optimization proceeds on nature's own terms. The ubiquity and repetition of the patterns considered here at many different levels of human behavior underscore the self-organizing nature of complex systems. It is a subject we will again encounter in chapter 12 (e.g., fig. 12.2).

The reason for the fifty-six-year periodicity lies beyond our comprehension at this time. One suspects that there is a biological driving force, a deep rhythm, that extends beyond the natural and human biological sphere to societal systems. Reverting to our analogy, in the introduction to part 4, of trying to steer a boat on a swift, turbulent, and unexplored river, we surmise that the stream has its own momentum. We have partial control, at best, and some of our steering efforts (or decision making) may prove quite ineffectual in the long term.

Let us consider the area of technology that will undoubtedly have the strongest influence on society in the coming decades, that of information.

INFORMATION TECHNOLOGY

For our purposes the term *information technology* embraces the gathering, processing, transmission (communication), storage, and display of information.* We should also include information at the molecular level, that is genetics and biotechnology. The post–World War II period has seen a revolution in information technology as stunning advances have followed one upon another. Consider the computer. The first computers were developed under U.S. Army sponsorship in the 1940s. They used large numbers of bulky, hot vacuum tubes. Soon there was a second generation with more efficient transistors, then a third generation with silicon chips, a fourth using very large scale integrated circuits, and now a fifth featuring neural networks for parallel processing.** Computing power has increased 100,000-fold in a mere three decades. Further advances, such as optical computers, are expected to raise computing power by another factor of 100,000 to 1 million early in the 21st century. In 1992 Japan's Ministry of International Trade and Industry (MITI) initiated a sixth-generation project, known as the Real-World Computing Program. It is a new ten-year quest to develop massively parallel computers and neural networks.[69]

Even so, there is still a long way to go before these advances approach the most efficient means of gathering, moving, processing, and storing information, namely biological systems. Consider transmission via genetic coding. A single human chromosome has an information content of $2 * 10^{10}$ bits,*** equivalent to 4000 books, of 500 pages each. Our growing understanding of this

---

*Sometimes the term *telematics* is used to encompass these functions.

**Japan's Ministry of International Trade and Industry (MITI) sponsored a ten-year, $400 million, fifth-generation project to produce a breakthrough in artificial intelligence, but it proved to be overly ambitious.[68]

***A bit is a unit of binary information, i.e., an answer to a single yes-or-no question. Thus, if we know whether a light switch is on or off, we have one bit of information.

genetic linkage is beginning to permit significant intervention, both therapeutic and preventive. Eventually this ability may facilitate enhancement of human capabilities directly. We must recognize, however, that the human brain has about 100 billion neurons, permitting a staggering number of possible states that far exceeds the 10^{78} protons in the known universe.[70]

Among the dazzling advances forecast in microelectronics and optics we have:

- New semiconductor materials that permit operation at higher speed and less power (e.g., gallium arsenide chips)
- Cheaper logic and memory components
- Software advances, such as new languages
- Computer-aided software engineering
- "Smart" neural networks
- Artificial intelligence, such as "intelligent" processes that solve problems in unstructured situations where there exists no universal solution program
- Advanced computer architecture, based on parallel instead of sequential processing
- Electron-beam lithography
- Fiber optics, a combination of lasers and low-loss silica-glass fibers; this approach permits much greater information movement at lower cost; also optical amplifiers
- Integration in digital form of high-resolution video, multichannel sound, and the desktop computer into the "telecomputer," with the aid of fiber-optic communications and high-density microchips
- Further miniaturization: "nanoelectronics," based on synthesis at the molecular level; nanostructures will be used for optical as well as electronic applications and will bring the computer another step closer to the human brain.*

---

*Nano is to micro as one-billionth is to one-millionth. It deals with dimensions ranging from the size of an atom to the wavelength of light (10^{-10} to 10^{-7} meters). We are by no means implying that the computer will replicate the brain. The immense neuronal connections in the brain, which vary from one individual to another, remain far beyond our capability.

A hint of the wide range of information technology applications is presented in table 7.2. Nanotechnology is of great interest because it undertakes the design of atomic- and molecular-scale materials and devices that can be implanted wherever needed. Thus everything becomes smarter. Today's automobiles and microwave ovens with microprocessors, as well as humans with pacemakers give but a foretaste of this.

The digitalization of electronic media is also accelerating their convergence. Today's fax machine combines the telephone and printer; interactive video combines the personal computer, television, and printer. Business already uses the telephone network more for nonvoice communication (data transmission, fax) than for voice communication.

A single fiber-optic thread may be able to handle 3,000 times the traffic over all current radio frequencies. In effect, the telecomputer provides every home with the power of a television station. The awesome combination of vastly more computing power and fiber-optic communications is leading us rapidly toward the concept of the true 'video computer'. It will accept input from voice, microphone, camera, remote sensor, vision system, and graphic processor. It will compute data, process images, recognize patterns, and act as video communicator. The challenge is to match the new computing power with the new communication power. Nicholas Negroponte, of MIT, sees the need for the following basic switch:

- what currently moves through wires, like voice, should move through air, for example, global cellular telephone networks replacing telephone lines
- what currently moves through air, like video, should move through wires—fiber-optic cables in a switched digital system for the video-computer.

The reason is that voice uses very little air, whereas the video-computer needs a far broader range of the electromagnetic spectrum and is severely limited by air, even with the expected increase in cellular bandwidth.[71]

The potential impact of the telecomputer and networking on the learning process amounts to a revolution that may drastically alter "education systems." The new approach, sometimes called

TABLE 7.2

A Sampling of Information Technology Applications

Function	Application Area	Information Technology
Data Collection	Weather prediction	Radar, infrared object detection, radiometers
	Medical diagnosis	CAT scanner, ultrasonic camera
	Fishing industry	Sensors to locate fish clusters
	Extractive industries	Remote sensors to locate likely deposits
	News media	Remote satellite imagery for news coverage
Data input	Word processing	Keyboards, touch screens
	Factory automation	Voice recognizers for quality control
	Postal service	Optical character reader
Storage	Archives	Magnetic tape, magnetic-bubble devices
	Libraries	Hard disks
	Ecological mapping	Charge-coupled semiconductor devices, video disks, satellite systems
	Scientific computation	Very high speed magnetic cores
Information processing	Traffic control	Minicomputers
	Medical Diagnosis	Expert systems
	Inventory control	Software packages for on-line management, multiuser super-micros
	Agriculture	Control systems for milking, feeding, harvesting
	Engineering	Computer-aided design and engineering

Function	Application Area	Information Technology
Manufacturing	Manufacturing	Computer-aided manufacturing, robotics
		Networks linking associated contractors
		Intelligent manufacturing control systems
		Rapid response product realization process
	Extractive industries	Robotics for extraction from hostile environments
	Construction industries	Expert systems for structural analysis, robotics
	Building maintenance	Climate-control systems, security control
	Utilities	Energy conservation and load management controls
	Banking	Automatic teller machines, global funds transfer
	Security	Voice- and image-recognition devices
	Learning	Computer-aided, active, "hyperlearning" systems
	Personal services	Interactive entertainment, high-definition TV
		Remote home-operations control ("smart TV")
	Health care	Implanted continual medical monitoring devices
	Defense	"Smart" guided weapons
		Battlefield vehicle, personnel identification systems
	Science research	Supercomputers
Communications	All-purpose	Worldwide digital cellular mobile telephone system
	Office systems	On-line work station network linkages
	Management	Coordination-intensive networks, audio computers
		Briefcase office (computer, modem, fax, telephone)
	Rescue operations	Cellular mobile radios

TABLE 7.2 (continued)

Function	Application Area	Information Technology
	Mail service	Electronic mail
	Translation	Multilingual translation systems
	Defense	Secure digital communications
	Political campaigns	Cable TV, interactive TV
Data output and presentation	Word processing	Personal computers, printers
	Management information	Computer graphics
	Pedestrian traffic control	Voice synthesizers

"hyperlearning," empowers the individual student by permitting both customized, interactive learning and linkage to learning resources throughout the world.[72] The same concept can also empower the patient by facilitating medical self-education and remote diagnosis by physicians.

In genetics we deal with the molecular level of information transfer and in biotechnology or genetic engineering with the application of such knowledge to practical concerns. The basic material of our genetic memory is deoxyribonucleic acid (DNA). The equivalent of a software package coded in the genes is activated in various ways to produce specific enzymes or complex protein catalysts on demand.

The range of molecular genetics can be grasped through a sampling of its applications:[73]

- pharmaceuticals—protein products: currently human insulin and growth hormone; in the future antibodies, vaccines, neurochemicals

- food processing—conversion of nonedible into edible products: single-cell proteins competing with soybean, fructose sweetener

- organic commercial chemicals—a widening range of products made by the organic chemistry industry that are producible by micro-organisms

- environmental processes—bacterial cleanup of oil pollution (see chap. 2), waste treatment

- agriculture—pest control by biological means (viral mutations), in effect, a plant producing its own pesticide

- energy from living trees—genetic splicing to induce growth of galls on trees that then act as machines to produce hydrogen or methane that is tapped for collection*

Future research is expected to focus on better understanding of the molecular architecture of proteins and, most importantly, lead to direct physical readout of the genetic code.

Information technology is so potent and pervasive in its impacts that Toffler[75] sees it as elevating "knowledge" to primacy

---

*The gall actuates a reverse of photosynthesis. A tree can thus produce about 1 kilowatt of power, enough to light a house.[74]

over "military strength" and "wealth" as a source of power in the 21st century.* It may become a production factor in industry more important than management, labor, and capital. It may induce a new age only dimly grasped today by most business and government leaders. This should not be surprising. Consider the fact that this is not the first technology-induced information explosion. The Gutenberg printing press revolutionized the spread of information five hundred years ago. It democratized knowledge and standardized education. Largely due to the political decentralization of Europe and its effective transportation systems at the time, the diffusion of the printing press proceeded with startling speed: only twenty-three years passed from the time of the single Mainz press of 1457 to the establishment of presses in 110 towns. Pocket books, "how to" books, and translations of Greek classics soon became available at low cost. The impacts ranged from a revival of intellectual activities distinct from manual ones to a weakening of religious authority. An early assessment of the impact of the technology would have found it exceedingly difficult to grasp the remarkable pervasiveness and depth of the changes. There were both advantages and disadvantages. Popular education enhanced human capabilities and raised the standard of living. But the dramatic acceleration of the pace of change also led to wrenching societal instabilities and revolutions.

Today's information technology is shrinking the world to a single megalopolis, creating an extraordinary borderless community with a global economy (see the discussion of "smart" capitalism in chap. 8). It makes possible *unprecedented redistribution and flexibility of control—concentration and diffusion of power, centralization and decentralization of decision making, homogenization and heterogenization of culture.* English has become the preferred language at the global level, while local languages such as Provençal, Catalan, and Welsh are enjoying a resurgence locally. CNN, the news network, is transcending national control over information and has become "the common frame of reference for the world's power elite." At the other extreme, E-mail and the proliferation of TV channels offer unprecedented possibilities for expressing personal uniqueness. Simultaneously, the

---

*We will use the terms *information technology society* and *knowledge society* interchangeably. Knowledge encompasses information, imagery, and other symbolic products of society.

significance of national media such as ABC, CBS, and NBC, as well as *Time* and *Newsweek*, is declining.[76] Thus *we attain a relative increase of power at the local and global levels, a decrease at the national level. It is a pattern of redistribution, a rebalancing of control, that augurs coming changes in both governance and business operations.*

The technology flattens organizational hierarchies and encourages "mosaic" enterprise structures that decentralize control. It favors smaller, more fluid, and flexible organizations with wide-ranging linkages. It permits de-massification and customization of products. Familiar concepts such as old-style 'factory workers', 'mass production', 'economies of scale', and 'vertical integration' are out of step with efficient operations in an information-technology era. At the same time the linkages and instant, quasi-universal information availability also permit much larger networks and groupings than ever before, creating a new kind of "global intelligence."

Computer-mediated communication (CMC) networks for commercial, academic, and proprietary use have already created dense coverage of North America, Western Europe, and parts of Asia. By April 1993, several million scientists across the globe were linked through 11,252 interconnected networks by Internet, a cooperative system. Indeed, its popularity and heavy use is now threatening data overload and psychological burnout. Modest networks, such as Relcom, exist in the former Soviet Union. With relatively low-cost technology, 391 organizations in seventy cities of the former Soviet Union have been connected by telephones and personal computers.[77]

Motorola is developing the Iridium global cellular communication system. By 1997, it proposes to provide the ability to connect individuals anywhere instantly by using seventy-seven satellites at 500-mile orbits. It will permit "virtual communities," providing electronically mediated relationships spanning the globe.

Interactivity is a key aspect of the technology. Video games are a primitive example; hypertext and hypermedia are evolving. They refer to the ability of the individual to access many kinds of text, video, data banks, and computer graphics as well as actively participate in group projects, electronic forums, and monitoring of government operations. Today's teleconferencing, information services (e.g., Prodigy), corporate satellite work centers (e.g.,

Pacific Bell), and electronic-format university degree programs (e.g., University of Maryland) merely hint at the future possibilities.[78] Every user can customize the videocomputer to his or her individual needs in a highly flexible, creativity-enhancing way. It heralds a fundamental change from the centralized television broadcasting system that made the user essentially passive, constrained to turning the dial to one of a few channels.

In the corporate environment, the technology permits top managers to obtain information more quickly and accurately but also allows middle- and lower-level managers to be better informed and make more timely decisions that are less parochial and suboptimal. Thus a highly centralized organization becomes more decentralized, and a highly decentralized one becomes more centralized. One major effect should be the decrease in the number of hierarchical levels in the corporation. Another concerns the organizational "memory": the computer provides more available data bases and in-house expert systems.[79]

In the military sphere, information technology is similarly revolutionizing combat operations, as the Gulf War demonstrated. "Smart weapons," technically skilled personnel, and superiority in communications/information over the enemy are recognized as critical factors. Simultaneous centralization and diffusion of control will differentiate the 21st-century military force from the 20th-century one just as it will change the corporation.

Value-added networks ("extra-intelligence") will revolutionize manufacturing and marketing operations in business as well as regulatory and intelligence operations in government. Already today more than two hundred corporations have Chief Information Officers (CIOs).[80] The quasi-instantaneous speed of movement of information calls for a responsiveness far beyond the abilities of traditional government and business operations. *It requires an entirely new organizational metabolism to maintain competitiveness and effectiveness.* High-speed information flow makes possible high-speed money flow through the global financial system. How fast can the money move through the system before control is lost?

Fear of losing control has already been in evidence as U.S. stock markets have proceeded to automate their operations. In 1975 Congress enacted legislation calling for linking all markets through communication and data-processing facilities. One aim was for "investor's orders to be executed without the participation of a dealer." This struck horror in the hearts of many exchange officials and their most powerful members, the broker/dealers, and implementation has been resisted. The technology is at hand and the pressure to automate trading and eliminate the dealer is expected to intensify.[81]

ON OTHER ASPECTS OF TECHNOLOGY

In the area of materials, lightweight strength, high operating temperatures, and materials for electronic components have been areas of primary interest to researchers for years. Much progress has been made and applications have been developed. Ceramics, composites, and polymers are already replacing metals. Composite fibers (such as boron and carbon), with properties of high tensile strength, high elasticity, and low density, are far superior to conventional materials in terms of their strength-to-weight ratio. Engineering ceramics (produced from compounds of silicon, carbon, oxygen, nitrogen, aluminum, and zirconium) are light, resistant to corrosion and friction, and have low conductivity—all desirable properties in engineering. In effect, we are moving toward custom designing new materials for specific uses by working at the atomic and molecular scale. Structure and function are being fused.

New energy-conversion systems are likely to move us away from the environmentally dangerous reliance on fossil fuels. High-efficiency, low-cost solar cells, or photovoltaics, that convert sunlight to electricity directly will provide a clean new energy alternative. Fuel cells that convert gas, such as hydrogen, to electricity offer another clean energy system.

The Apollo manned lunar landing project was motivated by the Soviet Sputnik and the thrilling fulfillment of age-old human dreams to travel to the moon. In the coming decades intense pressure to undertake super-engineering or megasystem projects will be the result of the combined population and technology explosions. The massive need for breathing and living space in the global village, for more land and new transportation modes, for resource extraction and processing, and for waste management will lead to innovative and mind-boggling proposals. Examples include:

- high-speed train networks in Europe and North America
- multipurpose offshore floating islands and vertical cities to create living space (housing several hundred thousand people in a single structure)
- high-speed coastal transportation systems using Maglev trains (connecting floating islands by submerged floating tunnels)[82]
- mega-airports
- hazardous waste storage sites
- mariculture facilities
- a water tunnel from Alaska to California
- the superconducting supercollider
- immense solar cell arrays in desert regions for clean energy

The high-speed train networks are under most serious consideration for Europe. The European Community is planning a rail network of 18,000 miles by 2000; on it, trains will be able to travel at speeds in excess of 200 MPH. The expected cost is in excess of $76 billion. Recalling the invariance of personal daily travel time discussed earlier in this chapter, it follows that this system could permit urban conurbations much larger than today's car city. The 34,000-square-mile Los Angeles metropolitan area may, in this way, ultimately become a true city.[83] The global shortage of capital is likely to be the major hurdle to all such development projects.

There will inevitably be much opposition to these proposals. The supercollider, for example, is being challenged as "big science" that gobbles up money that could be more effectively spent on "small science" and "intermediate technology" in the service of, say, waste management, pollution reduction, and low-cost housing. Similar arguments were, of course, invoked a generation ago against the Apollo manned lunar landing program.

One source of unpredictability is the interaction of innovations. Two historical examples illustrates this point. As noted earlier, the first information technology revolution was triggered by Gutenberg's printing press. But the success of this technological innovation depended on the availability of cheap paper, and that resulted from the innovative spinning wheel, imported earlier from China to Europe. The resulting boom in clothing production led to a cheap byproduct, linen, which in turn created a plentiful supply of linen rags for paper.[84]

In our times, the submarine-launched nuclear missile was made possible by the fortuitous confluence of several new technologies: nuclear warheads, nuclear propulsion for submarines, accurate inertial guidance for missiles, solid propellant rocket propulsion, and heat shielding materials for missile reentry. More such interactions yielding unanticipated innovations are among the "expected surprises."

FINAL COMMENTS

There are some certainties about the future. Demographic data indicate that the population in the affluent world is aging and that in the poor world there is a youth boom. By looking at the past, we have also detected regular patterns: apparent invariants, S-shaped curves, and cycles. The continuing dominance of information technology is also apparent, although we do not know what the overriding technology in the following (sixth) long wave will be (fig. 7.3).

Twentieth-century science has not only dazzled us with its remarkable advances; it has also shocked us with its limits. Einstein buried the concepts of absolute time and space, Heisenberg the faith in precise measurement, Gödel the idea of absolute proof and truth. The complexity of the human brain is such that we see no hope of simulating it, even though we can model some human cognitive processes. The question, Can computers think? remains a subject of unending, albeit heated, philosophical debate.[85]

John Casti concludes his search for certainty with these words:

> Modern science has redrawn the map of human knowledge so that it now shows potholes and detours not only along every side street and back alley, but on all the major highways and byways as well.
>
> It's in those areas of the natural sciences least susceptible to human influence that we have the best "programs" for prediction and explanation. As we move away from hard physics and astronomy and into the JELL-O-like realm of biology, our capabilities for prediction and explanation begin to deteriorate. And by the time we reach the almost totally gaseous state of econom-

ics and the other social sciences, there's far more "social" than "science" in our capacity to say what's next and why.[86]

Twenty-first-century science will present us with more brilliant advances but also undoubtedly with more frustrating awareness of constraints.* Among the achievements that may beckon our children and grandchildren are nuclear fusion, direct readout of the genetic code, and the beginnings of artificial life. Their understanding of complex systems will be deepened through the budding "science of complexity."[88] It is beginning to explore the spontaneously self-organizing, adaptive dynamics of systems that encompass both order and chaos (see also chap. 12). Such insight may facilitate the management of sociotechnical systems—governance, business, and ecological—in ways far beyond our abilities today.

Technology is one of the most powerful responses to problems that human beings have developed. It can be of great benefit, harm, or both simultaneously. On the negative side, we have the potential hazards in using the technology, the unchecked spread of advanced weaponry, and the widening gap the rapid pace of technology can create between knowledge-rich and knowledge-poor societies. On the positive side are the abilities of technology to relieve energy, food, and material resource shortages by developing substitutes, the abilities to improve health and education, and to achieve a narrowing of the rich-poor gap in the long term. *Whether the negative or the positive impacts dominate will depend on how well we organize and manage ourselves and our technology.*

---

*Cesare Marchetti provocatively speculates whether science and technology, seen as the new faith, may actually correspond to the third pulse or millenial cycle of Catholicism. With the canonization of saints forming the data base, he finds that the first Patristic cycle is described by a logistic or S-curve centered in the year 350; the second Tomistic one is centered in 1350. His proposed science-technology cycle begins in 1800 (1 percent of final saturation), reaches 5 percent in 2000 and 50 percent in 2350. The intriguing implication is that we are in a very early phase of this cycle![87]

CHAPTER 8

Linking Technical and Organizational Perspectives

We have begun to lose trust in our institutions . . . It is tempting to think that the problems we face today . . . can be solved by technology or technological expertise alone. But even to begin to solve these daunting problems . . . requires that we greatly improve our capacity to think about our institutions.

Our present situation requires an unprecedented increase in the ability to attend to new possibilities, moral as well as technical, and to put the technical possibilities in a moral context.

Today we must rise to a new level of economic sophistication and creative institutional imagination.

R. N. Bellah, R. Madsen, W. M. Sullivan,
A. Swidler, and S. M. Tipton, *The Good Society*

The interaction of the population and technology explosions is dramatically shaking the foundations of our institutions. A growing mismatch looms as societal change has great difficulty keeping pace with technological change. We are in danger of moving into the new era with 21st-century technology and 19th-century institutions, of being unable to manage ourselves and our technology.

In this chapter, we add organizational or institutional perspectives to the technical perspectives of chapter 7. After a brief discussion of the evolving global societal environment, we will focus principally on two T-O aspects: (1) the obsolescence of ingrained organizational assumptions in light of technological change and (2) the impact of information technology on organizational governance.

THE GLOBAL SETTING

Democratization—an Encouraging Global Trend?

Viewing the world from an American perspective, we begin on a positive note. Democracy seems to be spreading steadily across the world, although temporary setbacks occur along the way.* One recent study, involving a fascinating application of the S-shaped growth curve discussed in chapter 7, uses a data base from 1800 to 1990. In 1840 about 1 percent of the world's population lived under institutional democracies; today the total has passed the 40 percent mark (fig. 8.1). Extrapolation suggests that, by 2060, more than 70 percent of the world will be democratic.[2] In another study of the world's 186 nations by Freedom House, 99 are labelled in 1992 as "democratic" (of which number 75 are considered "fully free," another 23 partly free, and one not free). The same group had judged 44 as "democratic" in 1972 and 57 in 1983.[3]

Karl Marx expected the capitalist machine to spin out of control and self-destruct; instead, the communist system has collapsed. There are many indications that there is de facto an evolving "commonwealth of democracies" based on common values, non-military competition, and cooperative endeavors in many areas. It is interesting to observe that developing a market economy prior to attempting political democratization (the path followed by Taiwan, South Korea, and China) seems to work more effectively than does the reverse process (as attempted in Russia).

Today we hear virtually no discussion of plausible alternatives to a free-market economy and democracy. All but a minute number of libertarians accept the necessity of governmental involvement and the use of public money to avoid excesses and to smooth instabilities in the free-market system. The arguments are largely about matters of degree. The question no one seems to be addressing is this: are there limits to a global self-governing political-economic system, that is, to the democracy and the free mar-

---

*Admittedly, the definition of "democracy" is arguable. The term literally means "rule by the people," but throughout history, it has meant "some of the people." In ancient Greece women and slaves were excluded from participation. Of the Attic population of 315,000 at the time of Pericles, only 43,000 were citizens with the rights and responsibilities of democratic rule. [1] In the United States more than half of America's white men were excluded until 1828, women were excluded until 1920, and many blacks until the 1960s.

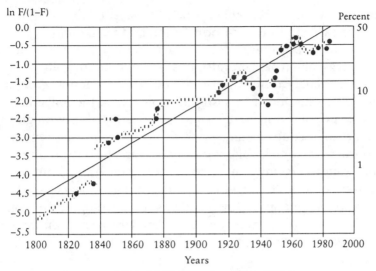

a. 1800–2000 (data for 1800–1986)

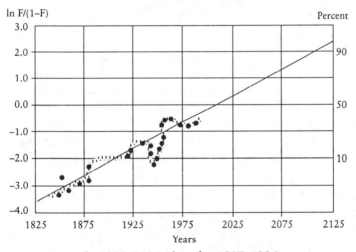

b. 1825–2125 (data for 1837–1986)

FIGURE 8.1

The Spread of Institutional Democracy

Note: F denotes the fraction of institutional democracies. The left scale is the natural logarithm of the ratio democratic to nondemocratic. If $F=1/2$, $\ln F/(1-F)=0$. The right scale is the percent democratic. Graph a focuses on the past, graph b projects the future.

Reprinted by permission of the publisher from "Democratization in Long Perspective," by G. Modelski and G. Perry III, *Technological Forecasting and Social Change, 39*, p. 31 (Figs. 2 and 3). Copyright 1991 by Elsevier Science Publishing Co., Inc.

ket concepts, in an overcrowded world stressed to the breaking point by the twin explosions of population and technology?

Clearly the opportunities for contact and conflict are multiplying. Just as an urban megalopolis accommodating in close proximity startlingly different ethnic and socioeconomic groups experiences a rising tide of tension and violence, so the polyethnic "global megalopolis" must anticipate growing instability ranging from terrorism to coups d'etat, separatism to irredentism, revolution to border war, and smuggling to holy war.[4] Just as urban life confronts vandalism, muggings, robberies, gang rapes, murder, and bombings, so we face globally increasing violence as the have-nots, often energized by religious or ideological fervor, struggle desperately for land, resources, and respect. Continuation of the democratization trend can thus not by any means be taken for granted.

Frustration, Fanaticism, and Fundamentalism

In the coming decades, frustration by billions in the Third World can vent itself with explosive force. High-tech weapons are widely available now, and exotic biological, chemical, and nuclear weapons may eventually reach the hands of fanatics. Charismatic leaders can arouse millions with the powerful help of information technology. While in exile, Khomeini and Abdel-Rhaman used audiotapes to transmit speeches to their followers at home; Saddam Hussein exploited television appearances effectively. Today's low-cost and high-speed telecommunications and transportation greatly facilitate the work of terrorists. The worst problems are internal to the Third World. Just as violence in the poor sections of America's multicultural cities (i.e., Washington, Miami, and Los Angeles) is endemic, so is strife and disorder in many of the world's poor regions.

Frustration provides fertile ground for *religious or ideological* movements. It is particularly significant that, of the five major world religious establishments, the youngest, Islam, appears to be the only one that has not yet entered the final stage of declining spiritual force; in fact, it is just beginning a robust reformation period.[5] Signs of pan-Islamic fundamentalism are seen in a wide belt encompassing the Middle East (Iran, Pakistan, Afghanistan, Jordan), the former Soviet Union (Tajikistan, Kyrgyzstan, Uzbekistan, Turkmenistan), Asia (Kashmir, Bangladesh, Burma), and

North Africa (Sudan, Algeria, Tunisia, Morocco, Libya, and Egypt). At least four of these states—Iran and Afghanistan in the Middle East, Sudan and Libya in North Africa—serve as bases that nurture global terrorist activity. Two events that strengthened Islamic fundamentalism were the end of the Cold War and the Gulf War. The former weakened secular forces in the Islamic world (including Afghanistan and Egypt); the latter embarrassingly displayed the impotence of both secular and traditional Arab governments (including Iraq and Saudi Arabia) vis-à-vis the West. Iran, a fundamentalist theocracy, is ready to fill the regional power vacuum; a hawkish Islamic Egypt is a growing possibility. A sheik-inspired "police state with Allah" or a "nuclearized caliph" conjure up nightmarish possibilities.[6] A nuclear war between India and Pakistan, triggered by fundamentalist Hindu and Moslem fanatics, is also readily conceivable. *The specter of a 21st-century militant Islam replacing a 20th-century militant communism as a global threat is casting ominous shadows.* Other radical ideological movements are finding fertile ground among the poor.

Frustration and a rush to radicalism in the developed world may result from a global economic collapse or extended depression. Germany, considered by some as the culturally most advanced country in 1930, is this century's most horrifying example of a fanatic ideology capitalizing on a severe economic depression. In Russia we have a nationalistic fundamentalism that focuses on the "threat" of non-Russian cultures, such as the Moslem republics and the depraved West that imposed on it Karl Marx and jeans. It glorifies the Russian peasant and yearns for the return of Russian greatness, a vision quite reminiscent of Adolf Hitler's call for a purified and reborn Germany. Spellbinding, paranoid leaders have created fanatical engines of apocalyptic destruction and even self-immolation in this century. Faced with economic hardships and millions of bitter, displaced communist bureaucrats, Russia is particularly vulnerable to such lures. In 1993 Vladimir Zhirinovsky garnered wide electoral support with his demagogic appeals to ultranationalism and in his writings Alexander Solzhenitsyn has urged the Russian leadership to turn away from "the false god of modern technology."[7] Even the United States is not immune from such seductive appeals in times of societal stress, as the rising tide of religious fundamentalism and the responsive chord evoked by American demagogues shows.

The worst scenario would be a new Dark Age engulfing a significant part of the earth, as barbarians and crusaders would practice genocide and wholesale pillage on a modest budget. The global spread of democracy (fig. 8.1) could well suffer a profound setback.

Economic Power and Prestige

A much brighter Third World scenario would envision the economic imperative displacing theology, tribalism, and nationalism as a motivating and unifying force. Success stories such as those of South Korea, Taiwan, and Singapore point the way. Economic reforms and downscaling of government are increasingly recognized as vital steps in Third World development. For example, since 1982 Mexico has sold off three-quarters of its state-owned companies, slashed its budget deficit to 4 percent of GNP, cut tax rates, and strengthened tax collection. Its growth rate for the 1988–90 period has been 3 percent to 4 percent and its inflation rate less than 20 percent. Its automobile production has doubled between 1988 (0.5 million units) and 1991 (1 million units) and is expected to double or triple again by 2000. All three U.S. automakers have factories in Mexico: General Motors in Ramos Arizpe, Ford in Hermosillo, and Chrysler in Mexico City. The uncommon combination of cheap labor and high quality give Mexico a substantial advantage. Chile, Thailand, Indonesia, and Morocco are others with an improving economic climate and, as a result, fast rising private investments. Countries such as these may displace South Korea, Taiwan, and Singapore as the growth stars by 2000.

The affluent world includes North America (headed by the U.S.), Europe (led by Germany), and the Asian Pacific Rim (headed by Japan). It is one of the great ironies of the twentieth century that Germany and Japan, the two Axis powers instigating World War II and decisively losing it, have nevertheless proven remarkably successful—Germany has seen the Soviet Union collapse and is the dominant power in Europe, while Japan has achieved its long-sought Greater East Asia Co-Prosperity Sphere.

Affluence must, of course, be seen in relative terms. In 1993 there is a serious global shortage of capital. The affluent countries are struggling to supply enough capital to finance perceived needs, such as

- reconstruction of the German *Länder* (provinces)
- huge industry debt liabilities in Japan
- aid to the Eastern bloc
- superengineering projects (see chap. 7)
- strengthening of economic competitiveness
- infrastructure rebuilding in the U.S.
- servicing of the enormous national debt in the U.S.—a debt estimated by the Congressional budget office at $6 trillion by 2002![8]

The United States has a more balanced combination of military, economic, and knowledge-based power than any other nation. But it cannot expect to continue exercising global leadership without deep internal changes. It must reverse the widening of its societal gaps and dispel organizational complacency. It must free itself of the tightening stranglehold that its huge debt exerts on the economy. It must greatly strengthen its infrastructure and work force skill level to compete successfully in the new 21st-century global-village environment. In view of the underlying cultural similarities, the German model of modern capitalism is particularly relevant for America. There is the well-trained work force, the social consensus through worker participation, the social health and welfare protection, the longer corporate planning horizon made possible by reliance on banks rather than stock markets in providing financing, and the fiscal conservatism of the government.

Three Scenarios

In an evolving global economy with intensifying competition, scenarios such as the following must be considered.*

A. Economic superpower struggles   There are three economic superpower blocs:

- North America, led by the U.S. and based on the North American Free Trade Area (NAFTA) agreement, possibly evolving later into an all-American bloc with the inclusion of much of South America

---

*These scenarios are based on the ideas and writings of Peter Schwartz, Alvin Toffler, G. Friedman and M. LeBard, and Harold A. Linstone.[9]

- Europe, led by Germany and based on the European Community, possibly drawing in Russia and other East European countries

- Far East, led by Japan, with Pacific Rim partners such as Singapore, Taiwan, Malaysia, Korea, Australia, and New Zealand. China is likely to challenge Japan as bloc leader by 2025. According to a 1993 World Bank estimate, China already matches Japan in gross domestic product, if that product is measured in terms of the purchasing power of its own currency at home. It appears to have abandoned communism for a corruption-ridden political authoritarianism with a market economy, and Singapore as the model to emulate.

These blocs could drift into a new kind of war, an economic confrontation of giants. Some examples: (A_1) a deadly competition between North America and the Far East, with the third bloc remaining neutral, (A_2) a combined "Atlantic bloc" North America/Europe versus the Far East, (A_3) a combined "Pacific bloc" North America/Far East versus Europe, or (A_4) a combined "Eurasian bloc" Far East/Europe versus North America. A_1 could conceivably involve a U.S. naval blockade of Japan to throttle its exports and its imports of raw materials that are vital in supplying its energy needs, specifically oil and plutonium.

A victory by the combined Atlantic bloc in A_2 over the Far East or by the combined Pacific bloc in A_3 over Europe could then be followed by another struggle pitting the victorious coalition partners against each other—North America against Europe in A_2, North America against the Far East in A_3. Case A_4 would probably be disastrous to the U.S. and would be followed by a struggle between the Far East and Europe. In one sense the A scenarios are surprise-free: they continue the focus on a new version of Cold War thinking, economic confrontation replacing the capitalism/communism confrontation. In these scenarios, the bureaucratic national and new regional organizations operate much like those of today.

B. New market economies:

(B1) "Smart" capitalism = competition + cooperation
There is a realization in the affluent world that the nation-state is no longer suitable to serve as societal power center. Local, regional,

and global organizations, in many cases nongovernmental, gain power at the expense of national entities. Global ethnic groups and corporations exemplify the new forces. In other words, considerable control is diffused from the national level both downward and upward. The world is seen as a global megalopolis, as one market where innovation and change are the rule. In this knowledge-based, free market society, the critical nature of superior education and training, as well as management of the endangered global environment, is fully recognized. Attitudes have changed about individual/societal responsibility-sharing and a longer-term perspective prevails. Cooperative international projects in science and technology are fostered, for example, the $8.2 billion American-Japanese superconducting supercollider project in Texas and a trans-European TGV rail system. This scenario extensively exploits the new capabilities afforded by information technology for institutional restructuring.

(B_2) "Dumb" capitalism = out of control

This is a dog-eat-dog world of self-centered individuals and corporations. The global shortage of capital has weakened national governments. There are no new loci of power as in B_1. Both wealth and poverty attain new extremes everywhere, in the affluent as well as Third World areas. The environment, already subject to severe degradation by the Third World population explosion, is further denuded by unrestrained industry in ferocious competition over, and ruthless exploitation of, the declining resources of the planet. Crises, including increasingly frequent industrial and environment-degradation-induced catastrophes, are handled by stopgap responses. Conflicts are very likely to arise over disputed resources. The global megalopolis becomes a global asphalt jungle.

   This power vacuum lends itself to exploitation by right-wing (neo-Nazi-type) extremists. In the U.S. it becomes easy to blame foreigners for America's ills. Well funded and armed vigilantes terrorize Mexicans, firebomb Japanese-owned plants, and hound Jews and environmentalists as the source of the society's economic ills. A populist leader gains strong support from growing numbers of disaffected middle-class whites and "give America back to true Americans" becomes a popular rallying cry. Police are generally sympathetic and moderate, mainstream politicians are afraid to take a strong stance of condemnation against an increasingly potent force. Powerful industrialists see the new populism (neofascism?) as a better alternative than the weak, discredited major party leadership. They want domestic order restored and national power regained to further their interests globally in this turbulent, dog-eat-dog world.

OBSOLETE ORGANIZATIONAL ASSUMPTIONS

National Security

An obvious indicator of organizational obsolescence or "organizational drag" in a rapidly changing world is the clinging or blind adherence to stale, outmoded assumptions. *National security* and *the threat* to national security in America have been defined over a forty-year period in terms of formal military capability in a U.S.-Soviet cold war environment. The U.S. became mentally, if not economically, locked into a kind of national security state.

The depth of the old thinking is well illustrated by the 1988 Navy-sponsored "Navy-21" report of the National Academy of Sciences. This study considered the U.S. Navy over the next several decades. It saw the Soviet Union as still the primary threat during this long period ahead and concluded that the "basic missions are not likely to change over *the next 30 to 50 years.*" No major organizational changes were suggested. And, of course, continuing high defense spending was taken for granted. The basic assumptions proved false barely three years into this long-term forecast. Despite the end of the communist threat, the U.S. government in 1992 still devoted $116 billion to contain the expired "evil empire." Of this amount, $95 billion came out of the $295 billion Department of Defense budget.[10]

This is not to suggest that there will be no military threat in the coming decades. Likely sources of trouble are "crazy states"[11] led by desperate Third World leaders who recognize that they can wreak considerable havoc with a modest, affordable nuclear, chemical, or biological arsenal. Recent years have produced nearly fifty ethnic conflicts, the majority being in Africa and Asia. Severe economic depression in developed countries can also produce tyrants who can unloose threats against the U.S., Germany, or Japan. Economic competition scenarios can transform themselves into military confrontations.

In view of such potential instability, how should the U.S. design its military capability? Maintenance of a nuclear deterrent force at a much lower level than in the Cold War is a reasonable and relatively inexpensive objective: it should deter an attack on this country by assuring annihilation of the aggressor. Of course, it cannot guarantee protection against random nuclear, biological, or chemical strikes. While the fear of communist subversion led this country to institute comprehensive internal security measures during the Cold War, there appears to be remarkable laxity vis-à-vis the new sources of terrorism. Easy illegal immigration and entry under the guise of "political asylum" create an open door for terrorists.

What should be the purpose of the remainder of U.S. military capability? President Bush, in his January 5, 1993, valedictory address on foreign policy, proposed a set of conditions for the use of U.S. military force:

> Using military force makes sense as a policy where the stakes warrant, where and when force can be effective, where no other policies are likely to prove effective, where its application can be limited in scope and time, and where the potential benefits justify the potential costs and sacrifice . . . [When the U.S. does intervene] it will be essential to have a clear and achievable mission, a realistic plan for accomplishing the mission and criteria no less realistic for withdrawing U.S. forces once the mission is complete.[12]

In practical terms, both Bush and Clinton administrations have agreed to a Pentagon force posture equivalent to the ability to fight two regional Gulf Wars simultaneously. However, the old military habit of planning to fight the last war over again with better weapons will prove ever more questionable, as entirely new kinds of action must be considered in the new century. For example, the threat of a pollution disaster of foreign origin or a tidal wave of immigration may be viewed as more serious than a remote military aggression and may require a very unconventional response. It is ironic that, while Americans are arming themselves privately with increasingly lethal weapons, the U.S. military services have begun to recognize the growing importance of non-lethal weapons. Examples are microwaves that melt electronics, lasers that destroy a tank's targeting optics, combustion inhibitors that immobilize engines, as well as advanced sensing and identification systems. Innovative regional police forces may prove as appropriate in the next forty years as NATO forces were for the preceding forty.

There is a basic question Americans must ask themselves: what does "the threat" really mean when 24,000 people were murdered within our own borders in 1991? What does national security really mean when the nation has become dependent on foreign financing? America's own future security depends first and foremost on the ability to deal with domestic violence, and with global economic and ecological "threats."

> Non-military concerns now affect the freedom, well-being, and physical safety of people more profoundly and more dramatically than any military development in any nation. The threat to human life posed by man's destruction of the environment far surpasses any threat posed by increases in the weapon stockpiles . . . As old enemies are transformed, new enemies float into view. National bureaucrats, no less than generals, tend to prepare for the last war, and so it is not surprising that they view the post-cold war world through a cold war prism . . . The challenge of the next decades is to make domestic institutions of nations reflect the political, economic, and ecological realities of the post-cold war world.[13]

How can deep shifts be effected by organizations (the services and defense industry) and their components when all are bent on self-preservation? Great value is placed upon traditional missions and "planning to fight the last war over again with better weapons." Each part of the military establishment is its own vested-interest group that defends its status quo. The organizational perspective becomes dominant.

As an illustration, consider the Army's drive in the 1960s to gain a high level of air mobility. When Robert McNamara, secretary of defense, heard the request by the concept's promoters, he essentially told the Army's Air Mobility Board he would consider the idea if the Army would also tell him what it would cut to pay for it. After arduous internal debates among its own vested interest groups, the board finally agreed to eliminate its narrow-gauge locomotives. They were the one group in the Army that did not have an effective lobby. We saw another example of that thinking in the recent Navy-21 report. The justification for this organizational approach is its value in maintaining internal morale or esprit de corps.

Furthermore, unneeded big-ticket military programs are defended against cancellation on the grounds of jobs rather than military need. In 1992, the Connecticut Congressional delegation campaigned to force the construction of two nuclear Seawolf submarines, $3 billion each, which the Department of Defense wanted to cancel. The 1993 long-range defense plan included a third Seawolf with the justification that it would be too costly to close the production line and possibly restart it in the future.[14] Similar pressure is exerted to build the unwanted vertical takeoff Osprey Army aircraft. The Department of Defense's plans to close old, superfluous military bases are fought in order to save jobs in small communities. Proposed reductions in the defense budget are seen in terms of serious defense industry job losses. The idea is hardly new. The provocative 1969 "Report from Iron Mountain"[15] boldly argued the case for the economic utility of a powerful military establishment:

1. Military spending furnishes the only balance wheel with sufficient inertia to stabilize the advance of the economy

2. Without a long-established war economy, and without its frequent eruption into a large-scale shooting war, most of the major industrial advances known to history, beginning with the development of iron, could never have taken place; the modern computer development was first supported in the 1940s by the U.S. Army and Boeing's 707 commercial airliner was an outgrowth of the Air Force's KC-135 tanker program

3. Military spending has been a consistently positive factor in the rise of gross national product and of individual productivity; in the words of a former Secretary of the Army, "If there is, as I suspect there is, a direct relation between the stimulus of large defense spending and a substantially increased rate of growth of GNP, it quite simply follows that defense spending per se might be countenanced on economic grounds alone as a stimulator of the national metabolism"[16]

With the end of the Cold War, arms sales to other countries have increased to take up at least some of the slack resulting from the cutback of U.S. defense needs. This "industrial policy" has helped to maintain the nation's status as the world's largest arms merchant. In 1991, it sold more weapons and related technology to the Third World than all other countries combined. (Our erstwhile Cold War enemy is trying a similar policy.)

But today a very different argument is heard: for the Japanese, "business is war." The organization of Japan's "economic war" machine, "Japan Inc.," shows striking similarities to America's defense sector (table 8.1). Japan's mobilization for global economic competition has much in common with the American approach to the Cold War. Government and industry work closely together for a common long-term goal. Maintenance of leadership in technology is a constant concern and risks are shared.

No such similarities exist between Japan Inc. and the nonmilitary U.S. industry sector. Where the Japanese are long-term goal oriented, American industry is short-term profit oriented; where Japan Inc. recognizes the entire society as stakeholders, American industry sees only stockholders and senior management as stakeholders. Where Japan Inc. takes collectively high risks, but provides a low risk, stable environment for the individual, American industry takes low risk, but maintains a high risk environment for its workers. Where Japanese companies subsidize intrapreneurs who then set up their own small companies, known as the "silver handshake," American companies provide the nonproductive "golden handshake" of huge executive job severance agreements.[17] Many American entrepreneurs wheel and deal in mergers and acquisitions, hiring and firing rather than building their work force competence.

One exception to the rule is America's Apollo lunar landing project, a nonmilitary example of highly successful, goal-oriented government-industry collaboration in the national interest. It is ironic that the National Aeronautics and Space Administration (NASA) is spending 94 percent of its total budget on space and only 6 percent on aeronautics. In fact, the U.S. is in danger of losing its preeminence in aeronautics, an important chip in the economic superpower competition between North America, Europe, and Japan.[18]

Rethinking of the threat from both T and O perspectives calls for a restructuring of the highly successful national security establishment to meet the economic challenge, a point not lost on President Clinton. It will not do to shift from the manufacture of titanium aircraft to titanium snow shovels and wheelbarrows (as a Russian factory has done). The current "streamlining" merely leads to slow dissipation of one of the nation's greatest resources. Los Angeles County, plagued additionally with particularly high costs of doing business, is expecting to lose 210,000 to 420,000 jobs by 1995 (in addition to 71,000 aerospace jobs already lost since 1989).[19]

A few aerospace companies are showing one way to the future. Hughes Aircraft Company, headquartered in Los Angeles, began, in the late 1940s, as a purely military contractor in the radar and air-to-air guided missile field. Its staff was led by Simon Ramo and Dean Wooldridge, two outstanding Ph.D.'s recruited from Bell Telephone Laboratories. It evolved into a remarkable high-tech, military-focused organization with an innovation-stimulating, campus-like atmosphere. Its research laboratory was a major contributor in the development of the laser in the 1950s.

At least four events combined to set the stage for its weaning itself from the Pentagon: branching out from its military reconnaissance satellite projects to commercial telecommunication satellites, gaining commercially oriented General Motors as a corporate parent, purchasing a small company that builds video equipment for jetliners, and bringing in a top executive of commercially focused IBM as its chief executive.

Hughes is counting on internal flexibility and retraining of its highly skilled employees. Thus, Hughes Research Laboratory engineers redesigned a fire-suppression system meant for Army tanks to be used on municipal buses, resulting in an order for 202 such systems worth a total of more than $1 million. Hughes is working on the electric-powered Impact car for General Motors, specifically, it is developing the inverters needed to convert direct electric current from the car batteries into alternating current to run the engine. It has built a plant to produce 10,000 or more inverters for the car. The development work is being done by engineers in the Radar Systems Group who previously worked on military projects. Hughes Communications is investing $500 million to develop a 150-channel system to directly broadcast TV programs and movies on a pay-per-view basis; this would use small home satellite dishes and its own powerful HS-601 satellite. In addition, Hughes is seeking to develop its overseas commercial business.[20] In these ways a valuable national technological asset is being preserved. More often, however, restructuring will require decentralization, leading to many small high-tech enterprises.

TABLE 8.1

Japan Inc. vs. the U.S. Military-Industrial Complex

Japan Inc.	U.S. Military-Industrial Complex
Goal, rather than profit, oriented (economic strength)	Goal, rather than profit, oriented (military strength)
Long-range planning done: implications of new research results and future industries identified early	Long-range planning done: new threats and their technological capabilities identified early
High-tech emphasis	High-tech emphasis
Good intelligence gathering and analysis: Science and Technology Agency of Japan does technological forecasts; environmental scanning using MITI, JICST, JETRO, company subsidiaries	Good intelligence gathering and analysis: Defense Intelligence Agency (DIA) does enemy capability forecasts; environmental scanning using CIA, corporate planning groups
Coordination of funding and intelligence for development of new products: needs analyses	Coordination of threat evaluation and formulation of new system requirements: needs analyses (e.g., MIRAGE studies)

Japan Inc.

State-industry collaborative R&D arrangements: risk sharing; R&D support by MITI

Vertical and horizontal information flow structures

Intrapreneurs and spin-offs encouraged (the "silver handshake")

Parent-subsidiary and subcontractor arrangements

U.S. Military-Industrial Complex

Dept. of Defense-industry collaboration: risk sharing; R&D support by Defense Advanced Research Projects Agency, Office of Naval Research

Matrix organizations

"Greenhouse" projects often lead to spin-offs

Prime and subcontractor arrangements

Power and Growth; or, Is Bigger Better?

Power has been defined in modern times largely in terms of a nation-state's military or economic strength, and we have talked about superpowers in precisely those terms. However, such separation is becoming a dangerous anachronism. Furthermore, the terms themselves are questionable when knowledge becomes the qualitatively dominant source of power.

By the same token, powerlessness must be viewed differently in a knowledge society. Often impressive-sounding institutions are not connected to anything and are thus quite powerless. However, they may be effective by playing two power centers off against each other. Many countries used the superpower Cold War to their advantage in this fashion, for example, gaining aid from one superpower by threatening to become associated with the other. In the future, there will be new ways of managing powerlessness, and they may involve effective manipulation of information.

Prior to modern times, power was wielded by groups and entities that were not nation-states. The Roman Catholic Church and the Hanseatic League are examples of power based on religious and economic linkages, respectively. Today, geographically dispersed, but culturally or economically based, networks can and do exercise power. Communication and transportation technologies vastly increase the potential of such power.

What constitutes a knowledge superpower, if such a thing is possible? It has recently been suggested that *information* might be defined as the power to exercise control over the disposition and behavior of matter and energy by goal-oriented systems. Information is seen as involving a relationship between a source and a recipient. The reasoning implies that information does not exist until it is used to control matter and energy. Thus an unread book is powerless, but the Bible, which has literally influenced the lives of millions over centuries, is a very potent source of information. It is, in this context, irrelevant whether the Bible is true or not. Its power lies in the fact that people believe it and act upon that belief.[21]

Wealth has been traditionally defined in monetary terms, such as dollars. But we must in the future deal with three aspects of wealth: matter, energy, and information. For example, matter and energy can be combined in a measure of wealth defined by Odum as "emergy." It expresses raw materials, finished goods, energy, and human services on a common basis in terms of the energy needed to generate them. It has been used to calculate the losses caused by the Alaska oil spill (chap. 2). Emergy is inverse to dollar price. When a resource is abundant, its price is low, yet its contribution to an economy is great. Money buys more and overall living standards are high. When a resource is scarce, its contribution to an economy is small. Scarce resources require more energy to collect, concentrate, process, and transport per unit; therefore their net contribution to the economy is much reduced, yet their prices are high.[22] Using the definition of information as power, it may even become possible to combine it with matter and energy to measure wealth.

Growth has, in Western culture, been interpreted as material or economic growth. "More leads to more" and "bigger is better" have formed the popular gospel for generations. It is time to question this assumption.

Let us begin by considering the growth curve *AB* shown in figure 8.2a. Anywhere from *A* to *B* "more leads to more," that is, "bigger is better." More or greater input leads to more or greater output, although at a diminishing rate the farther one moves up the curve. Analogous to this phenomenon in the organizational realm is the notion of "economies of scale": the more units are produced, the cheaper the per capita cost or the greater the profit. However, we find that, from *B* to *C* to *D*, more or greater input now leads to diminished output or "diseconomies of scale." If at some point *C* the output diminishes catastrophically, then the curve *CE* represents the case "less leads to less." At some point, if the catastrophe is not too great and hence is one from which one can recover, one hopefully moves from *E* to *F*, or "less leads to more." Ideally, one would like to sense the point B before one gets to point *C*. If one could do this, then one could switch from *B* to *G*, or "less leads to more," without having to go through catastrophe.

Notice that whatever one does, in principle the whole process repeats itself endlessly. Thus *F* goes to *H*, and so on, because the phenomena with which we are dealing are dynamic. They never stay put or remain the same; they are always developing. The best one can do is cope with them continually; one can never solve them completely. Finally, it must be noted that figure 8.2a represents only one pattern out of many that exist for combining these four basic processes.

Four scenarios for the U.S. economy can be developed that illustrate these processes (fig. 8.2b):

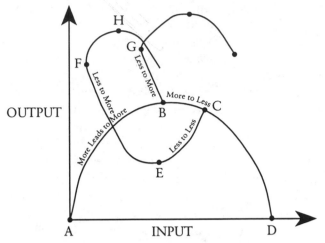

Pattern combining the four basic processes (a)

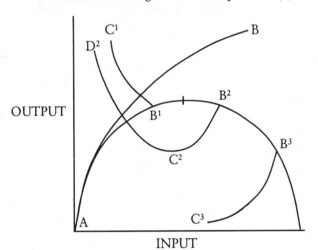

INPUT

Four scenarios for the U.S. economy (b)

FIGURE 8.2
Growth Patterns

Source: Reprinted by permission of publisher from "Slip, Sliding Away: Are We Any Closer to Understanding the Stuff of Which Social Reality is Made?" by I. I. Mitroff, *Technological Forecasting and Social Change*, 36, p.73 (Fig.1) and p.74 (Fig. 2). Copyright 1989 by Elsevier Science Publishing Co., Inc.

1. Continually increasing prosperity without substantial changes or dislocations from the past (curve *AB*): Optimistic—no change in thinking is needed. What succeeded in the past will succeed in the future: "more leads to more."

2. Continued prosperity with substantial early adjustment (curve $AB'C'$): Optimistic—U.S. industries are highly adaptive. Having seen clear signs of the decline of basic industries, a shift to new patterns is begun. Less bureaucratic, smaller, and more autonomous companies are seen as more flexible and effective in global competition. "Less leads to more."
3. Late and slow recovery only after substantial pain (curve A B^2 C^2 D^2): Mildly optimistic—U.S. industries adapt to change only after suffering serious losses in the global economic competition: first "less leads to less," later "less leads to more."
4. Catastrophic decline after severe pain (curve AB^3 C^3): Pessimistic—by the time the need for change is recognized, the economic competition has been lost. The poorly skilled work force and inferior infrastructure make an unattractive setting for corporations that have superior options for their location.

Let us look at growth from another point of view. On a finite earth, growth in all matter and energy use must be limited (see chap. 7). Most of the nation's business establishment has implicitly assumed that exponential growth (fig. 7.2a) will continue indefinitely. Under such assumption, a huge national debt is no cause for concern. Future generations will be so much richer that the debt will prove of little consequence. However, an S-shaped growth curve (fig. 7.2a) totally invalidates such reasoning. A mushrooming debt now inexorably leads to a financially induced national crisis. While a nation cannot be placed in bankruptcy like a private enterprise, the resulting upheaval can certainly endanger the democratic system.

Growth has been measured in recent times by quantities such as gross national product (GNP). GNP was created for an industrial society, and its usefulness as a scale is questioned by some even in that role. For example, it omits public assets, such as the infrastructure, intellectual property, and human capital. It is even less suitable for a pre-industrial (agricultural) or for a knowledge- or information-technology era (see chap. 7). We stress that it would be misleading to call the new period a postindustrial era. There will be strong industrial production in the knowledge society just as there has been strong agricultural production in America's industrial society. But in both cases we experience a shift—doing more with less—that implies the need for a new measure of growth compatible with sustainability and a society that is less consumption-oriented.

In hunting societies, the animal population may have served as a meaningful variable of a society's economic status. In agricultural societies, manpower and arable land were the critical resources and farm production measured economic growth. In industrial societies, skilled manpower, energy, machines, and materials are essential elements. The three kinds of society have often been depicted as a sequence of growth curves like that shown in figure 7.2b. What we are suggesting here is a radical departure, and figure 8.3 makes the point schematically. When faced with a limit to the growth of V_1, say agriculture, humankind has shifted to an entirely new growth curve, industrialization, which requires a different measure (V_2). This curve has a limit also, of course, but we will shift again to a new curve with a new measure. Information is not the same as matter or energy. Therefore, the knowledge society does not have the limits of the industrial society. In other words, *we can still talk about contin-*

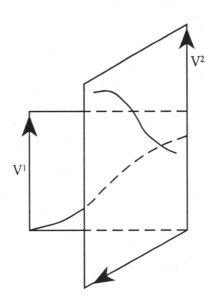

FIGURE 8.3
Three-Dimensional Presentation of Growth Curves

ued growth, provided we define it in a new way. Growth for a child has a strong physical component; growth for a mature adult may have a strong intellectual component. Economic growth in the familiar mode (certainly "dumb capitalism" [B_2]) is not likely to be sustainable in the 21st century, but the new growth definitely will be. How do we measure the growth of a supersymbolic economy? And what will be its limits? How can we grasp the economics of information? Thoughtful answers to these questions may do much to lift the cloud of pessimism that has engulfed many in the developed world.

Forward-looking industry is providing us with some clues, yielding interesting success stories based on "less is more." Consider Quanex, an old (1927) steel company led since 1971 by mechanical engineer Carl E. Pfeiffer.[23] It has taken the efficient mini-steelmill technology and upgraded it to make high-quality steel for specialized applications such as ball bearings and camshafts. Now it is broadening out to become a specialty-metals company and is already supplying $25 million of titanium annually for aircraft engines. The modus operandi is unorthodox: diversified customer base, virtually no inventory, production to order with bar coding of the steel with the customer address while cooling, lean headquarters (down to thirty-two from ninety-four people in the early eighties), worker bonuses based on defect-free steel shipped, and use of profits for growth rather than leveraged buyouts. Its mini-mill efficiency and high profit margins have vaulted Quanex into the *Fortune 500.**

Lockheed provides another case in point. About 450 engineers participated in that company's P-3 Navy aircraft development. However, the technologically far more sophisticated high-altitude U-2 and SR-71 surveillance aircraft produced by Lockheed were designed by a team of only about fifty engineers. The latter aircraft, flying at an altitude of 100,000 feet and at three times the speed of sound, holds the Los Angeles-Washington speed record: sixty-eight minutes. The P-3 was developed by the rules of the procurement process originated in World War II, when 50,000 aircraft had to be built annually. That system was designed to be

---

*Quanex presents a stark contrast to the more typical American steel companies like Homestead Steel Works. A pioneer of the open-hearth steel process, it grew mightily from 1879 to 1950. But size and success brought on inflexibility and lethargy. Austria developed the basic oxygen furnace and substitutes, such as plastics, provided new competition. By 1986, the steel mill closed down.[24]

"idiot-proof" and achieved its purpose. The technically far more advanced surveillance aircraft, as well as the more recent F-117A stealth fighter, were built in Lockheed's "skunk works" (named after the foul-smelling factory in the comic strip *Li'l Abner*).

The organizational concept was the innovative idea of Clarence (Kelly) Johnson, vice president of Lockheed and one of those rare executives who had a vision and a reputation for excellence. A brilliant aeronautical engineer, his greatest talent was his ability to push his team and technology to their limits.[25] He not only had the title of chief engineer but also functioned as such. There was a minimal hierarchy, and Johnson personally selected and supervised the team. Thus mistakes were caught early and corrected promptly. "Analysis paralysis" and budget overruns were avoided. (In France, Dassault's Mirage development team operated similarly.) The result was "less is more."

One important contributing factor was the engineering culture of the Lockheed Corporation as a whole. It was ready to innovate and tolerate the coexistence of advanced and conventional subunits. Equally important was the client, the U.S. Air Force. As the newest of the military services, it was least hamstrung by tradition and most willing to bypass normal project control mechanisms, such as frequent progress reports, briefings, and inspections. Thus there was a fortuitous confluence of circumstances.

It has been demonstrated in many contexts that, as a small organization grows, it often becomes more effective, but when a certain size is exceeded, its effectiveness begins to diminish. The paperwork, meetings, and other bureaucratic activities throttle the substantive work. Admiral Hyman Rickover once proposed to a congressional committee that the effectivenss of the Pentagon could be raised considerably if half the employees were sent home to stay there with pay but without a way to generate memos. The other half would continue working and the result would be a net increase in the Pentagon's cost effectiveness.

IBM was founded before World War I and grew to 407,000 employees in 130 countries by 1986. Long admired as a paragon of the modern corporation, it is now struggling with reorganizations to make it less bureaucratic and more flexible for an information technology era that it was instrumental in creating. As is typical with large, aging organizations, middle management has been particularly resistant to change. Only after its worldwide computer market share declined from more than 50 percent to 35 percent did the impact of the rapid technological advances and effective competition become obvious. IBM recently eliminated 9,500 management jobs, reduced the number of hierarchical levels from ten to seven, and announced plans to eliminate another 25,000 jobs. It is significant that its best-selling product of the last decade was developed by its Boca Raton, Florida, version of Lockheed's "skunk works," that is, entirely removed from its normal slow product development system. Whether the current reorganizations and decentralization efforts will be thorough enough to lead to marketing of new products rapidly and successfully remains to be seen.[26]

It is ironic that, in the 1970s, the U.S. Department of Justice made strenuous efforts under the Sherman Antitrust Act to break up two giant corporations, IBM and American Telephone & Telegraph Corporation (AT&T). In 1982 it gave up on the highly resistant IBM, but AT&T agreed to divest itself of its regional operating companies. A decade later, IBM is in dire straits and AT&T is flourishing. In January 1993, IBM reported a quarterly loss of $5.46 billion, while AT&T had record earnings of $1 billion. Thus, history has proven IBM's victory to be a Pyrrhic one.

The giant Japanese companies are beginning to encounter similar size-related problems. Japan's automobile industry is now finding that an optimum scale of production is one-quarter of that in a mass production operation.[27] There is an increasing effort to reduce company size by underwriting employees as internal entrepreneurs so that they can then leave to create their own small enterprises. This previously noted "silver handshake" clearly implies "less is more."[28]

Southern California Edison Company has learned that "less is more" in a different way. The experience of electric utility planning over the past twenty-five years has taught it a vital lesson: unforeseen events over which the company had no control negated the assumptions of every single business plan. Examples of such events were the 1973 Arab oil embargo, the California Environmental Quality Act, the National Energy Act of 1978, the Iran-Iraq War, the economic slowdown of 1979–82, and the 1983 Washington Public Power Supply System bond default. The recognition dawned that huge new energy projects with long lead times were inappropriate in this uncertain setting. The company changed its strategy drastically to maximize flexibility. Plans for large, long lead-time, capital-intensive resources were replaced by relatively small, modular, short lead-time supply options that can be brought on-line in three years or less.[29]

The situation at General Motors echoes that of IBM. Its success has made it large, conservative, inflexible, and bureaucratic. As with long-established churches or armies, its size inhibited needed strategic change for at least a decade. Although the automaker's U.S. employment dropped from 612,000 in 1978 to 368,000 in 1992, its strategies in the 1980s under Roger Smith, the chairman of the board, and his docile board of directors were deeply flawed. For example, a massive new layer of bureaucracy was created between corporate headquarters and the automaking divisions.[30] Even the giant U.S. airlines—American, United, and Delta—are recognizing that "bigger is better" is not likely to be the preferred strategy for the 1990s.

According to Peter Drucker, a management consultant: "The advantages of smaller size are becoming very great . . . Some businesses will have to be very big, but I see more and more businesses where medium size is much better and where it simply diffuses results and destroys profitability to try to be big."[31]

The bloated size of the U.S. government suggests that "more is less" applies equally to nonprofit organizations.* Factors that have contributed to the awesome growth of the federal government include the population growth, the power centralization trend apparent throughout this century,** and the "lawyerization" of the country after World War II. The last led to an incredible growth of rules and statutes. The federal civilian workforce has attained elephantine proportions: it now exceeds three million. The bureaucratic arteriosclerosis and profusion of superfluous jobs boggle the mind. The Congress (535 members) has a staff of more than 31,000 people. The White House staff has increased from 35 in Theodore Roosevelt's time to 3,366 (including 800 Secret Service and 1,300 military) under Reagan, whose stated aim it was to reduce the size of the federal government. There are researchers who study the shape of pickles and the flow of catsup, elevator operators who operate automatic elevators,

---

*Bureaucratic resistance to change established a kind of record in 1992 when the Vatican finally admitted that Galileo was right 359 years earlier in claiming that the earth revolved around the sun.

**See the discussion on governance later in this chapter.

and job slots without any tasks (these are designed for whistle blowers so that they can be neutralized).

Governmental institutions have become so entrenched that their relevance and legitimacy for the coming century is rarely questioned. The U. S. Assay Commission, formed in 1792 to ascertain that U.S. coins contain as much silver or gold as promised, a chore routinely duplicated by the Bureau of Standards, was finally abolished only in 1980. The Rural Electrification Administration long outlived the electrification of America's farms.

"Bigger is better" becomes "bigger is worse" in the modern metropolis, as well. As a community grows, more job opportunities and more cultural facilities make it more attractive. Up to a certain point, "bigger is better." Beyond that point, the tendency seems to reverse, and "bigger is worse." The quality of life deteriorates as fear of attack, tension due to congestion, and disgust with ugliness and vandalism take hold. The vibrant city becomes an urban jungle.

For a nation that is maturing, quality becomes more important than quantity. We need not more schools but better schools, not a bigger work force but a better-trained one, not more land but a cleaner environment. For the long haul, incentives for negative population growth will appear increasingly attractive as a means to assure a good life for future generations of Americans.

In the past, "more leads to more" or "bigger is better" has governed much of our thinking. In tomorrow's society the pattern of thinking will increasingly shift to "more leads to less" or "smaller is better."

Other Parameters

Productivity is a well-established industrial manufacturing measure of worker output for a given level of input. In an information-technology era, the traditional means of measuring productivity become inappropriate. For example, additional information provided may improve the quality, while its increased speed of availability should raise the timeliness of decision making. The technology will permit greater variety of products and previously unattainable customization of products. How are these benefits to be measured? *Production factors* traditionally encompass management, labor, and capital. In the next decades, knowledge will come to be recognized as an even more important production factor.

In economics, the term *externalities* is hallowed. It defines aspects of a system viewed as outside some arbitrary system boundary and therefore to be omitted from economic consideration. For decades, the waste and effluents from an industrial enterprise were considered externalities in studying the costs and returns on investment in doing business. Someone else, usually the taxpayer, bore the costs of cleaning up the pollution. Economists have long resisted recognition that "everything interacts with everything" as an inconvenient complicating factor in their work.

The same problem afflicts many other disciplines. Engineers assume that system management is not part of system design. Energy producers take energy conservation as an externality. Agricultural technology focuses on increasing food production ("the Green Revolution") but ignores the food distribution problem.

Traditional personal status demarkations no longer apply. In recent times, the individual's lifetime was divided by the society into three very distinct states: childhood (learning), active societal participation (work), and retirement. Both childhood and retirement were brief periods; children started working while still children and the short life span precluded lengthy retirement. In 1900, American teenagers worked to support the family, and life expectancy (at birth) was 47.3 years. In other words, roughly three-fourths of life was filled with responsibility and work. Today that mainstream phase covers only about one-half of life. Young people are in school much longer and old people live in retirement much longer. We expect the segment of the population aged fifty-five and older to increase by 62 percent between now and 2015. Initial planning for the Social Security System assumed that a retiree at age sixty-five would be paid for only a few years of remaining life; the fallacy of that assumption is endangering the future fiscal soundness of the system.

With today's extended youth and longer life, these three divisions are losing their meaning. It is becoming foolish to identify school with the young and retirement with the old. Education is no longer just a preparation for a job and career. It is a lifetime pursuit as an individual changes careers or requires retraining and frequent educational updating, even in the same career. The traditional implication that the individual is part of the mainstream only during the adult working period gives the young and old an intolerable sense of being unneeded. It is a situation that is unhealthy for the individual and unsound for the society.

Similarly, the demarkation among "poor," "middle class," and "wealthy" is increasingly unclear and subject to argument. It is a popular, albeit not very useful, pastime today to debate what income range defines the middle class. Sometimes a range of $20,000 to $50,000 per year has been used (this accounts for 34 million individual and family tax returns, or 35 percent of all tax returns in 1990). An extended range might be $15,000 to $75,000 (53 million tax returns or 55 percent of the total number of returns). In Washington the upper limit of middle-class income is often equated with the salary of members of the House of Representatives, $125,100. Many families earning $80,000 consider themselves middle class, although that income places them in the top 6 percent of American households.[32] It is to be expected that, in the knowledge society, the use of income as a defining measure of "class" will become increasingly questionable.

Traditional scientific discipline boundaries are obsolete. The scientific establishment has evolved very successfully by specialization and compartmentalization. Most members' vision is narrow; they know more and more about less and less. Even in a single field such as mathematics, workers in one subarea often have great difficulty communicating with those in another. They speak different languages and read different journals. They all focus on research that will produce academic papers to help their career advancement. Like Ph.D. dissertations, they are of interest only to a small coterie of fellow academics. Little of the work addresses the critical concerns of humanity in a global village where everything is interacting and problems do not fit such discipline compartmentalization. Little of the work goes beyond theory to implementation. Little ventures beyond the technical perspective. We are dealing in the real world with increasingly complex systems that cannot be managed by separation and compartmentalization, by divorcing theory and praxis, by relying on the technical perspective.

Traditional industry and work force assumptions are obsolete. Old business labels are misleading. The fusion of digitalization of electronic media and fiber-optic cables is blurring the boundaries between computer hardware and software companies, cable companies, TV networks, telephone companies, and consumer electronics manufacturers. Already new linkages are being created to explore the melding of video, cable, computer, shopping services, and entertainment for the home. One example is the recent linking of Toshiba and Time Warner, another is that of Apple Computer, Sharp Electronics, Pacific Bell, Bellcore, and Motorola.

Japan's Toyota and Honda produce automobiles in America, and America's General Motors and Chrysler sell Japanese-built cars. Honda is also building engines in Ohio with American and Japanese parts for car assembly in Canada. Among 1992 cars, the Ford Probe is basically identical to the Mazda MX-6; the Chevrolet Geo Storm is the Isuzu Impulse; the Dodge Stealth is the Mitsubishi 3000GT. The typical "American" car is now an amalgam of components from several countries.

General Motors itself has 800,000 employees in thirty-eight countries. Its new Program for Improvement and Cost Optimization of Suppliers (PICOS) promotes global sourcing, that is, international coordination in purchasing. U.S. suppliers will be able to sell parts to overseas G.M. operations, but they must compete with foreign suppliers selling to G.M. in North America.[33] The Airbus and the Channel Tunnel are European transnational projects. New regional and global linkages between manufacturers and their subcontractors, branch companies, marketing and venture groups are being formed at a quickening pace. At the same time we are seeing corporations evolve smaller in-house cores and use more employees in peripheral modes.[34]

In a global village the work force is global, yet we expect large wage differences for performance of the same work in rich and poor areas. General Motors finds that its Mexican workers produce the same quality of cars as its U.S. workers, but at a fraction of the wages. Institutional resistance to the changes wrought by the population and technology explosions remains strong. Jobs are artificially maintained even as the work force becomes increasingly irrelevant. Instead of constantly training staff for new functions, employers protect obsolete operations. Consider as a case in point the banking business, a prime example of the impact of shifting boundaries. It is grossly overstaffed in an era where:

- Electronic banking is today's state of the art and requires minimum manpower
- The public is putting savings into mutual funds rather than banks
- Functions traditionally performed by banking are now performed by other businesses (e.g., corporations are raising

money by issuing commercial paper bought by other organizations and auto loans are extended to buyers by auto
manufacturers); commercial banks now hold only 30 percent of corporate debt (among nonfinancial firms) and 44
percent of auto loans[35]

Powerful unions and industry interest groups have placed obstacles in the way of badly needed change to insure their survival.
One result is that U.S. government policy protects old, declining
industries, in contrast to Japan, which protects young, rising ones.
Dirty, old industries are permitted to be more polluting than
clean, new ones. Stricter environmental standards are placed on
new technologies than on old ones.[36]

The familiar concept of the infrastructure is no longer adequate. Roads, bridges, and airports may be vital for the movement of people and goods, but the movement of information will
be at least as important in the knowledge society. Global cellular
telephone systems and fiber-optic cable networks exemplify
tomorrow's "communication highways" and must receive priority in a competitive global economy. It is also becoming clear that
traditional means of infrastructure improvement smother functional innovation. The emphasis on refinement of existing infrastructures is based on rigidities in the supply system, product
standardization, and the consequent social contracts oriented to
rights to these products. In air transportation, for example, the
DC-3 essentially molded the entire system structure in the 1930s
and jet aircraft stretched it in the 1960s. Pliable initially, the system becomes hardened and static. Dysfunctions multiply, but
there is no flexibility. New procedures, such as hub creation and
noise amelioration, simply shift rigidities or create new ones.
"Demand management" focuses on low-risk incremental
improvements and resists functional rethinking, experimentation,
and innovation. The bundling of air and ground transportation,
operational in Switzerland but only fitfully attempted in the
United States, would constitute an innovation.[37]

The education system from kindergarten through high school
(K–12), as well as college, is being thrust into total obsolescence
by information technology. Computer power, fiber-optic telecommunications, sophisticated software, and neural network and
artificial intelligence concepts are fusing to make the cumbersome
education system with its entrenched, multilayered bureaucracy

an expensive dinosaur ($400 billion annually). The technology places the emphasis on learning, rather than on education. Lewis Perelman calls the approach "hyperlearning."[38] The school becomes a learning center, with telecomputers linking students to the home, libraries, learning modules, great teachers, and it makes possible participation in individualized learning games and large group projects. Educational software will permit each child to learn at the pace he or she finds most comfortable; networking will allow interaction with peers in the same neighborhood and in other societies anywhere in the world. Thus we have simultaneous decentralization and globalization of learning.

Traditional national governmental regulation of business is no longer effective. Transnational corporations faced with strict regulations in one country simply move their base to another that has loose regulatory mechanisms or where bribery easily avoids their enforcement. They can shift operations among countries to take advantage of the lowest tax rates. Thus it is often more advantageous in terms of federal taxes to operate in the U.S. as a subsidiary of a foreign corporation than as an American one. Havens such as Luxemburg (banking), the Cayman Islands (banking), and Liberia (shipping) are well known. Carnival Cruise Lines is incorporated in Panama and, thanks to U.S. tax laws, paid no U.S. corporate income taxes whatsoever on its $502.5 million profits from 1985 to 1988. At the time the corporate income tax rate for U.S. corporations ranged from 46 percent in 1985 to 34 percent in 1988. Thus the Panama home basing saved Carnival $200 million in taxes.[39]

Today, billions of dollars can be transferred halfway across the globe in seconds. Criminal enterprises operate with great complexity internationally, making their apprehension and prosecution exceedingly difficult. The Colombian drug cartel thrives with relative impunity on several continents. The seizure in 1991 of the global Bank of Credit and Commerce International (BCCI) for massive fraud revealed that it had been operating in more than sixty countries with $20 billion in assets. The case gives a foretaste of the future possibilities. Corruption is one of the few truly global afflictions; we find it rampant from Japan to Germany, from the former Soviet Union to Australia, and from sea to shining sea in America. In this new environment, even companies facing bankruptcy now shop around for the country in which the process is most advantageous for them (for example, in 1991,

Maxwell Communication preferred the U.S. over the U.K. for this step).[40]

Not surprisingly, computer crime has already reached staggering levels, with annual embezzlement losses measured in billions. Experts believe that most computer crimes go totally undetected. As one observer has noted, "Regulators are always behind the curve, always correcting last year's folly while people are finding new ways to beat the system."[41] The confluence of computing power and fiber-optic communications requires total rethinking on the part of the Federal Communications Commission (FCC). Current regulatory policies severely constrain the development of cellular telephone networks as well as fiber-optic systems for computers. U.S. telephone companies are precluded from laying fiber-optic cables to homes and from operating information-based businesses. Their Japanese competitors are under no such restrictions and are thus likely to gain a strong edge in optoelectronics for consumer use (as they did with the VCR a generation ago).[42]

The line between private and public functions is blurring. United Parcel Service, Fax services, and E-mail compete with the U.S. Post Office. Private security services compete with public police, private schools with public schools.

Information technology knows no national boundaries. CNN makes the same information available at the same time in all countries. It is highly significant that the network has banned the use of the word *foreign*, using instead *international*. It has become almost impossible to isolate a nation from information. In August 1991, the coup plotters in the Soviet Union could not even isolate Gorbachev for seventy-two hours from world information sources. By the end of the 21st century, information networks may replace the territorial state as the basic organizing unit of human society.[43]

It is evident that the time has come for challenging and revising many of the obsolete assumptions, including definitions and boundaries, to which we have long been accustomed.

INFORMATION TECHNOLOGY AND GOVERNANCE

The Public Sector

The impacts of the many facets of information technology are complex indeed. They subject us to information overload, and we often fail to see that accumulation of data is not the same as knowledge or insight. Rogers cites Japanese research showing that information supply, measured by the volume of words flowing through the media and available to an individual, increases faster than consumption, the volume of words the individual absorbs by reading or listening.[44] Inevitably, there is less time for thought and analysis. The torrent of raw, undigested information leads to confusion and overreaction. In other words, information becomes as much the problem as the solution.[45]

The passage of time can have a beneficial calming effect. The further an event is in the past (or in the future), the more we tend to downgrade or ignore it.* In earlier times a crisis was often over by the time the information was widely disseminated by word of mouth or print, and thus it did not create strong waves of global reaction. Today the instant visual transmission has an immediate, virtually undamped effect across the world. Often impressions are instantly created that subsequently prove totally false or misleading:

> TV's runaway instant speculative reporting is often outrageously over the top. When stories break, for example, morning TV becomes a vat of steamy, gaseous verbiage, continuous emissions of disposable talk that choke and pollute the airwaves, then vanish. Inevitably, these sprouting wild words of morning become the wilted weeds of afternoon, which by evening's end usually have been supplanted by different words.[46]

The rise of information as the primary source of power naturally places a premium on this asset. It is profoundly altering presidential politics in the United States. First, media experts became key figures as campaigns were transformed into a kind of show business. The birth of media politics can be traced to the 1934 California gubernatorial campaign, which featured muckraking novelist Upton Sinclair, running as a populist on the Democratic ticket against lackluster Republican Frank Merriam. One group of leading Los Angeles businessmen hired an advertising agency and the first campaign management consulting company entered the fray. National fund raising, local press manipulation, and direct mail campaigns were initiated. The movie industry helped to produce anti-Sinclair radio dramas and propaganda shorts. The effort was successful, and Sinclair was soundly defeated.[47]

---

*See the discussion on discounting in chapter 9.

The wordsmiths have become adept at transforming unpleasant words into more satisfying ones. The War Department became the Department of Defense; a depression is now "a life-style downsizing opportunity"; taxes are camouflaged as "user fees."

In the 1980s correspondents and journalists declined in importance as visuals were set up by the campaign managers and distributed directly to the networks.[48] The image of a candidate was fashioned by photo opportunities, thirty-second advertising spots, and one-minute news items on the evening television news programs. The 1988 presidential campaign has been described as "a triumph of the Bush image makers employing modern political technology . . . 'We are running a campaign that is designed for network television,' says Roger Stone, a senior Bush adviser. 'That means only one message a day, and getting it out early enough to get on the networks and major media markets that night. It means not allowing anything unplanned [to happen].'"[49]

The Nixon campaign's suppression of the Watergate scandal until after the president's reelection and both parties' suppression of the magnitude of the savings and loan scandal until after the 1988 election illustrate the focus on the media. It is a far cry from Winston Churchill's broadcast exhortation of his countrymen to fight on the beaches in the early days of World War II, which he began by announcing "the news from France is very bad."

Russell Baker was moved to offer a witty rephrasing of Lincoln's adage, "You can fool some of the people all the time, and all the people some of the time, which is just long enough to be President of the United States."[50]

The 1992 presidential campaign produced further evolution of the role of the media. Outsider candidates Ross Perot and Jerry Brown struck a responsive chord with many voters by blasting campaign professionals as well as special interest lobbies. They scorned handlers and connected directly with the citizenry. They innovated using free (800) telephone numbers to collect pledges for donations and making free appearances as guests on radio and TV talk shows. Perot purchased television time for plain substantive chart talks whose popularity dumbfounded the media political advisers. He also espoused the idea of an "electronic town hall," and Bill Clinton used this concept with startling effectiveness to establish direct dialog with citizens and cut out mediating journalists.

The Economic Summit Conference after the election in December 1992 moved the concept a step further. It brought together more than 329 leaders representing business, labor, academia, and other areas, in two full days of open, televised discussions moderated by the president-elect. There were also telephoned questions from citizens across the country. Here was an example of leadership to focus the public on (1) the long term as well as the short term, and (2) the interrelatedness of the problems facing the society. It was a remarkable learning experience for those who availed themselves of the opportunity to see and hear. Such interactive radio/TV use has a profound potential: building a real citizen-governance connection and diminishing the power of the traditional vested interests.

Reagan was nurtured in the film industry, whose products were designed for a national audience and a one-way flow. Clinton, by contrast, likes to listen to people and enjoys unmediated, interactive communication with them. In Portland, Oregon, Clinton listened to regional and local concerns about timber/forest use in the Northwest; in talks with California reporters, he focused on the military base closings in that state. Citizens can use E-mail (address: "clinton pz") to contact the White House and obtain press releases and documents instantly on computer networks. The shifting balance in media power between national media (White House reporters, national networks) on the one hand and local/global media (E-mail/CNN) on the other is a pattern symptomatic of an information technology era that we shall encounter again (see table 8.4).

In the Soviet Union the media also played a key role in Gorbachev's drive to awaken his people. Movies and television were in the forefront as they drew audiences of 80 million and 150 million, respectively. In Hedrick Smith's view: "If there was one movie that changed the climate and the popular political mind, it was *Repentance*, a powerful surrealist allegory of Stalinist terror . . . [it] became the cinematic flagship of *glasnost*."[51]

In the battle between radical advocates of change and hardliners, documentaries became effective weapons, changing the nature of public dialog through their shock effect. The most sensational object lessons for the Soviet people were the nationally televised sessions of the nineteenth All-Union Conference of the Communist party, which held a truly open debate for the first time in seventy years of communism. As Smith observes, "[it] is a classic lesson in the power of television to change the political dynamics of a society and alter the mindset of a nation . . . that conference . . . opened up the country to debate and ultimately to breaking the power monopoly of the Communist Party."[52]

Similarly, the subsequent televised Supreme Soviet sessions were avidly watched by millions, often until 2:00 A.M., and their open debates presented riveting lessons in the operations of democratic governance. The nationalist movements in the Baltic republics, in Georgia, and elsewhere were practically born on local and regional television shows.[53] We expect the roles of information security, industrial intelligence gathering, disinformation, and intervention in computer programs and communications to be magnified. Economic intelligence may become as much a concern of the federal intelligence community in the next half-century as military intelligence has been in the last.*

Efforts to manipulate neural network and artificial intelligence systems will not lag far behind their introduction. Some governments, such as those of Thailand, Singapore, Guatemala, and Israel, have already begun to use computers to develop national data bases or dossiers that permit "Big Brother" control of their residents.[55] The staggering increase in crime may even make frightened and desperate citizens call for the use of such technology to reestablish personal security. Cybernetic surveillance may bring Aldous Huxley's "Brave New World" closer to reality.

On the positive side, the Freedom of Information Act has proven to be a landmark in countering to some extent the misuse of secrecy. It is also becoming apparent that individual citizens can make use of information technology to check on misuse of government power. An example is the use of the camcorder by a citizen to expose police brutality in the Los Angeles Police beating of Rodney King in 1991.

As the value of information as power rises, the gap between the information-rich and the information-poor grows. The elite consisting of those who have access to information and are able to use it most effectively becomes more sharply defined. This applies not only within the U.S., but also globally. Consider that today one-quarter of the world's adult population is illiterate and three-quarters of the world's telephones are concentrated in only nine countries. The creation of a true global telecommunications

---

*However, we must note that Russia's entry into the information technology era is severely impeded by its backward level in computers. In 1987, when the U.S. had 25 million personal computers, the Soviet Union had about 150,000.[54]

city brings its billions into very close proximity. And the gap between the information technology haves and have-nots can become a cultural chasm. On the other hand, the same technology also has the potential to close such gaps.

Information technology is already able to "amuse us to death" by creating artificial worlds that appear "more real" than reality itself:

> Television . . . renders diaphonous the division between objective and subjective realities. The switch gives you at one moment "hard" news of earthquakes, murders, and assassination attempts, and at the next, the doings of federal agents on other star systems. The blend and flow of fantasy and fact is so swift that consciousness loses its sense of absolute determination of what is real and what is imaginary.

> Television and media simulate personal fantasies . . . There is a growing primacy of secondary realities: styles of clothes and fashions offer fantasy identities, media and advertising flood the mind with dreams so that the technology of secondary reality has become central to marketing, to production, to economic health itself.[56]

In Brazil, where over half the population is chronically hungry, more homes have televisions than have refrigerators. Immensely popular television melodramas, *telenovelas,* totally replace reality in the lives of both rich and poor from 6 P.M. to 9:30 P.M. every evening. The breathtaking corruption scandals that burst on President Fernanco Collor de Mello in 1992 intertwined these melodramas tightly with real life. One miniseries fueled the mass demonstrations of outrage which helped to drive the president from office and another was quickly created from the rich flow of real-life revelations. Thus reality and fantasy were woven into one fabric.[57]

In Paddy Chayefsky's satiric motion picture *Network,* the overwrought TV news anchorman chides his audience: "You people sit there, night after night. You're beginning to believe this illusion we're spinning here. You're beginning to think the tube is reality, and your own lives are unreal. This is mass madness."

The competition for the attention of the consumer forces the producers to make their output as undemanding, emotionally appealing, and entertaining as possible. CNN is eminently successful because of its ability to present news as living theater. It does not merely report events that have happened but is often a participant in making them happen. Ronald Reagan played the role of president to popular acclaim, even as his White House chief-of-staff Donald Regan warned that "a destructive American tendency to trivialize the nation's business [has been carried] very close to the pathological . . . The government . . . cannot design its operations with theatrical effect as the primary objective."[58]

The intimacy of the television setting and the small screen favor video-scale, visualizable, and dramatic events. The shooting of Kennedy assassin Oswald, President Yeltsin speaking on top of a tank in Moscow, the rescue of a child fallen into a well, a few whales caught in an arctic ice pack, a lone protester facing tanks in Tiananmen Square, NASA's Challenger explosion, the quasi-courtroom drama of the Congressional Watergate and Clarence Thomas hearings all provide memorable, real-time images. The oft-repeated video footage of Los Angeles Police batons battering Rodney King fueled nation-wide viewer outrage that, after the verdict of not guilty sparked the 1992 Los Angeles riots (and more vivid images of mindless violence).

The Gulf War was not only a spectacular military success but also a smash hit on television. Before the attack, television served the administration as "our chief tool . . . in selling our policy."[59] The administration wanted to present a clear, black-and-white, hero-villain image. Once the fighting began, the emphasis was on clean, technologically impressive images of pinpoint strikes on bridges and other military targets taken at long range. In the view of Pentagon controllers, the only good war is one where the cameras are far distant from the targets. Images of Scud/Patriot encounters reminded one of Nintendo games. Televised military briefings were highly professional and polished. The administration was very anxious to avoid the impact that televised images of collateral damage might create. In fact, television admittedly influenced the calculation that led to the abrupt halt to the fighting. Of particular concern was the fallout the images of American forces marching to Baghdad would have on Arab viewers.

Small-scale events are also easily magnified and distorted by the TV screen. Organizers of demonstrations are keenly aware that shots of small groups can be made to appear to viewers at home as close-ups of vast crowds. Selective shots of the very limited damage in the recent San Francisco earthquake suggested massive devastation. This characteristic of TV has been exploited effectively in trials by defense lawyers to counter strong, prima facie visual evidence introduced by prosecutors in police brutality cases. Radical minority groups recognize that they can vastly amplify their influence by use of the media. Demagogic radio and TV personalities can instigate a massive volume of audience telephone calls or letters. This form of corrupted participative democracy or "electronic populism" can frighten politicians and generate political power. Anti-abortion demonstrations exemplify this strategy, which enhances the potential to thwart the will of the majority, that is, it permits "tyranny of the minority."

The various facets of information technology have already increased the choices available at any time of day or night on the home screen through videocassette recorders and the profusion of cable channels catering to every taste in spectacular fashion. And this is only the beginning! The result may well be a society hooked on massive doses of unreality and functioning on the basis of trivializing the complex world through entertainment.[60] Researchers are already working toward developing devices to create "virtual reality," which will even make it possible to see through computer-display eyeglasses and wear clothing with built-in sensors to create the illusion that one is in a real world. The transformation of political conventions, terrorist actions, and the Gulf War into television spectacles serves as an augury of this new "unreality world."

Indicative of the pervasive and complex impact of this technology is the fact that, while tying the individual to the world through networks of all kinds, it also effectively promotes isolation. The individual is presented with the choice of operating in a global family, that is, linked in a "virtual community" with colleagues and strangers through globe-spanning networks or in virtual isolation from the world. Through the personal computer and telecommunications, the individual gains independence, working and playing in antiseptic solitude. If the latter implies the absence of social support, it may have serious negative impacts on the individual's health.[61] It is evident that the simultaneous operation of such trends confounds the forecasting process.

The ability of information technology to improve system coordination has profound implications for the structuring of governance. Simultaneous globalization and localization in governance has the potential of restoring effective operation to governments mired in bureaucratic impotence. This possibility should resuscitate debates that occupied America's Founding Fathers and is mirrored in the fascinating Federalist Papers.

James Madison carefully studied the Greek city-states and confederations for "experimental instruction" about the balance between local and federal governments.[62] The precedents of the Amphyctionic Council and the Achaean League of cities were particularly illuminating, showing the value of an effective confederacy in contrast to cities singly exercising all the prerogatives of sovereignty; the German Empire and the Dutch Confederacy also offered insights:

> In the Achaean League, it is probable that the federal head had a degree and species of power, which gave it considerable likeness to the government framed by the convention. The Lycian Confederacy, as far as principles and form are transmitted, must have borne a still greater analogy to it . . . The powers delegated by the proposed constitution to the federal government are few and defined. Those which are to remain in the state governments are numerous and indefinite.[63]

> The general government is not to be charged with the whole power of making and administering laws. Its jurisdiction is limited to certain enumerated objects, which concern all members of the republic, but which are not to be attained by separate provisions of any. The subordinate governments which can extend their care to all those other objects, which can be separately provided for, will retain their due authority and activity.[64]

> The federal Constitution forms a happy combination . . . ; the great and aggregate interests being referred to the national, the local and particular to the state legislatures.[65]

According to its Preamble, the constitution enables the federal government to "establish Justice, insure domestic Tranquillity, provide for the common defence, promote the general Welfare, and secure the Blessings of Liberty." According to the Tenth Amendment, all powers not delegated to the federal government by the Constitution or prohibited by it to the states are reserved to the states or to the people. In the first century of the republic, the federal/state power balance was severely tested by the Civil War. With America soon taking on England's mantle as foremost industrial power, the balance between federal and state power tilted increasingly to the former. In its second century the U.S. became a true world power. The balance between the three branches of government also tilted: the executive branch steadily grew and became dominant. In other words, America was relatively decentralized in its first century and centralized in its second. It is interesting to note that complex system evolution generally appears to proceed by periodic swings between centralization and decentralization (see fig. 12.2).

A challenge for the third century of the country is to create a new "happy combination," balancing global/regional, national, and local governance levels in a way appropriate to the shrunken global village and the coordination-intensive structure made possible by information technology.

How much system decoupling is feasible, while maintaining both national cohesion and effective regional and global linkages? History provides fascinating clues. The vast Roman Empire could operate only with considerable decentralization, even though it had outstanding communications for its time. It was physically impossible to exercise centralized day-to-day control. *The key to effective decentralization was simultaneous effective centralization.* It worked as follows. Great care was taken at the center to appoint a provinvial governor or military commander. He was highly trained in the way the Roman system operated, and Roman policies were deeply ingrained in him. He had to hold high office in Rome before being appointed to a distant post. Thus coordination between the provinces and Rome was assured.

Of the various power centers at the beginning of the modern era (ca. 1500)—Ming China, the Ottoman Empire, Muscovy, Tojugawa Japan, and the cluster of states in west-central Europe—all but Europe "suffered from the consequences of having a centralized authority which insisted upon a uniformity of belief and practice, not only in official state religion, but also in such areas as commercial activities and weapons development."[66] It was decentralized Europe that benefited from a stimulating competitive entrepreneurial environment as well as from the coordination made possible by effective communications and transportation. As a result, it was able to overtake all the other powers.

Today we see forces at work pulling simultaneously in both directions: separation and integration, tribalization and globalization. Tribal and ethnic boundaries conflict with established national boundaries, and there are widespread moves toward decentralization. More than a hundred new nations have been formed since World War II, and the trend continues. Recent pressure points have included the Soviet Union (all republics), United Kingdom (Northern Ireland, Scotland), Spain (Catalonia), Canada (Quebec), Yugoslavia (Slovenia, Croatia, Bosnia-Herzegovina), Israel (West Bank), Ethiopia (Eritrea), Iraq (Kurds), India (Kashmir), and China (Tibet). In Italy, the northern, wealthy "Bavarian" provinces with Milan as their center, see Rome and the "Third World" southern provinces as separate, at best a source of cheap labor. Unmanageably large nation-states, such as China and India, may, in the next century, follow the Soviet Union into dismemberment, making their components more viable in the competitive global village.

New transnational communities are also appearing in Europe and stirring in the Arab world. If the Maastricht Treaty is ratified, Western Europe will be moving toward partial integration of independent nations in a European Community. Certain functions will be the responsibility of the common government and the remainder left to each participating state. The desire to maintain cultural distinctiveness confronts the combined effects of information, communications, transportation, and commerce to break down all barriers. Barber labels these opposing forces "Jihad versus McWorld."[67]

Coordination-intensive structures can help to develop effective new arrangements, confederations that tie semi-autonomous communities smaller than nation-states together into regional economic systems larger than nation-states. The most natural coordination-intensive structures are, in the words of Joel Kotkin, the geographically dispersed "global tribes" that combine strong ethnic self-identity and values with a passion for knowledge and adaptability to technological change. They seem to have a particular affinity for dealing with multiple perspectives, their own and that of the local society, as well as the technical perspective. Examples in today's world are Anglo-Saxons (British/Americans), Japanese, Chinese, Indians, and Jews.* Their cultural bonds or tribal intimacy and rapid transportation/communication make for effective global networks. Nodes are multicultural cities like New

---

*In the century beginning in 1832, the United Kingdom accounted for over one-third of all European emigration. In the case of the Jews, three-fourths lived outside Palestine even at the time of Christ. The growing importance of the 20 million of Overseas Indians and 55 million Overseas Chinese is evident. Singapore, Taiwan, and Hong Kong are striking examples of world class financial power centers created by these Chinese. Potential future global tribes include Middle Easterners (Lebanese and Palestinians), Armenians, and Mormons.

York, Los Angeles, San Francisco, London, Hong Kong, Singapore, and Sydney. These tribes contribute much to cross-fertilization of commerce, technology, and culture. Their networks have had a global impact in areas such as finance (British/Americans, Chinese, Japanese, and Jews) and science and technology (British/Americans, Japanese, Indians [software], and Jews).[68] The relevance of these "global tribes" to one of the scenarios sketched earlier in this chapter, "smart" capitalism (B_1), is evident.

A coordination-intensive network linking the global private and public sectors is being developed by corporations to influence regulations that affect them in their transnational operations. International lobbying is now a soaring industry. In Brussels, for example, more than a thousand lobbyists or "agents of change" were at work in European Community matters in 1992. This is one way to disseminate the results of regulatory experiments and innovations across the globe. In the same year, daily private foreign currency exchange trading reached the level of the worldwide foreign currency reserves held by governments ($1 trillion), demonstrating a new global private/public financial interdependence fashioned by electronic technology.[69]

There are already seventeen common markets or free trade areas; some of the principal ones are shown in Table 8.2. There is also a growing recognition that governance is possible without formal government. Social institutions and informal organizations can be created to deal collectively with specific issues, developing rules of the game and settling conflicts.

There are physical and biological systems that lie outside the jurisdiction of any one national government, such as Antarctica, the oceans, the electromagnetic spectrum, the geostationary orbit, and the global climate system. Such "international commons" may be managed by a world government, extension of national jurisdiction to encompass the system, or restricted common property (such as common economic zones). Joint management regimes are feasible and have been demonstrated in Europe (Rhine River) and Antarctica.[70]

Other environmental problems suitable for new forms of governance include the sharing of resources (such as stocks of fish and oil pools straddling national boundaries) as well as industrial catastrophes (see part 2) with transnational impacts. The concept is an effective answer to Hardin's Tragedy of the Commons (see chap. 6).

TABLE 8.2
The Rise of Regional Economic Linkages

Name		Nations	Population (in millions)
EEC 1957	European Economic Community (incl. Germany, France, Italy, Britain)	12	345
ASEAN 1967	Association of Southeast Asian Nations (incl. Singapore, Malaysia, Thailand)	6	333
Andean Group 1969	(incl. Colombia, Venezuela, Ecuador)	5	92
ECOWAS 1975	Economic Community of West African States (incl. Nigeria, Ghana, Liberia)	16	206
LAIA 1980	Latin American Integration Group (incl. Argentina, Brazil, Chile, Mexico)	11	380
GCC 1981	Gulf Cooperation Council (incl. Saudi Arabia, Kuwait)	6	23
ECO	Economic Cooperation Organization (incl. Iran, Pakistan, Turkey, Azerbaijan)	8	250
NAFTA 1993	North American Free Trade Area (incl. U.S., Canada, Mexico)	3	365

Source: *Los Angeles Times*, Mar. 22, 1992

Issues that have to be addressed include effective administration of such governance systems, use of feedback loops to permit adaptability to change, robustness, and institutional bargaining. Introduction of "good" uncertainty serves as an incentive to maintain bargaining and settle on provisions that seem equitable to all. Thus uncertainty helps to assure the legitimacy of the governance system (see chap. 6). Multiple perspectives are vital in this process.

The individual may simultaneously want to live in a nation-state, enjoy a culturally compatible or ethnic community, work in a professional community, and participate in virtual communities for avocational activities. A living community should be of reasonable size, permitting meaningful personal association, not a megalopolis where the individual is an unidentifiable statistic. Such a community is easily linked to every other community and has access to all the same resources. If the individual feels secure enough, he or she may prefer to live in a nonethnic, multicultural neighborhood and be tied to a particular cultural community by a network. A characteristic of this restructuring is maximum flexibility and coordination; an aim is national unity with diversity.

The further we move into the global knowledge society, not just more, but also more frequent, restructuring is to be expected, and the technology will provide the means. For the United States, possible restructuring for its Third Century suggests functional shifts such as are shown in figure 8.4. Many functions now centralized in Washington can be devolved to the states and local communities. The federal government must keep responsibility for certain parts of the infrastructure, such as air-traffic control and "electronic highways," as well as manage national security and monetary policy. New ways of funding such shifts will be necessary. The deputy director of the federal Office of Management and Budget has proposed that taxes be centrally collected and distributed to the states in proportion to their population.[71]

It is time for a serious reexamination of the Constitution on a scope not undertaken since the days of the Founding Fathers. This document is remarkably flexible, but it was created in a setting of 18th-century technology and world environment. It would be a very appropriate to ask, What modifications are desirable for an America in a global village and networked computer era of the 21st century?

The fifth article of the Constitution states that amendments can be proposed in two ways, by a two-thirds vote of both houses of Congress or by a convention requested by two-thirds of the state legislatures. Ratification requires approval by three-quarters of the state legislatures or by conventions in three-quarters of the states. Why should the citizens not use this power? Thomas Jefferson wanted to hold a Constitutional Convention at least once every generation. As he put it, "no society can make a perpetual constitution, or even a perpetual law. The earth belongs always to the living generation."[72] The longer Americans refuse to deal with the needed coordination-intensive restructuring, the weaker they will be as they enter the competitive world of the new century.

In many ways relentless, uncontrolled growth has made government increasingly dysfunctional. With the 1992 election, a beginning was made in raising the question, How can "quality management" reshape the structure to operate effectively, fully attuned to the information-technology era?

An overhaul and streamlining of U.S. governance is an essential task in adapting the nation to the information-intensive environment of the 21st century. The technology will make "unity with diversity" a realistic objective, but its achievement demands innovative leadership and institutional flexibility.

The Private Sector

The debate between centralization and decentralization is an old one in business. It is reflected in the changes in organization shown in figure 7.3; these changes follow the same cyclic pattern as that of technological innovation clusters. Management experts draw lessons for the private sector from the balance between centralization and decentralization in successful empires in history.[73] We observed the Roman solution in the preceding section. In chapter 7, it was noted that today's information technology can fine-tune the balance. A centralized organization readily becomes more decentralized and vice versa. Indeed, the technology can effect stunning improvements in the corporation: fewer hierarchical levels, a more knowledgable, responsible, and productive work force, as well as more participative management.

New technologies that have altered industrial processes have had a particularly profound influence throughout history. Consider the typical factory of the nineteenth century. It used steam boilers or water wheels as power sources. They were connected to rotating drive shafts that, in turn, powered individual machines through belts. The power source was centralized, and the need for many machines forced construction of multistory buildings to permit connection of all of them to the same single drive shaft. In other words, the power source dictated centralization. With the introduction of electricity, the drive shaft could be electrically driven. However, the improvement in productivity was minimal for some time. Only when it dawned on factory managers that machines could now be powered individually by their own motors did it become clear that totally different, more decentralized factory layouts were feasible and could increase productivity

FIGURE 8.4
U.S. Governance: Illustration of Functional Restructuring in the Information Technology Era

A possible shift in power balance facilitated by the evolving telecommunications, information processing ability, and coordination-intensive networks: *MORE local and global, LESS national.*

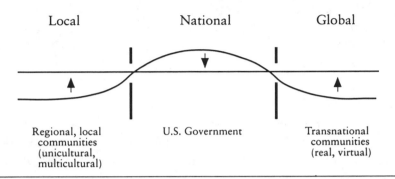

Local	National	Global
Regional, local communities (unicultural, multicultural)	U.S. Government	Transnational communities (real, virtual)
Public and Private Function Providers	*Public Function Providers*	*Public Function Providers*

personal safety
  police, fire protection
 	
infrastructure upgrading
	national security
	  armed forces
	  national crime control
	  criminal justice information network
		global security
		  peace-keeping forces
		  terrorist surveillance network
		  transnational crime network (including drug traffic)

local services
  utilities, transportation, welfare, waste disposal, youth and elderly activities
	national services
	  immigration and border patrol
	  monetary system
	  air traffic control
	  tax collection and monitoring
		global services
		  resource monitoring network
		  conservation policies
		  population monitoring network
		  family planning assistance

learning and skills training
  high-tech learning centers, apprenticeship systems
	education standards monitoring
	  national service program (aids infrastructure revitalization)
		education/training networks
		  (satellite systems, video)

creation of the wired city
  community-home links, networks linking com-
	the wired nation
	  "electronic highways"
	  coordination of subna-
		the wired world
		  global information networks

Public and Private Function Providers

munities, both real and virtual, such as similar cultural centers, common interest groups, learning centers

science/technology cities

health care

environmental protection
pollution control and cleanup
urban water purification

Public Function Providers

tional and transnational networks

economic competitiveness support
R&D incentives and technology transfer

dismantling of part of federal bureaucracy

health care umbrella protection

environmental protection coordination

modernization of Constitution
accountability of corporations
limitations on lobbies
election campaign reform

Public Function Providers

regional data banks
conferencing systems (experts to aid Third World)

regional economic compacts
(European Community, etc.)

global health monitoring

global environment monitoring
pollution control and cleanup

coordination of international projects (space, ocean, etc.)

Private Enterprises

new entrepreneurial arrrangements
R&D consortia
innovation facilitators ("intrapreneurs")
information networks
small manufacturing company networks

common interest virtual network activities

autonomous internal enterprise units in corporations

ethical management

Private Enterprises

crisis management training for industrial and natural catastrophes

Ethical management

Private Enterprises

"global tribes" to promote finance, commerce, technology, and information transfer

regional trade systems

global coalitions for macroengineering projects

alliances among internal enterprise units of different corporations

ethical management

significantly.[74]

Now let us turn to the present. Federal Kemper Life Assurance Company introduced a computer system in 1974, but the assembly-line policy processing operations were carried out as before. Stacks of paper moved from one work station to the next, with the computer simply providing faster access to information at each work station. It was only when the entire assembly-line concept was abandoned and a radical new mode of operation was introduced and productivity jumped that the value of the new technology became evident. Employees were grouped into self-contained, three-person teams, each handling all the functions involved in the issuance of policies.[75] In the same way introduction of robots on the factory floor should be viewed as merely the first step in rethinking and reorganizing the entire manufacturing process.

Aetna Life and Casualty Insurance Company also radically "reengineered" its office operations. Full-service claim offices were reduced from sixty-five to fewer than thirty. A 24-hour hot line replaced a maze of local agents and adjusters, as each claim was assigned to one person for immediate processing from beginning to end.[76] *Information technology is creating the most profound change in corporate organizations since the Industrial Revolution. Whereas the earlier revolution was driven by changes in production, the present one is driven by changes in coordination.*

Consider the impact of an earlier technology: transportation shifting from horse to train and automobile. The first level of impact was technological substitution—the replacement of the horse by railroad and automobile. The second level involved an increase in travel frequency and distance. The third level of impact produced new transportation-intensive social and economic structures such as suburbs and shopping malls. Information technology is having analogous effects in coordination capability. First there is substitution of information technology for human coordination. Data processing systems replace clerks, and middle management becomes superfluous. The second-order impact is an increase in the overall amount of coordination. Automated reservation systems permit fine-tuning of airline fares to maximize profits. The third-order impact is a shift to *coordination-intensive structures.*

An example is Frito-Lay Inc. Its 10,000 route salespeople record all sales of 200 grocery products on hand-held computers as they deliver goods to stores. Each night the information is transmitted to a central computer. Changes in pricing and product promotions are sent back to the hand-held computers for use the next day. The main computer summarizes the centrally stored data each week and combines it with information about competitive brands. The senior executives then have access to this information through an executive information system.[77] This permits rapid response to changes in customer demand and flexibility in strategies. Other companies, such as United Parcel Service and Federal Express, are also using computers to coordinate and track parcel deliveries with remarkable efficiency.

Networks may link companies within an industry and may even create "virtual corporations." These may be of limited duration and reinvent themselves periodically. Small independent companies are linking themselves into "flexible manufacturing networks" with the aid of "network brokers." Each company makes a part of a series of joint products. Research and development, training, quality control, marketing, and accounting services are shared to create cost-effective common capabilities. In 1993, 150 such manufacturing networks were identified in this country.

In the garment industry, sales of sweaters in stores may be linked instantly to all parts of the production chain, thus permitting minimal inventories. Such coordination capability also makes it in many instances more desirable to purchase components rather than make them in-house. The result is less vertical integration and smaller, more flexible corporations. Increasingly, they are also creating "internal markets" with quasi-autonomous "intrapreneurs" and internal quasi-market economy structures. In other words, the functions of external markets are replicated in-house, that is, internalized.[78]

The manufacturing process itself is transformed through information technology. The availability and integration of information on product design, process flows, marketing, reliability, productivity, and costs permits "intelligent manufacturing control" (IMC). Manufacturing processes can be controlled using a distributed, hierarchical computer-based system that has the ability to learn from experience. The computer work station forms the tool that networked peer-to-peer business teams can use to create a highly responsive product realization process (PRP) that evolves into a new style of manufacturing. Complete product images are produced that encompass data, algorithms, and mathematical and multimedia models to capture the various activities composing the product development cycle. These images reside primarily in computers and can be marketed just like the physical products they represent.[79]

Japanese industry recognized the value of coordination-intensive structures early, and they proceeded to implement them even before the potential of computer networking was fully grasped (table 8.3). The coordination that may be said to define Japan Inc. is shown in figure 8.5. The synergism and interdependence, the simultaneous centralization and decentralization, make the whole more than the sum of its parts, and therein lies the core of its success. The loose coupling of the components of the hierarchical system permits each to operate with much autonomy. In systems terms, the total system has "nearly completely decomposable" subsystems.[80] At the same time, information flows smoothly both vertically and horizontally through the entire system. Once the networked computer is fully exploited, the system should reach a much enhanced level of effectiveness. Japan has also pioneered in extending its networking beyond its borders. The lavishly financed and highly successful lobbying campaigns by Japanese companies in our Congress and state legislatures and trade associations illustrate the new trend.

An examination of one industry, data entry and processing, sheds light on another aspect of the simultaneous centralization-decentralization trend: the spatial shifts in job locations. For this U.S. industry, it will be reflected in losses to rural America and gains for large urban areas, both American and offshore.[81] The industry has concentrated its nonroutine, knowledge-intensive, and rapid turn-around services such as development of internal communication networks for business and tax return processing in urban areas. It has turned to rural areas and offshore for routine, labor-intensive, low-skill data-entry tasks such as keying of mailing lists and credit card receipts. Workers in locations such as Barbados, Jamaica, Ireland, and Thailand are replacing the rural U.S. work force due to their lower wage levels. American Airlines has created a subsidiary in Barbados that receives about 900 pounds of used airline tickets daily for sorting and data entry into its computer terminals. Total processing costs including salaries are about $6 per hour lower than they would be in the U.S.

In terms of work locale, information technologies such as product specialization, optical scanners, and point of transaction

TABLE 8.3
Examples of Japanese Information Flow Structures

Horizontal

1. Plant level:
   - Quality circles, job rotation, "kaizen" (see chapter 11)
   - Flow between supervisors and shop floor operators
   - Production development teams, training

2. Industry level:
   - Industry associations ("keidanren"); policy making for industry sector
   - Joint R&D projects, such as the Supercomputer Project and the Integrated Services Digital Network Project
   - Industrial Technology Council; industry structure and policy decisions
   - Corporate groups based on "keiretsu" (financial conglomerates)

3. Technical support level:
   - Japan external trade organization
   - Ministry of International Trade and Industry (MITI)
   - Japanese Information Center for Science and Technology (JICST), industry information centers (such as the center set up by the Japanese Iron and Steel Industry Federation)

Vertical

1. Parent/subsidiary hierarchy: parent company is minority owner of autonomous subsidiary
2. Prime contractor/prime subcontractor/secondary subcontractor hierarchy
3. Banking hierarchy
   - The Bank of Japan
   - Private and government financial institutions
4. Trading firms
   - "Sogoshosha": large multinational trading firms
   - Large "senmon-shosha": national information gathering, consultation
   - Small "senmon-shosha": marketing, regional networks
   - Mini "senmon-shosha": limited regional networks

Source: B. Bowonder, T. Miyake, and H. A. Linstone, "The Japanese Institutional Mechanism for Industrial Growth: An Analysis," *Technological Forecasting and Social Change*, forthcoming, 1994.

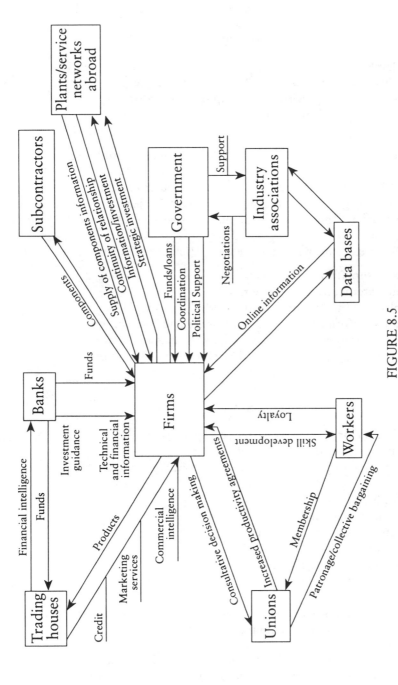

FIGURE 8.5

The Coordination that Characterizes Japan Inc.

Source: Reprinted by permission of the publisher, "The Japanese Institutional Mechanism for Industrial Growth: An Analysis," by B. Bowander, T. Miyake, and H. A. Linstone, in *Technological Forecasting and Social Change*, forthcoming, 1994. Copyright 1994 by Elsevier Science Publishing Co., Inc.

data entry favor *centralization* in urban areas in the developed countries:

- product specialization—increasingly specialized accounting software production, sales, and service for tax accountants who now use their personal computers leads to more intense firm-client linkages
- optical scanners and bar coding—advances will reduce scanning error rates and reduce the need for manual scanning, a current low skill rural area job opportunity
- point of transaction data entry—merchants can enter their sales, doctors their claim forms, and court reporters their legal records directly into computers, facilitating work centralization

Advances in imaging, fiber optics, and satellite transmission favor *decentralization*, specifically offshore movement. Combined with investments in state-of-the-art telecommunications, improved imaging will permit high speed, accurate transmission of pictures over long distances. Processing delays, costs, and turn-around times will be drastically cut. Several Caribbean countries are making the necessary telecommunications investments to gain a major market share. The industry decentralization trend is strengthened by the rising skill levels available offshore. Singapore and Taiwan illustrate highly successful strategies to upgrade offshore work-force skills.

In the U.S., the jobs lost in using the increasingly skilled, low-cost global worker pool and downsizing large enterprises is not balanced by the needs of newly rising high-tech industries. Computer manufacturing is already being automated and protein production by biotechnology techniques needs minimal facilities. Physical-production and physical-service jobs are declining while managerial-administrative and technical-professional jobs are rising.[82] Manufacturing jobs have declined from 22 percent of the work force in 1980 to 17 percent in 1992. Reich talks about one-fifth of America's workers as prospering "symbolic analysts," that is, people whose skills involve manipulation of data, words, and images, while the other four-fifths, comprising routine production and low-level "in-person" service workers, is increasingly struggling.[83] Businesses with "smarter jobs, dumber workers" and with obsolescent forms of organization will not be able to compete globally in the coming decades. The 1993 Clinton economic program, emphasizing intensive manpower training and incentives to small, innovative enterprises, is clearly a step in the right direction.

Finally, we observe the pervasive effect of information technology in rebalancing organizational structures, a thread that has run through our discussion in different guises and is reflected, for example, by the localization/globalization trend (table 8.4).

LESSONS ON TECHNICAL-ORGANIZATIONAL LINKAGES IN THE THIRD WORLD

Vertical Linkages: A Himalayan Example

Two systems analysts, Michael Thompson and Michael Warburton, were commissioned by the United Nations Environmental Program to construct a systemic framework for the severe environmental degradation in the Himalayas (Nepal). Their first problem, very typical for complex systems, was to understand what the problem really was. In the case of the Himalayas there is

TABLE 8.4
Examples of the Localization/Globalization Trend

Localization	Globalization
The media	
local E-mail	global CNN
Languages	
local Provençal and Catalan	global English
Corporations	
decentralizing information services (American Airlines)	global sourcing (General Motors) top-level executive information system (Frito-Lay)
internal markets	global alliances of internal enterprise units
Governance	
decentralization (Soviet Union)	integration (European Community, NAFTA)

remarkable uncertainty in the physical facts (values of variables such as per capita fuelwood consumption) as well as in the causation (interactions among variables, i.e., the system structure). However, there is reasonable certainty about the institutions. Each has its own perception of the world, its own O perspective and its own definition of the problem.

The analysts saw the value of shifting from *product* thinking to *process* thinking, a characteristic difference between the T and O perspectives already noted in chapter 6. In Nepal they realized that they needed an approach that

> relegates to the wings the alarm bellringers and their immaculate prescriptions . . . Grand designs are appropriate only when there is a shared understanding of "the problem" and complete knowledge of the causes of the problem. . . .
>
> [T]he process of institutional development is inherently unplannable. But it does, at certain places and at certain times, offer points of leverage—localized opportunities for facilitating and integrating the development of institutions in the desired direction.[84]

Technical fixes imposed from the top (the provider level with usually much T input) that have no support at the bottom (the O/P-oriented village/individual delivery level) will fail. Similarly, projects that enjoy support at the village level are often strangled at birth because they lack support at the top. The situation facing a proposed project is summarized in table 8.5. Many projects fall into category 3, few into category 4. Thompson and Warburton conclude that top-down development is in the nature of a project (an intervention), bottom-up development is in the nature of a process. The project comes down from the top, but must mesh with the process ascending from below.* They also find that the meshing of top-down and bottom-up requires constructive intervention at the "right" points of leverage; it may be described as *tinkering* in contrast to a grand-design approach. Finding the "right" points is not easy, indeed, it is an art, for they may be few and far apart, invisible to the unsensitized viewer.

---

*In the Soviet Union, Gorbachev recognized that *perestroika* also needed both: "Press on, comrades—we from above, you from below. This is the only way *perestroika* can happen. Just like a vise. If there's pressure from only one side, it won't work.[85] Unfortunately for him, the pressure from below was directed more strongly toward ethnic nationalism than a new Soviet socialist democracy.

TABLE 8.5
Planning Alternatives

Institutional Support: *Match or Mismatch*	*Probable Result*
1. Neither top- nor bottom-level support	Dead loss
2. Bottom-level support only (grass roots enthusiasm but no funds or power to make major changes)	Strangled at birth
3. Top-level support only (top-down planning, creation of new structures)	Difficulties in implementation
4. Top- and bottom-level support	Potential success

Source: M. Thompson and M. Warburton, "Decision Making under Contradictory Certainties," *Journal of Applied Systems Analysis,* vol. 12 (1985), p. 24.

The authors illustrate the point with an anecdote. A motorist has tried in vain to get his car started. He pushes it into a nearby auto repair shop. The mechanic lifts the hood, looks at the engine for a while, and then, selecting a large hammer from his tool chest, gives it a hefty clout. "Try it now," he says to the owner, and it starts immediately. "How much do I owe you?" asks the delighted owner. "Ten dollars," says the mechanic. "Ten dollars," says the owner, his face dropping, "ten dollars for just hitting it with a hammer?" "Oh, no," says the mechanic, "fifty cents for hitting it with the hammer, but nine fifty for knowing where to hit it."

Of course, if the automotive problem is serious enough, for example, an engine or transmission breakdown, the hammer won't do any good. The example shows that a rethinking of linkages may be critical even with implementation of moderate system improvements.

Horizontal Linkages: A Chinese Example

For a long time the Chinese used a centralized top-down approach as did their political mentors, the Russians. However, they now fully recognize the need for meshing top-down and bottom-up approaches. In their efforts to balance the interests of the state with those of local collectives and individuals, they use a "two top-down and one bottom-up" planning method. But in China we encounter a different linkage problem: a striking imbalance between vertical and horizontal linkages.

In an application of multiple perspectives to the technology-society gap in a northwest province, we found that top-down information flow was good, but horizontal flow was quite poor.[86] For example, there are agricultural institutes directly linked to the central government while others belong to the provincial government. There are no direct, horizontal links between them. At lower levels, many tasks are duplicated by agricultural colleges, prefecture, and county government bureaus. Each office or department head presides over a "power pyramid" and hoards information, recognizing that it is useful in maintaining power.* The Chinese newspaper *Liberation Daily* noted "impermeable information barriers" of epic proportions, resulting in enormous redundancy.[87] Techniques such as desktop publishing may ultimately propel a major breakthrough in institutional effectiveness. And linkages with technical experts outside China would have a similar impact.

Both these examples merely hint at how much needs to be, and can be, done to improve coordination. Without wise application and management of information technology, the gaping chasm between the First World and Third World will further expand. There are already signs that the technology is widening the gap between rich and poor within the Third World. The elites buy small electric power generators and cellular phones, while the poor must rely on the failing power grids and telephone systems of the national infrastructure.

This chapter has tried to show how *technology and social structures are intertwined, how change in the former must be accompanied by thorough rethinking of the latter. This linkage is equally strong in the private and the public sectors, locally and globally.* In the following two chapters we shall find that other linkages are similarly critical.

---

*It is intriguing that this ancient Chinese concept is highly relevant to the "knowledge society" of the 21st century (chap. 7), where knowledge will be a source of power as never before.

CHAPTER 9

Linking Organizational and Personal Perspectives

Ask not what your country can do for you; ask what you can do
for your country.

John F. Kennedy, January 20, 1961

Greed is all right . . . I want you to know that. I think greed is
healthy. You can be greedy and still feel good about yourself.

Ivan Boesky, May 18, 1986, to the graduating class,
University of California

What kind of ethical social system takes as its fundamental
precepts the words "I," "me," and "mine"? Our two-year-olds
start like that and we spend the next twenty years trying to teach
them there's more than that to life.

Scott Turow, *Pleading Guilty*

We began this exploration with the technological perspective
(chap. 7) and proceeded to draw in the organizational perspective
(chap. 8). Now we focus on the relation between the organizational perspective and the personal one. In effect, it must deal
with the social contract, the nature of the ties between the individual and the society.

The American Revolution, unlike the French Revolution, was
a middle class revolt. The struggle in France pitted downtrodden,
despairing, and hungry sans-culottes against a callous, despotic
aristocracy. The rebellion in a rapidly developing America was
fueled by taxpayer anger and merchant unhappiness with British
colonial economic policy. The Founding Fathers created, in the
Constitution and Bill of Rights, a sound and flexible basis for a
balance not only between local and national governance (see
chap. 8), but also between the individual and the society. This lat-

ter balance has withstood repeated wars, depressions, and, in its second century, transformation into a melting pot of diverse cultures and a global military superpower. However, entering its third century, the relationship between individual and society appears under increasing tension. We focus on four issues: the widening gaps among members of the society, the balance between individual rights and responsibilities, the tension between democracy and freedom, and the balance between short-term and long-term concerns.

WIDENING GAPS

When statistical averages are used, this phenomenon is easily masked. Polarizations are not detected by describing the "typical American"—the white, married, Protestant, suburban homeowner (family size 3.17), with some German, English, or Irish blood, who owes $2,317 on her credit cards as of Dec. 1991, likes steak and hates liver.[1]

Rich and Poor

The gap between the rich and other Americans has widened enormously. In 1976, the wealthiest 1 percent held 17.6 percent of the national wealth, the lowest share since the 1770s. By 1989 the figure had soared to 36.3 percent, reflecting a concentration of wealth exceeded only during the stock market boom of the Roaring Twenties.[2] In 1989, the top 1 percent of U.S. households (834,000) was worth more than the bottom 90 percent (84 million)! During the Reagan years, the top income tax rate was slashed from 70 percent to less than 30 percent. In the 1980s the salaries of people earning more than $1 million soared 2,184 percent while those of people earning between $20,000 and $50,000 rose only 44 percent, less than the 71 percent inflation rate for the decade.

Today, the chasm between rich and poor is truly stunning. There are two million millionaires, while one-fifth of the under-eighteen population lives in poverty, often without the advantage of the sturdy support provided by a traditional family. In 1991, 35.7 million Americans, or 14 percent of the population, fell below the government's poverty line, exemplified by a four-person family income of $13,924 or less. The U.S. now contains its very own Third World. Nowhere is the stark First World/Third World contrast more obvious than in cities like New York. In Manhattan, multimillion-dollar high-security condominiums with doormen can be seen in the same field of view as the homeless living on the street, and the condominium owners pass by condemned tenements cocooned in their chauffeured stretch limousines. Piled-up trash and walls sprayed with graffiti make some parts of the city look like a war zone. In 1991 America's black teenagers had an unemployment rate of 37 percent. For them prison is no longer a place to be feared but one to be accepted as part of normal life, even a badge of manhood.[3] While the affluent enjoy the benefits of a luxurious style of health care, and the occupancy rate of community hospital beds in 1991 was only 64 percent, more than 30 million Americans have no health insurance, and the poor are often turned away from medical facilities or given minimal care. Health care is rationed de facto by income despite surplus capacity.

We have a society strikingly different from that observed by Alexis de Tocqueville in 1830:

> During my stay in the United States, nothing struck me more forcibly than the general equality of conditions. I readily discovered the prodigious influence which this primary fact exercises on the whole course of society . . . The more I advanced in the study of American society, the more I perceived that the equality of conditions is the fundamental fact from which all others seem to be derived.[4]

Top Management and Workers

The gap between top management and workers has widened. In the period 1973–75 the after-tax pay of chief executive officers of ten major companies averaged 24 times that of the average U.S. manufacturing worker; by 1987–89 that differential had widened to a factor of nearly 160.[5] Overall, from 1977 to 1987, corporate profits remained about the same, the consumer price index rose about 85 percent, average worker wages 80 percent, and CEO salaries rose 320 percent.[6] In 1990, amidst recession, rising layoffs, and falling corporate profits, the wages of America's top executives rose 12 percent to 15 percent, more than the wages of any other group of salaried American workers. Our CEOs today earn almost twice as much as their counterparts in Germany.[7]

The Generation Gap

The generation gap was noted in chapter 8. The pace of technology is also exacerbating the difficulties associated with this gap. Communication between parents and children has become increasingly difficult when input received by the young from school and the media differs radically from that stored in the minds of their elders. Computer literacy and interactive video games create a cultural separation frustrating to many parents.

Another impact of technology is the mobility made possible by transportation and communication technologies. The nuclear family with its physical separation of children, parents, and grandparents has replaced the mutually supportive extended family. Generation gaps are intensified as the sense of family history and continuity is shattered.

The Cultural "Salad Bowl"

The United States, a country known for most of this century as a "melting pot," is in danger of becoming a "salad bowl" or Tower of Babel, a collection of disparate societies. During the 1980s, 8.6 million immigrants arrived, almost as many as in the previous record decade 1900–10. In terms of ethnic origin, the largest increases were a 135 percent rise in the Vietnamese, 126 percent in the Asian Indian, 125 percent in the Korean, and 104 percent in the Chinese. In 1990, the largest immigrant population in the U.S. by a wide margin was Mexican, comprising 13.5 million.

Unlike the earlier waves of immigrants, much of today's vast body of Hispanic immigrants, as well as the Muslim and Asian influx, remains completely unassimilated. The ease of making illegal border crossings, low international air fares, and direct international telephone dialing help to avoid the severance of social ties with the homeland that encouraged earlier immigrant masses to eagerly seek assimilation into the mainstream.[8] Long distance telephone company revenues from international calling are expected to nearly double between 1991 and 1996. Major marketing campaigns are targeting Asian immigrants.[9]

The proportion of the population speaking a language other than English at home rose in the 1980s from 6.6 percent to 13.8 percent, or nearly 32 million people.[10] There is insistence, often strident, on bilingual education and rejection of homogenization. Ethnic groups are increasingly moving to elect one of their own to represent them. As noted in chapter 1, three minorities will jointly form the majority in the America of 2030.

The Result

The combined effect of these widening gaps is a fragmentation of society, a loss of a sense of community. As cohesion fades, the solid historical core of middle America feels increasingly battered and disarrayed. One is reminded of the words of poet William Butler Yeats in his poem "The Second Coming":

> Things fall apart; the center cannot hold;
> Mere anarchy is loosed upon the world;
> The blood-dimmed tide is loosed, and everywhere
> The ceremony of innocence is drowned;
> The best lack all conviction, while the worst
> Are full of passionate intensity.

Can these trends be reversed, the fragments fused once again to form a unique and productive American amalgam with the whole greater than any of its parts? Or are they symptomatic of a new, more globalized *and* decentralized society as discussed in chapter 8?

The Individual versus the Society

The widening gaps among subgroups of the society do not constitute the only reason for societal fragmentation. Since World War II there has been a subtle, but profound, change in the relation between individuals and their societal institutions, such as governments and corporations. It is closely linked to the uncontrolled growth of the number of lawyers in America—the U.S. is now a litigious, adversarial society like no other. It has 281 lawyers per 100,000 people; Great Britain, which operates under the same common-law system, has only 82, and Japan has 11.[11]

The subtle change to which we refer is the shift of *responsibility* from the individual to the system and a shift of *rights* from the system to the individual. The 1961 plea of John F. Kennedy in his inaugural address, quoted at the head of this chapter, has been largely ignored for the past thirty years. The Peace Corps has been a notable exception. The same plea as that of Kennedy could be made if *state, city,* or *neighborhood* were substituted for the word *country.*

The individual's shirking of responsibility is illustrated by the amount of delinquent child-support payments owed by fathers today, an estimated $16 billion. Easy divorce and unwed mothers, as well as absentee fathers, have seriously weakened the family structure. The self-centered desires of today's men and women increasingly breed tragic results for their children.

Elsewhere, the individual's concern with rights or entitlements rather than duties or responsibilities is reflected in the power of the thousands of lobbies—labor, corporate, and single-issue—that constitute de facto a hidden branch of government in Washington. This power is exemplified not only by legislative and regulatory obsequiousness to industry associations (e.g., agriculture, automobile, petroleum, savings and loan) but also by the intimidating pressure the National Rifle Association and the anti-abortionists can exert on elected officials.

The typical voter wants his or her senator and congressperson to gain special benefits for the home district, no matter how unsuitable for the nation, but views every other elected official as a scoundrel who wastes the taxpayer's money. An example of the reasoning: cut the defense budget, but keep the military base in my district open, even though the Department of Defense insists it does not want it. A member of Congress who places the national welfare above the short-term interests of the constituents is endangering his or her own short-term goal of reelection.

Single-issue politics is particularly destructive because it distorts the democratic process and places the will of the special interest group above the common welfare. Individuals prevent society from action through endless litigation, appeals, and the exercise of other "rights." One need only look at prison "death rows" to verify the effectiveness of such delay procedures. In fifteen years only 121 of the 2400 people sentenced to death have been executed.[12]

"I am not responsible" has become the operative maxim for many. If a long-time cigarette smoker dies of lung cancer, his or her family sues the tobacco company. If a child does poorly in school, the parent blames the school. Fired incompetent employees sue employers, prisoners sue jailers, and patients sue doctors. Suits against doctors and hospitals have increased 300-fold since the 1970s, steeply driving up doctors' malpractice insurance costs. "Ghost riders" board buses after seeing them involved in minor accidents to be able to file personal injury claims.[13] Companies need large legal staffs to protect themselves against legal harassment, and civil liberty advocates thrive on ever more arcane and convoluted argumentation in defense of the most outrageous bigots.

Personal sacrifice for the common good has become decidedly unfashionable. There is now a weak personal commitment to collective needs such as high quality education and health care for all, as well as other improvements of the infrastructure. When a bow is made to collective needs, it is likely to be transitory. The brief Gulf War was popular, the long Vietnam War was not. The same passions that were unleashed by the Gulf War have seemed much harder, if not impossible, to arouse with regard to the tremendous social problems Americans face. The quasi-fanaticism that galvanizes the personal issues represented by the gun lobby or the anti-abortionists has been singularly lacking when it comes to education, the federal budget deficit, widening societal gaps, or massive corruption. Antisocial attitudes thrived during the 1980s: America's prison population doubled and now, with more than one million people in prison, the U.S. leads the world in incarceration.

Most seem to have forgotten the old business axiom, "there is no such thing as a free lunch." Reaganomics assiduously cultivated the fantasy that the collective treasury can be filled by cutting individual taxes without significantly reducing services to the individual.* It seems at times that the American Revolution's slogan of "no taxation without representation" has been transformed into "representation without taxation." The federal government is spending more annually on each person for benefits, services, and protection than it collects from that person in taxes. No wonder so many Americans have faith in "the free lunch"!

---

*A particular favorite of the Reagan administration was the Laffer curve. John Kenneth Galbraith—that rare type of person, a witty economist—writes:

> This economic formulation . . . held that when no taxes are levied, no revenue accrues to the government. An undoubted truth. And if taxes are so high that they absorb all income, nothing can be collected from the distraught, starving and otherwise nonfunctional citizenry. Also almost certainly true. Between these two points a freehand curve, engagingly unsupported by evidence, showed the point where higher taxes would mean less revenue. According to accepted legend, the original curve had been drawn on a paper napkin, possibly toilet paper, and some critics of deficient imagination held that the paper could have been better put to its intended use.[14]

Half of the entire federal budget now is allocated to entitlements, with programs benefitting primarily the elderly population (social security and medicare) alone accounting for over 30 percent.[15] The affluent elderly constitute a potent political force. Consider the well-intentioned efforts some years ago to provide senior citizens under Medicare with catastrophic health insurance. This was designed to prevent their impoverishment in the event of a severe medical crisis entailing huge medical bills beyond those covered by Medicare. The cost would have been $4 per month and, for the more affluent pensioners, a 15 percent income tax surcharge with a cap of $800 per year. The proposal was passed in Congress but raised great concern on the part of the beneficiaries—it was, after all, not quite a free lunch. The fear on the part of the elderly was fed by questionable fund-raising advocacy groups, such as the Taxpayers Education Lobby and the Seniors Coalition Against the Tax. Through massive mailings, which implied that all retirees would have to pay sizeable new taxes, they gathered millions of dollars in small donations.[16] The merchants of fright were effective: television news showed Dan Rostenkowski, Chairman of the House Ways and Means Committee having his car nearly tipped over in his district by angry elderly people.[17] Congress, mindful of the voting power of senior citizens, quickly reversed itself and killed the program.

The U.S. is the lowest taxed among affluent nations! Of the twenty-four wealthiest nations, only Turkey collected a lower percentage of gross domestic product in taxes in 1989 than did the U.S.[18] This country has been described as a debtor nation that lives like a creditor; Japan, as a creditor nation that lives like a debtor. The credit-card principle embodies "pleasure now, pain later." For example, this land of instant gratification offers its people houses and cars in return for much lower down payment than do countries like Germany. A return to fiscal responsibility implies an unpopular alternative: "pain now, pleasure later."

As noted in chapter 8, the unstated assumption justifying the rising national debt has been that U.S. economic growth will be so strong in the coming decades that today's enormous debt will pale into insignificance. With this approach, once correctly termed "voodoo economics" by George Bush, politicians have had no trouble getting votes. The Reagan era offered the ultimate extension of the "buy now, pay later" philosophy.

In the 1980s individualism was carried to new hedonistic

heights. Bloom reports that egocentrism appeared to be the norm of the young generation: "Their primary preoccupation is themselves, understood in the narrowest sense . . . There seems to be no reason to be a conscious part of civic existence . . . America is experienced not as a common project but as a framework within which people are only individuals, where they are left alone."[19] *This statement goes far in explaining the rise of crime, drugs, gangs, and general decay of American communities.* Even the older generation, the senior citizens, now reflect this trend: many refuse to vote for school bonds in their localities because they have no children of school age.

Plato said of the Athenians: "Love of wealth wholly absorbs men and never for a moment allows them to think of anything but their own private possessions."[20] These words were echoed in the American 1980s as money came to be considered an end in itself: the commandment was "enrich thyself." Ethical qualms evaporated while greed, ostentatiousness, and excess were "in." Where once Americans admired authentic heroes, such as Washington, Jefferson, and Lincoln, for their accomplishments and deep values, the eighties produced dubious media stars such as Malcolm Forbes, Michael Milken, and Donald Trump. Grave damage to the country was inflicted by venal savings and loan and bank executives, corrupt Wall Street traders, and taxpayer-subsidized entrepreneurs, as well as by regulators and politicians bought by domestic and foreign corporations. For example, the failures of savings and loan institutions have cost the taxpayers about $200 billion. Michael Milken developed and controlled a junk bond market that bore a strong resemblance to the classic Ponzi scheme.*

Executive salaries and bonuses have come to bear little relation to corporate success. American managers of troubled corporations average nearly three times the salary of the heads of healthy Japanese ones. For example, Roger Smith, the CEO of General Motors in the 1980s, made more than four times the salary of his counterparts at Toyota and Nissan. In 1980 GM held 46 percent of the U.S. car market. By the time Smith retired in 1990, the share had dropped to 36 percent. His retirement pay is $1.2 million annually, twice the $550,000 salary of Honda's cur-

---

*Apparently crime still pays. A court-endorsed February 1992 settlement of more than 150 law suits appears to leave Michael Milken with a family fortune of nearly a half-billion dollars after he completed serving a forty-month sentence at a minimum-security "country-club" jail for his conviction for massive fraud.[21]

rent CEO, Nobuhiko Kawamoto.[22] Top executives usually receive large stock options in addition to their salaries and other fringe benefits. If these options were to be counted as costs, they would totally wipe out the 1991 profits shown by companies such as Eastman Kodak and Federal Express.[23]

The case of Ross Johnson, the chief executive officer of RJR Nabisco, offers a particularly egregious example of the 1980s attitudes. After he assumed office in 1986, he moved the headquarters from Winston-Salem, where the tobacco company* was created by R. J. Reynolds in 1874, to Atlanta, which he found less dull. He formed a corporate "air force" of ten jets and thirty-six pilots, providing them with airport facilities that included a building with Italian marble floors, a three-story atrium, $600,000 in new furniture, $100,000 in objets d'art, and $250,000 landscaping. He used the jets to fly his German shepherd from Palm Springs to Winston-Salem and to transport sporting friends and politicians. One baseball star was ensconced for a year in a company apartment, complete with a company car, and received a personal-services fee of $400,000 annually. A professional golf star was allocated a $1 million annual fee for a half-dozen appearances at social events.[24]

The tax laws make the interest paid on corporate debt fully deductible, whereas stock dividends paid on earnings are not deductible. Wall Street and top corporate management suddenly realized that there was a whole new way to make money quickly by leveraged buyouts, that is, buying companies with borrowed funds, slashing expenses, and selling their assets at a huge profit.

The stampede began in 1981 with the purchase of Conoco, the ninth largest U.S. oil company, by du Pont for $7.8 billion. Before the bidding for Conoco began, its stock price was below $50 a share; the bidding raised the price to $98. Arbitrageurs such as Ivan Boesky buy stock on early knowledge of a potential takeover, then sell the stock when its price has escalated. In this single case, Boesky made a profit of $40 million.[25] The enormous corporate borrowings of the 1980s—$1.3 trillion in new debt— being fully deductible, resulted in more corporate taxes lost to the government ($92.2 billion) than paid ($67.5 billion). The share of

---

*The tobacco companies also have disturbing ethical problems as corporate citizens. Through their Council for Tobacco Research (*sic*) and Tobacco Institute, they have effectively fought for a generation to discount scientific evidence of the danger of smoking.

U. S. taxes paid by corporations has shrunk from 39 percent in the 1950s to 17 percent in the 1980s, while the individual tax-payer share has risen correspondingly.[26]

While leveraged buyouts in the eighties provided fabulous short-term windfall profits for key executives, lawyers, and arbitrageurs, they proved ruinous to thousands of employees and extremely harmful, if not destructive, to the long-term health of many companies. The huge interest payments forced drastic cutbacks in plant modernization as well as in research and development. Not surprisingly, we now find that Japanese industry has invested substantially more in nonmilitary research and development than have American businesses.

Of the thirty largest bankruptcies in U.S. history, all but five occurred since 1987. Just as federal tax laws have proven highly advantageous for leveraged buyouts, new laws make bankruptcy highly lucrative for the scavenger specialists. They "reorganize" bankrupt companies, often liquidating them while pocketing huge fees. A case in point is that of S. Sigoloff, who "restructured" L. J. Hooker Corporation (which owned Altman, Bonwit Teller and other retail stores). He cut the number of employees from 12,000 to sixty-two, earning $6.5 million in fees for the first year of his stewardship.[27]

Even nonprofit charity and religious organizations played the game. United Way of America paid its longtime president $463,000 annually in salary and benefits and generously subsidized other amenities.[28] The president of the National Geographic Society received $344,000 and the chief executive officers of twenty charitable foundations were paid more than $300,000 in fiscal year 1991.[29] Evangelist Jim Bakker used his large flock's church donations to finance a luxurious lifestyle. It included enormous salaries, chandelier-lit closets that were fifty feet long, and air conditioned dog houses.

The odor of corruption in government was matched in pungency only by the Harding presidency in the 1920s. Indeed one begins to wonder if the long-wave cycles (fig. 7.3) also apply to this domain. According to a *Time* survey, more than a hundred Reagan administration officials have faced allegations of questionable activities, including favorable treatment of corporations in return for stock or money and personal use of government funds, property, and services. For example, Melvyn Paisley, assistant secretary of the Navy, was the center of defense-procurement scams during and after his service in the Pentagon.[30] The military-industrial complex, awash in billions of dollars, has long been particularly vulnerable to overcharging, test rigging, and other forms of creative bookkeeping.[31] The quixotic crusades of New York police officers David Durk and Frank Serpico to expose pervasive police corruption and its toleration at high levels illuminate another facet of the plague.

The United States is, of course, by no means unique in this regard; one need only look at the 1993 Italian governmental corruption scandals. Indeed, history suggests that such democracies are following in the footsteps of the first democracy. Historian Will Durant observes that, in Greece's Golden Age, "there is hardly a man in Athenian public life that is not charged with crookedness; an honest man . . . is considered exciting news."[32] Incidentally, Athenians also loved to litigate.

Egocentrism means a focus on "me" and "now." It goes hand-in-hand with a disinterest in the future or the welfare of others. It knows no socioeconomic limits, infecting both billionaires and those at the low-income level. Wealthy executives have taxpayers subsidize their meals with wives and girl friends as "business" expenses. Doctors and medical laboratories collect millions for unnecessary procedures and tests. Worker's Compensation and Welfare programs are flooded with fraudulent claims; "stress" and problems unconnected with work are freely blamed on the workplace. Fraudulent "petition mills" help tenants delay evictions and withhold rent, costing landlords (and renters) thousands of dollars. Egocentrism also knows no age limits. Many elderly have no compunction about depleting the social security reserves available for the next generation; many young long ago abdicated any personal responsibility for the welfare of their parents. There is even talk of a "generational war" to divide up the "free lunch."

American citizenship is viewed solely in terms of benefits for the self rather than as a privilege with responsibilities.* Some cities in the U.S. have now extended voting rights to noncitizens. Illegal immigrants are legalized, and the status of English as the language binding the society together is threatened. Thus U.S. cit-

---

*The new citizenship test for immigrants asks: "Name one benefit of being a citizen of the United States." The three possible "correct" answers: to obtain a federal job, to travel with a U.S. passport, and to petition for close relatives to come to the U.S. to live.[33]

izenship has been devalued. In chapter 8 we suggested the rising importance of local, regional, and global entities. The devaluation of the nation-state may well be symptomatic of its relative decline as the dominant societal entity to which the individual feels a strong attachment.

Physical security has always been a foremost right demanded by the individual from the society. As egocentrism has abetted civic disintegration, personal physical security has also declined. With more weapons in private hands than there are in the military forces and cities becoming battle zones, physical security can hardly be taken for granted any more. The year 1991 saw one murder every twenty-one minutes; in Los Angeles alone, there were 2,355 bank robberies. According to the Federal Bureau of Investigation, the annual violent crime rate per 100,000 population rose from 597 in 1980 to 758 in 1991.

Job security has declined, and the impact of the technology and population explosions is increasingly felt. Just as technology has enabled 2 percent of the American labor force today to out-produce the 50 percent of the labor force engaged in agriculture a century ago, so the nature of future technology, combined with the global labor market, instills fear in American workers. A "social health" index, which integrates sixteen measures of over-all social well-being, including unemployment, poverty, homicide, drug and child abuse, shows a 44 percent decline from 1970 to 1990.[34]

Participation of the individual in a society must be based on mutual benefits. The individual seeks personal safety, material welfare, and the freedom and opportunity for self-fulfillment in return for contributing his or her labor (and sometimes life itself) for the common good. *We have observed growing signs that the individual has failed society and that society has failed the individual.*

When individuals feel that the society has abandoned them and they see their personal security being eroded, they become willing to trade their freedom for that security. During the Great Depression, Germany's desperately hard-pressed middle class willingly surrendered its rights for prospects of economic improvement and security. The result was an authoritarian and repressive society. The foremost responsibility of the individual was full support and obedience to leaders (regardless of personal ethics). Millions were slaughtered by obedient Germans who claimed they were "just following orders."*

In the U.S., the popularity with young people of the Rash-neesh, the Moonies, and ultra-orthodox Hassidic groups attests to the attraction of sects that absolve individuals of responsibility for making decisions in return for providing them with a strong communal structure (see chap. 8). Gangs offer another kind of "community," albeit a rather lethal one—there were more than 700 gang murders in Los Angeles in 1991 alone.

In the Soviet Union, the supremacy of the collective over the individual undoubtedly contributed to the collapse of the system. Its poet Vladimir Mayakovsky celebrated the ideal, *edinitsa nul'*, translatable as "the single unit is zero":

> The individual: who needs him? . . .
> The Party is the all-encompassing hurricane . . .
> The individual is nonsense,
> The individual is nothing.[35]

The Third Reich lasted only twelve years, not long enough for old habits of self-reliance to be forgotten. A good balance between the individual and the society was quickly reestablished after World War II. But in Russia the system lasted more than seventy years, and individuals now free to take risks for the first time in their lives find the challenge quite uncomfortable. Freedom means freedom to fail as well as to succeed, and the accompanying loss of personal security frightens many Russians. Not surprisingly, it provides fertile soil for demagogues.

It is a common forecasting fallacy to assume recent trends will continue. Indeed, the 1990s may see the beginning of a reversal of the "me-now" trend in the United States. The pendulum may be starting to swing back toward a better balance between individualism and societal cohesion. One recent survey suggests an

---

*Germans have a long tradition of surrendering individual rights for the perceived common good through societal regulation (*Ordnungsgesetze*). Rules still forbid residential noise above twenty-five decibels from 1:00 P.M. to 3:00 P.M.and from 10:00 P.M. to 6:00 A.M. (e.g., loud music and lawn mowing). The *Laden-schlussgesetz* places strict constraints on retail shopping hours: 8:00 A.M. to 6:30 P.M. except Thursdays to 8:30 P.M., no shopping from 2:00 P.M. Saturday until 8:00 A.M. Monday (exempt are bakeries and flower shops, also in recent years specified Saturday openings).

encouraging rise in voluntarism.[36] The 1992 groundswell of support for Ross Perot and the defeat of Bush by Clinton suggest widespread individual awareness of the need to redress the balance. Clinton's program, emphasizing reduction of the deficit, infrastructure improvement, education, and retraining, clearly showed a longer planning horizon. As the new president stressed, "the test of this plan cannot be what is in it for me. It has got to be what is in it for us."[37] Initial popular reaction to this call for shared sacrifice and more of a "we-future" attitude was encouragingly positive; polls indicated a 60 percent approval level.[38]

If the balance between rights and responsibilities, between individual and societal perspectives, is not restored, the 21st century will see in America a society playing out the "Tragedy of the Commons" (chap. 6). Individuals striving to maximize their personal gain will have precipitated their collective ruin.

With the 1992 Los Angeles riots a vivid memory, one is reminded of the sickness of social life in the late days of the Roman Republic:

> The rule became "everyone for himself" . . . the emphasis was increasingly put upon private wants, private ambitions, private possessions, personal enjoyment, and ease of life, on things to divide instead of unite one man with another . . . They set the poor against the rich, for in an age when all the emphasis is upon wealth, great is the frustration of those forced to remain poor.[39]

It is instructive to recall lessons from even earlier times. In the sixth century B.C. the Athenian reformer and lawgiver Solon enacted a law that any citizen who did not participate in "politics," the affairs of the *polis* (city) in times of crisis lost his citizenship. The great leader Pericles used the term *idiotes* not to denote mental incapacity, but as a disparaging label for citizens who took no part in public affairs. The Athenian community's business was considered to be every citizen's business.[40]

DEMOCRACY VERSUS FREEDOM?

Athenian democracy flourished because the moderation of leaders like Pericles minimized the war between the social classes. He wished not to destroy the rich but to preserve them by easing the condition of the poor. Every Athenian was at heart an individualist and loved private property. A balance was attained between individualism and moderate regulation of business and wealth. Wealth spread sufficiently so that private property could remain secure. The number of citizens with a comfortable income doubled during the years 480–431 B.C. Public revenue and public expenditures grew, yet the treasury was fuller than at any time in Greek history. Thus the Golden Age came about.[41]

It should serve as a warning to us that Athenian democracy suffered when the balance between the individual and the community was seriously impaired. Greed and the drive for power on the part of the *demos* (citizens) led to the defeat of Athens by Sparta. Democracy and individual freedom are related in complex ways; they may be mutually reinforcing as well as mutually counteracting. The organizational and personal perspectives illuminate the connections.

The Population Explosion as Threat

The biological sex drive has successfully overcome all attempts at demographic rationality. There seems to be no correlation between ability to nurture and the number of births: the poor deepen their poverty by creating the most children. In much of the Third World, including its U.S. branch, "the right to life" often means "the right to death," as poor children face a bleak future of miserable poverty, street life, gangs, drugs, and disease.*

Unrestrained population growth is certainly an expression of freedom, but it means increased crowding and economic strains. The consequences are hardly conducive to the flowering of democracy. In a democracy freedom is circumscribed by rules. There is not the freedom to disregard traffic signs and laws that constrain the movement of each driver for the good of the community. The more cars are on the road, the more are restrictive rules necessary to prevent gridlock. While initially the car offers a great gain in freedom of movement, continually increasing traffic subsequently throttles that freedom of movement. Does not the

---

*By 2000, Asia (including Oceania, but not Japan, Australia, or New Zealand) will have 3.6 billion people and 50 million HIV infected cases. Africa will have 814 million people, with 14 million adults and 4 million children infected with HIV. AIDS is a true family disease in Africa, with roughly equal numbers of male and female victims. Throughout South Asia and much of Africa, 65 percent to 80 percent of women do not receive adequate prenatal care.[42]

same concept apply more generally? The United States has doubled its population since the 1930s, drastically reducing the amount of space per person. Using the analogy, *the more crowded the society, the more rules are needed and the more challenging is the task of maintaining a true democracy. One way to confound the analogy is to argue that human beings can alter their ethics to match the needs of a crowded society. A change in human behavior obviates the need for heavier regulation. Greater self-restraint makes it possible to maintain freedom. However, the evidence to date hardly justifies reliance on greater self-restraint.*

The more elements a system has, the more complex it is (see chap. 12). The more complex it is, the more difficult is democratic governance. We see that many cities, formerly reasonably well governed, have reached a size where they are in perpetual crisis and a state of near-anarchy.

A View from the Organizational Perspective Migration from Eastern Europe and the Third World represents a natural release of the steepening pressure gradient between the exploding poor population and the nongrowing affluent population. Although many immigrants make a significant contribution, they also place a heavy burden on the local community for services, such as health care and education. If the flow continues unabated, the affluent areas may become, at least in the short-term, less affluent. In view of the limited perspective of most people, the mounting economic burden borne by the "haves" will create a rising wave of resentment against the "have nots." Aggressive antiforeigner feeling is growing in most Western European countries (for example, Austria, France, Germany, Belgium, and Russia). It is also increasingly noticeable in the United States. One may expect pressure to halt the flow until those already moved can be adequately absorbed. Balancing of immigration and emigration would ease the strain. Currently one million illegal immigrants cross the Mexican border annually, while total annual emigration from the U.S. is approximately 200,000 (chap. 1).

In contrast, Japan has avoided this problem by development of a self-centered, cohesive island culture, strict immigration controls, non-absorption of foreigners, and low birth rate.

A View from the Personal Perspective Unrestrained reproduction is taken to be one of the few inalienable rights of the individual and, amid the many "don'ts," is one of the few "do's" encoded in the religious gospel. Many individuals burden the society by reckless breeding. Some women on welfare have one child after another. Many of them live in abject poverty and become addicts (alcohol, drugs) who, as wards of the society, exact a further toll from the community.

By far the potentially most effective personal approach in a free society is a change in individual attitudes.* Lowering the fertility rate from 2.1 to 1.5 children per woman, combined with balanced emigration/ immigration, would go a long way to create an easier-to-manage, sustainable U.S. economy. A greater sense of personal responsibility is vital, and it can be achieved with the aid of the Three E's: ethics, education, and economic gain.

It may seem puzzling that the highly personal question of abortion is seen as a valid political issue, whereas that of the population bomb is not. From the P perspective, abortion is clearly a short-term, personal issue, whereas population growth is perceived as a long-term societal issue. Thus population raises an issue not only of the individual versus the society but also of short- versus long-term concern. We shall soon return to this point.

The Extra-constitutional Corporation as Threat

Corporations are "persons" under the U.S. Constitution. They have rights but are not recognized as parts of the society that wield considerable power. As we shall see, this power is exercised through lobbies to assure a favorable operating environment. The power is also seen in their position in communities and their transnational modus operandi. As constitutional law expert Arthur S. Miller observes:

> America's political economy is dominated by business enterprises not even remotely in the contemplation of the Founding Fathers. Some supercorporations overshadow in economic importance most of the 50 states and many nations of the world. Both domestic and transnational corporations . . . have in significant respects taken over some of the substance of sovereignty, and also are warping original constitutional structures. Decisions made in corporate boardrooms, on such matters as

---

*In an authoritarian society such as China, a one-child-per-couple policy is enforced in urban areas by social controls. There are female block wardens in cities that monitor women's compliance.

investment, prices, and plant location, directly affect (and thereby govern) Americans as well as people elsewhere . . . Corporations are private governments.[43]

Yet many corporations show concern neither for the local community in which their business evolved nor for the nation whose soil nurtured them.

> The governing power of corporations should be legitimated. That is two-pronged: first, corporate decisions in some way should further the interests of both the entity *and* the larger corporation called society; secondly, members of the corporate community who are directly affected by corporate decisions should be protected against arbitrariness . . . In a democratic society, power, to be legitimate, must be responsible, i.e., "accountable."[44]

As with natural persons, the corporate persons must assume responsibilities to the society as well as accept rights. One effective step in this direction would be mandatory jail sentences for the top executives for major corporate crimes. It is hard to rationalize jail sentences for the petty theft of $100 and only wrist-slapping fines for billion dollar corporate frauds. The Catch-22 is that, as shown in the next section, corporations have ample power to resist imposition of meaningful corporate responsibility to the society.

The Free-market Economic System as Threat

A pure free-market economy can lead to the scenario described as "dumb" or out-of-control capitalism (B_2 in chap. 8). We have pointed to the danger signals of excessive egocentrism in this chapter. Indeed, the Thatcher-Reagan-Bush policies of the 1980s in particular pushed both U.K. and U.S. well along this scenario path. A dog-eat-dog society with ever wider gaps between rich and poor induces system crises and instabilities that undermine democratic governance. As population increases, a redefinition of what constititutes the "free-market economy" may be necessary to preserve democracy. It should not surprise us that, in its transition phase, post-Communist Russia is intensively struggling with exactly the same dilemma.

Government of, by, and for Money as Threat

We urgently need to evoke "new thinking." The free market is eminently desirable, but its excesses pose a subtle danger to democracy. Indeed, we have here another case of "more leads to less."

Using the 1990 election as a base, we can calculate that it costs $4 million to be elected U.S. senator. A senator expecting to be reelected must thus figure on raising almost $13,000 a week for the six years of his term. It is obvious today that, at all levels, we have less a government of, by, and for the people than a government of, by, and for money. Money buys power, and a prime conduit is the special interest organization or lobby. In the period March 1991 to June 1992, political action committees (PACs) contributed $73.4 million to congressional candidates. The 1992 presidential campaign raised a staggering $400 million in contributions. Some examples: an agricultural products company, Archer-Daniels-Midland, and its affiliates gave $1 million to the Republican campaign; and an insurance company, American Financial Corporation, donated $715,000 in the period Jan. 1991 to August 1992; the American Medical Association contributed $1.1 million over the period January 1, 1991, to June 30, 1992.[45]

This perfectly legal dispensation of bribery serves special interests in a number of ways.

1.   A lobby can paralyze the lawmaking bodies and prevent enactment of legislation that will benefit the majority but is opposed by the special interest group. Recent examples: meaningful gun control thwarted by the National Rifle Association and effective health care reform by the insurance industry.

2.   A lobby can push through the enactment of legislation that benefits the special interest group but hurts the vast majority of citizens. An example is the support provided to farmers for tobacco and milk production. In 1992 the federal government stored 488 million pounds of unneeded butter that it purchased from dairy farmers who could not sell the butter elsewhere. Farmers also get paid for not planting, with those grossing more than $500,000 annually receiving more than half of the annual price support money distributed by the federal government.

Congressmen from Pennsylvania's coal country have been instrumental in persuading the Department of Defense to buy 10 percent of the nation's total high-sulfur anthracite coal production; it has now amassed a ten-year supply. U.S. bases in Europe are barred from buying gas or coal for heating from foreign sources, so that they must ship in anthracite coal from the U.S.[46]

Junk-bond king Michael Milken created the Alliance for Capital Access lobby solely to block any curbs on the deductibility of interest on corporate borrowing, the engine that propelled his highly profitable strategy. It was a boon to clever financial manipulators and ruined many corporations and their work forces.

3.   A lobby can get its supporters to fill regulatory commission positions, that is, have the regulated industries furnish the regulators.

4.   A lobby can operate in hidden ways, for example, by masquerading under a deceptive title. Oil companies and real-estate developers are financing the National Wetlands Coalition to defeat wetlands protection legislation, and auto manufacturers are financing a Coalition for Vehicle Choice to defeat fuel-economy legislation. Taking a leaf from the environmental movement, the timber and coal-mining industries, agribusiness, and large land developers are financing the "Wise Use" antienvironmental "movement."[47] President Bush had a "Team 100" of wealthy campaign contributors ($100,000 minimum). These "soft money" contributions skirt the federal campaign contribution limits by being laundered through designated party organizations rather than going to a specific candidate.* According to Common Cause, a watchdog group, one-fourth of the "team," which has raised $6 million, consists of real-estate developers, who were interested in tax relief. A single 1992 "President's Dinner" raised more than $9 million in contributions, with gifts ranging in size from $1,500 to $400,000.[49]

5.   A lobby can push for lax enforcement of undesirable legislation that somehow escaped its pre-enactment efforts (e.g., installing costly environmental pollution controls). In the Bush Administration, lobbies could appeal to Vice President Quayle's Council on Competitiveness to ease or circumvent rules and legislation. (The council was disbanded by President Clinton.)

6.   Lobbyists can produce a direct-mail or telemarketing campaign to quickly generate 100,000 messages to Congress

---

*A 1979 change in the federal law removed restrictions from "soft money" contributions. They are supposed to go for "party building" activities, but are routinely diverted to help candidate slates. In 1991 the Republican National Committee received $17.1 million ($12.8 million from businesses, $3.9 million from individuals), the Democratic National Committee $7.1 million in soft money donations ($4.3 million from businesses, $1.5 million from individuals, $1.2 million from labor unions). An example of individual contributions: the U.S. Ambassador to Austria gave $198,000 to the Republican Party.[48]

"from the grass-roots voters" across the country for or against any issue.

7.   A lobby can operate transnationally and exert influence to subvert other governmental systems and regulations. It can essentially focus on the weakest links in global networks.

These seven functions suffice to show the central role of the 80,000-member lobby "industry." In part, the effectiveness stems from its recruitment of former members and staff of the Congress and the executive branch of government. Their contacts translate into power and earn them highly lucrative jobs upon leaving government service. Lobbies are now so entrenched and well financed that they can successfully fight any reform that would doom their effectiveness—an instance of the notorious Catch-22. Elimination of the throttling power of lobbies is badly needed, but this is exceedingly difficult to achieve as long as they are still there, richly financed to assert their power.* Remember that this would include not only big business but also labor unions, professional associations (e.g., of lawyers, physicians, and educators), environmental organizations, retirees' groups, black and Hispanic groups, war veterans' organizations, foreign governments, religious foundations, and so forth. Furthermore, we still must contend with large political payments from corporations and wealthy individuals. A corporation may disguise its largesse by asking its upper-level employees to give specified amounts to specified candidates in an election campaign.

All such special relationships facilitate the free-market economy but they hardly strengthen democracy. As the old saying goes, "money talks," and it drowns out the barely audible voices of the many that lack financial clout. It thus distorts the concept of 'democracy'.

How can sufficient counterpressure be built up to at least achieve some major reforms? One proposal is a massive grass-roots voter revolt, threatening to oust elected politicians who accept special-interest support. This might clean up the Congress and the top level of the executive branch. That is not enough. It would leave untouched most of the gigantic executive-branch bureaucracy, including regulatory enforcement agencies. The new

---

*A very mild reform bill, the Lobbying Disclosure Act of 1993, is under current consideration. It requires all lobbyists to register, identify their clients, and the payments received from them. The lawmakers will not be identified, however.

Congress and elected executives would soon be subject to the same lobby pressures. Another idea is the enactment of term limits for the members of Congress, comparable to the one that already exists for the Presidency. Proposals to date typically involve three terms for representatives and two terms for senators. But this would still leave those elected forced to plead for financial support to finance reelection and eager to make connections to cash in on their term of office by landing lucrative lobbying jobs after their term is completed. Public financing of political campaigns has also been proposed. This would simply shift the pressure points to indirect channels and to incumbents. Finally, we must always remember that the lobby power is exercised not only through elected officials and through appointments to regulatory commissions but also by "revolving door" personnel movement between the executive branch and the private sector (see the Department of Defense and the aerospace industry or the chapter 3 example of the U.S. Coast Guard and the oil industry).

Many voters seem to feel that, in Hamlet's words, we would "rather bear those ills we have / than fly to others that we know not of." The U.S. Congress has been as hard to change as the entrenched bureaucracy: 98 percent of those who ran for reelection won in 1986; the figure was 99 percent in 1988. Between January 1, 1989, and December 31, 1990, 406 Congressional incumbents raised $88 million from Political Action Committees (PACs), while 331 challengers raised only $7 million. Overall campaign resources in the same period split $240 million for incumbents to $37 million for challengers.[50] A close affinity between incumbents and special interest groups is evidently mutually beneficial, minimizing chances for real reform.

However, by 1992, public dissatisfaction with "Washington" reached a threshold of concern. Several highly respected legislators retired in frustration at the gridlock that throttled meaningful system reform. Disclosure of legislative and executive "perks" or perquisites, such as House Bank overdrafts, legislative boondoggle overseas travel, and top administrative golfing trips funded by the taxpayers, finally fueled a slightly higher personnel turnover in the Congress. There were more retirements, and "only" 93 percent of incumbents running for reelection kept their seats.

The linkage between organizational and personal perspectives has made a point of considerable importance for our future: democracy and freedom have an increasingly delicate relationship in our crowded global village of tomorrow. Jean-Yves Calvez, a Jesuit author, isolates

> the two main conditions for freedom in societies of growing complexity. First, there is the necessity for true self-government in the decentralized social bodies which make up a great society. Second, there is the equal necessity of a personal participation of each member within each body, each social network, each organization. One must participate with full responsibility and let others participate with full initiative. These are some of the main aspects of the social ethics required by socialization, if man is to take advantage of the possibilities of freedom in tomorrow's complex societies.[51]

Facilitating Participation

Participation of the individual in governance is vital for a healthy democracy. Let us consider organizational steps to increase the participation of the two age cohorts that American society tends to keep outside the mainstream: the young and the old. Past innovations provide some clues. The 1930s produced the Civilian Conservation Corps and the 1960s the Peace Corps and Vista. The 6,000 Peace Corps volunteers serve two years in grass-roots development overseas; the 3,300 Vista volunteers serve one year in low-income communities in the U.S. In 1991 the Peace Corps received 13,735 applications for 3,800 trainee positions. About 30 percent of its field positions are filled by people twenty-three to twenty-eight years old; 45 percent are age fifty or older. In 1993, there were plans to open 250 to 500 field positions in the former Soviet Union.[52]

These concepts have inspired proposals for a mandatory "national service" by America's young people. Thoughtful individuals ranging from Robert McNamara, a liberal, to William Buckley, a conservative, have favored such a concept. It would concretize the idea of the obligation of the individual to give something to the society that gives so much to him and her. Like the earlier corps concepts, it could engage young people in constructive work involving environmental restoration, slum rehabilitation, health care, and other socially beneficial activities. The 1993 Clinton proposal of a National Service Program was designed to involve youths in just such activities, but it diluted the concept of obligation by offering them college or vocational-training grants of $10,000 in addition to pay as an inducement for their two-year participation.

The increasing pool of older, retired persons also constitutes a valuable resource. As with young people, retirees could gain feelings of self-esteem, of being useful, even needed, by the society that should not be underestimated. The pool of older persons should be tapped much more effectively in such activities as teaching and tutoring, caring for neighborhood children, working with the disabled, beautifying the community (flower gardens outside, artwork inside buildings), as well as providing cultural, legal, and health-care services to the poor. Once they gain familiarity with computers, they can participate in community projects even if confined to their homes. As Betty Friedan points out, senior status is a time for the assumption of meaningful new roles, not the time for desperate denial of age.[53]

There are many ways participation at the local community level can bridge the gap between the citizens and the state. Examples are farmers' alliances, labor unions, church groups, and electronic town halls.[54]

Unfortunately, token and symbolic forms of participation are all too often in evidence. The sense of frustration, powerlessness, loss of control, and cynicism by the individual citizen is palpable today. But leadership exercised at the level of ordinary citizens has galvanized Americans in the past and can do so again. The Civil Rights movement and the anti-Vietnam struggle arose from the bottom and slowly reached the top. They represent examples of effective citizen-propelled change and a reaffirmation of democracy.

Information technology can do much to make citizenship active rather than passive. Localization, information networks, and communications have the potential to make the exercise of citizenship as real in the 21st century as when it was envisioned by the Founding Fathers in the 18th century (see chap. 8). Access to information technology empowers the citizen. Today the computer-literate citizen is already able to engage many others in computer network discussions about major issues. Desktop publishing and the fax machine offer the individual powerful communication tools. Any possessor of a camcorder can document perceived government malpractices with stunning effect. The concept of coordination-intensive structures has the potential for facilitating participation in entirely new ways.

Participation requires an ability to communicate across ethnic, socioeconomic, and gender boundaries. We constantly become aware of differences in perspectives that impede such communication. Group A perceives a problem differently than does group B or C and makes assumptions about B and C that vary significantly from the views actually held by B and C. Whites see blacks in a different way than blacks see themselves. According to a 1992 *Time*/CNN poll, 43 percent of whites feel that the criminal justice system favors whites over blacks while 84 percent of blacks do.[55] The angry response of women to the Clarence Thomas nomination hearings by a Senate Committee demonstrated the different sensitivities of men and women about the sexual harassment issue.

One constructive step in bridging these differences and facilitating communication is assumptional analysis. It is designed to clarify the assumptions made by each group, such as the certainty and importance attributed to them by the others. Another is development of the ability on the part of individuals and organizations to be comfortable with, and open to, multiple perspectives. (For more details, see Appendix A.)

THE SHORT-TERM VERSUS THE LONG-TERM VIEW

The *Exxon Valdez* case (chaps. 2–5) revealed how the short-term and long-term views become the basis for conflict. For the oil company and most Alaskans, the short-term view dominates. Environmental damage is seen as requiring an intensive short-term cleanup effort. For the environmentalists the major concern is the long-term impact. For the oil company high near-term profits call for maximum oil sales and minimum regulatory requirements; for the state's citizens uninterrupted oil drilling means jobs, no state taxes, and an annual check—all immediate benefits. The environmentalist is concerned with the adverse global impact of fossil fuel use (specifically carbon dioxide emission) and the long-term need for nonpolluting energy sources.

The *Exxon Valdez* incident is just a microcosm of the universal conflict between short-term and long-term concerns. More than 2000 years ago, Aristotle defined two distinct kinds of economics:[56] *chrematistics*, the manipulation of property and wealth to maximize the short-term return to the owner; and *oikonomia*, management of the household to increase its value to all its members over the long run. The population and technology explosions amplify the effect of the *chrematistic* priority on the near term. The desire for immediate sexual gratification and the need for jobs and profit seem obvious to all. The potentially catastrophic long-term effects of overpopulation on the *oikonomia*, the resulting environmental degradation and decline in quality of life, are not apparent to most people.

Belching smoke stacks mean work; protection of the forests means closure of lumber mills and stagnating mill towns. At this writing the U.S. government, bowing to business pressures, is the principal stumbling block to a global treaty for effective reduction of carbon dioxide emissions that constitute the foremost cause of atmospheric warming. As an old adage has it, "all politics is local." Utilities and automobile manufacturers cite the near-term costs to them and, like tobacco companies investigating cancer, dismiss the scientific evidence of long-term harm on the grounds of insufficient proof. In the case of Chernobyl, chronic monitoring will be needed for decades, not years.

In all these cases, the conflict seems rather one-sided, because (1) the seriousness of the effects cannot be proven beyond a reasonable doubt, (2) corrective action will hurt those who exercise political power today, (3) effects may be harmful to a future generation that has no voting power today, and (4) human beings resist changing their habits.

Dow Corning knew in the early 1970s that some silicone gel would seep out of the breast implants' envelopes but saw no crisis in its short-term planning horizon. Its management downplayed the clear trouble signals it had, in the form of internal reports of leakage, until a fullblown crisis was at hand in 1992.[57] Now the company's future is endangered. This general myopia is also reflected in the corporate financial machinations of the 1980s discussed earlier and in the priority of near-term profits over research and development, which has a long-term payoff.

A 1992 study of 300 companies shows a stark contrast between long-term-oriented European and Japanese companies and their short-term-focused American counterparts. In the study, "long-term" was defined as profits not expected in the first five years of an investment project. Japanese companies considered 47 percent of their projects long term, European companies 61 percent, and American companies 21 percent. The difference in attitude is related to the large, long-term institutional stakeholders of the overseas companies, the exception rather than the rule in America.[58]

The lack of concern with the long term is even reflected in the prospective Year 2000 studies dealing with state and local governments. A 1985 review of thirty such studies showed that they "had a clear, sharp view of the trees, but proved weak in seeing either the leaves or the forest; that is, in seeing the micro or the long-term macro factors."[59] They shied away from the potential for long-term structural changes and radical changes in infrastructure technology, such as the wired city and mass networking, focusing instead on near-term strategies that emphasized continuity and financing.

The bias is reflected in the heated debate on "the right to life," a near-term personal issue, while the population explosion, a huge long-term societal issue, is largely ignored. The same bias motivates many of our self-centered youth and senior citizens. A recent survey found that "the typical young person doesn't want to hear about it 'unless it's knocking on my door.' According to a 22-year-old entering law school: "people in that age group are only concerned about issues that affect them. When the drinking age went up, quite a few people were upset."[60]

Most Americans have moved light years away from the old adage that "a penny saved is a penny earned." In this respect, they are evidently unlike citizens of their principal economic competitors, Japan and Germany. Those countries demonstrate that capitalism does not make a short-term view inevitable—on the contrary, their longer-term view seems to facilitate success. Can such deeply ingrained American attitudes be changed?

In discussing the differences among perspective types (chap. 6), one parameter was the planning horizon (table 7.1). It tends to be far for the T perspective, intermediate for O, and short for P. This at once explains a source of the difficulty of communication between the analyst and the lay person. Forrester's engineering perspective made it seem reasonable to run his world dynamics computer model out to the year 2100. The O and P perpectives are most unlikely to do so.

The "fly now—pay later" philosophy is evident among young people. About half of the 3.5 million unplanned pregnancies annually are attributed to the lack of any birth control. The immediate "benefit" of spontaneous pleasure outweighs the "cost" of pregnancy later.[61] A "fly now—pay later" philosophy is even promoted commercially at the high school level. Mastercard International has proudly proclaimed that 10 percent of high school juniors and seniors now use credit cards regularly. At the other end of the age spectrum, more than 55 percent of doctors' patients over sixty-five fail to keep taking prescribed medicines vital to their long-term survival. A frequently cited reason is the disappearance of the immediate symptoms. Failure to follow doctors' preventive or therapeutic orders results in 125,000 preventable deaths annually at all ages.[62]

It is evident that this myopic philosophy is deeply imbedded in the American psyche and can ultimately prove exceedingly hazardous. It is analogous to driving at high speed at night on a dark, unfamiliar mountain road with low-beam headlights. The impacts of infrastructure decay and educational system deterioration become apparent only many years in the future. They do not enter the field of view of O and P until they become near-term crises and reach a catastrophic stage. We observed in chapter 1 that the destruction of commercial fishing in the Great Lakes from 1939 to 1949 could be traced to the construction of the Welland Canal in 1829.[63] Today's carbon dioxide emissions may not increase global warming with any catastrophic effect until the middle of the 21st century (chap. 7). The population explosion, particularly in the world's tropical areas, may not be reflected in catastrophic effects due to deforestation and water pollution for decades.

Consider a comparison of two types of risk: *(a)* oil spills and waste dumps containing dangerous chemicals (pt. 2), and *(b)* ozone-layer depletion and the greenhouse effect. Many scientists view *(b)* as being of higher priority than *(a)*; the public views *(a)* as being of higher priority than *(b)*. The scientists see in the long term far more human lives endangered by *(b)* because of the global pervasiveness of the hazards. The public focuses on *(a)* because these hazards are "here and now."

A 1990 survey of 1,000 Maryland households posed the following choice: Unless controlled, a certain kind of pollution will kill 100 people this year and Y people T years from now. The government has to choose between two control programs costing the same, but there is money for only one. Should it initiate program A, which will save 100 lives now, or program B, which will save Y lives T years from now? Y and T were varied, with T taken as twenty-five years and 100 years. The results showed a strong preference for the present. For example, 68 percent preferred saving 100 lives today over saving 200 lives in twenty-five years. If the saved future lives remain the same, but the time T is stretched to 100 years, the figure becomes 85 percent. Consistently, only 10 percent to 15 percent consider saving future lives to be at least as important as saving lives now.[64]

O and P look at the future through the wrong end of a telescope—the distant objects look smaller than they actually are. The same applies when we look at the past. We refer to this ingrained human habit as "discounting."

The word *discounting* is used because the tendency we have described is analogous to the practice in business and economics to discount future dollars relative to present dollars, that is, to determine the present value of future dollars. Zero inflation is assumed in this discussion. A contract giving us a million dollars ten years from now is worth less than a similar contract providing the million dollars up front. The reason is that we can work with the money received now to earn more with it for the next ten years. Thus a positive discount rate means that the present value of the future dollar is less than a dollar.* Often the discount rate used is the cost of capital, that is, the rate of interest at which a company can borrow money.

If a discounting factor is imposed, as O and P perspectives are likely to do, the future crisis appears to shrink in significance compared to an equivalent near-term crisis. The higher the discount rate used, the greater is the shrinkage. We are aware, for example, that the poor generally discount time more heavily than the affluent. They are concerned with the near-term crisis of their own survival, not with long-term problems such as environmental degradation.

We have already observed that a temporal discount rate can be applied to parameters other than money. The pattern of wars in U.S. history (fig. 7.1) suggests that a war is discounted com-

---

*Correspondingly, a negative discount rate signifies that the present value of the future dollar is more than a dollar. In this case, *with no inflation*, today's dollar buys less in the future. In our discussion of sustainable growth (chap. 7), it was noted that nonrenewable or slowly renewable natural resources are being depleted and thus their cost will rise. Future dollars will buy less and less. This is equivalent to a negative discount rate.[65] Instead of assuming zero inflation throughout this discussion, we can equally well interpret the same discount rate as "after inflation" (often labelled the "real discount rate").

pletely in the span of about one generation. It takes about that long to regenerate enthusiasm for another war. Figure 9.1 illustrates the perception of the future global population problem in discounting terms, pointing up the analogy of looking at the world through the wrong end of the telescope. Using the global population in 1970 as a base, Forrester's world dynamics model calculated its behavior to 2100.[66] The zero discount rate case shows a catastrophe in the middle of the 21st century. The population reaches three times the 1970 level and then plummets in an ecological system collapse. However, if a 5 percent annual discount rate is applied to the population increases, the crisis appears to shrink to total insignificance from our vantage point. No dramatic worsening of the current situation is perceived by today's observer.

The discount rate found in the Maryland survey is 7.4 percent for T = 25 years and 3.8 percent for T = 100 years. That means six lives would have to be saved in twenty-five years or forty-four lives in one hundred years to be equivalent to one life saved now. A commonly used real (after-inflation) "social discount rate" is 3 percent. That means a 3 percent economic growth rate per year, equivalent to a doubling of the economy in one generation or a tenfold growth in a human lifetime, is needed simply to assure no decline in the standard of living. Such considerations lead T-oriented analysts to question the entire discounting concept when intergenerational time spans are involved. They see in discounting no satisfactory way to protect the rights of future generations.*

In economic policy making, the discount rate can have a startling effect on decisions. Consider, for example, the choice between twenty-year coal and solar energy programs that faced the Congress some years ago. If the after-inflation costs are compared using zero discounting, the solar energy program has more expensive initial investments, but very low later costs; the coal program has fairly constant costs over the same period. The solar program, measured by the area under the curve, appears to be cheaper (fig. 9.2a). However, if a discount rate is used, the high initial costs of the solar program far outweigh the subsequent

---

*Rothenberg proposes that time discounting be restricted to a single generation at a time and that, beyond the present generation, the discount be considered as a very slowly decreasing step function so that costs and benefits beyond the present generation are not strongly diminished.[67]

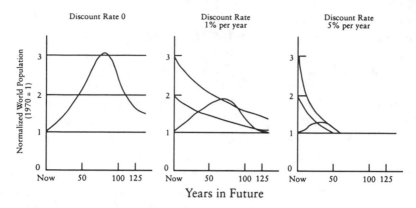

FIGURE 9.1

Discounting the Population Explosion

Source: H. A. Linstone, *Multiple Perspectives for Decision Making* (New York: North-Holland, 1984), pp. 21, 322 ©1984. Reprinted by permission of Prentice Hall, Englewood Cliffs, New Jersey.

higher coal costs and the economic decision favors the coal program (fig. 9.2b).

The preceding discussion strongly suggests that discounting occurs in space as well as time. In figure 9.3, the heavy shading denotes the primary focus of concern, the "me-now" (*a*). Point *x* denotes a future time point, *y* a distant place, and *z* a distant time and place. To O and P perspectives, what is far away in time and space is normally of less interest than what is near. They are more comfortable with human-scale dimensions. On the other hand, T perspectives move easily from subatomic to astronomic levels, from nanoseconds to light years.

For O and P, an event that happens thousands of miles away is of less concern than the same event in our neighborhood. This is the famous NIMBY syndrome (Not In My Back Yard) and it is central to environmental issues. Santa Barbara citizens become concerned about offshore oil drilling and its pollution of the Santa Barbara Channel; they show little interest in the same problem in the Gulf of Mexico, thousands of miles away.

The space-time discounting tendency probably has its origin in human evolutionary development. It was undoubtedly vital for survival to concentrate on the "me-now"; in this constant struggle, there was not the luxury of ruminating about distant space and time. O and P perspectives evolved before T perspectives.

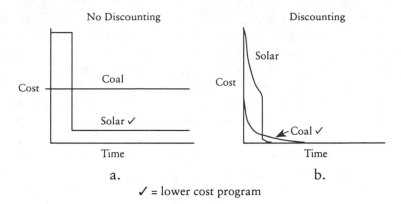

No Discounting

Discounting

Cost

Coal

Solar ✓

Time

a.

Cost

Solar

Coal ✓

Time

b.

✓ = lower cost program

FIGURE 9.2
The Effect of Discounting on Energy Decision

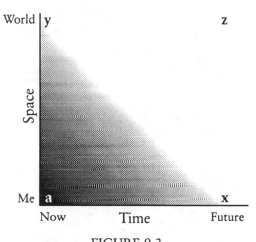

World y

z

Space

Me a

x

Now

Time

Future

FIGURE 9.3
Discounting Time and Space

It is even today possible to make a potent argument for discounting. Consider a current issue or problem X. Suppose we can conceive of three alternative solutions, S_1, S_2, and S_3. Each of these is likely to create impacts or new problems, say S_1 leads to X_{11}, X_{12}, and X_{13}. New problem X_{11} in turn will suggest new solution possibilities, and so on. We see that X leads to an ever growing sequence of problems and solutions. A historic example of X is the problem of epidemic diseases such as bubonic plague, cholera, and polio. Public health measures solved the problem, cutting the death rate without affecting the birth rate. The solution helped to create an overpopulation crisis and massive starvation. The Green Revolution and other measures alleviate this problem, but cause ecological imbalances and worsen environmental pollution. This reasoning may well lead to a Hamlet-like paralysis of action: why treat X when that leads only to unending new problems that may be even more difficult to solve? The answer: discount, so that the subsequent new problems can be ignored and action focused on dealing with the near term.* Many corporations also discount both space and time heavily. They ignore the world beyond the immediate corporate activities, a potentially dangerous weakness in a global village economy where everything interacts with everything. But discounting certainly makes the decision process easier; unfortunately it does not make it better.

It comes as no surprise that *the tendency to use differing discount rates or planning horizons seriously impedes the interaction that is so vital among T, O, and P.* This will be an increasingly critical matter in the coming decades.

The space-time discounting problem also makes us acutely aware of a profound ethical dilemma: *inter*generational versus *intra*generational equity.

In an affluent society, how should we balance our concerns for our grandchildren (x) with those for the poor in our world today (y)? Which is more important, dealing with the inexorable carbonization of our grandchildren's atmosphere or with today's water contamination afflicting billions in poor areas? Will our grandchildren benefit more if we focus today on environmental protection at the expense of economic growth or on economic growth at the expense of environmental protection? Should we move along the diagonal in figure 9.3, that is, toward z, keeping a balance between x and y? We shall confront this ethical dilemma once more in chapter 12.

Finally, we observe that *discounting leads to a failure to recognize overarching societal transformations.* Myopia keeps our eyes focused on the trees as we fail to see what is happening to the forest.

---

*In mathematical terms, discounting is equivalent to converting a divergent infinite series to a convergent one.

America's Constitution seeks a delicate balance among the Federal executive, legislative, and judicial branches of government. Here, too, we see highly significant differences in discount rates. The judicial branch has a much longer horizon or lower discount rate than the other two branches. This results from the lifetime tenure of judges as well as the judicial philosophy. As many presidents have recognized, the impact on American society of their judicial appointments is likely to be more profound and lasting than most of their other activities.[68]

The wisdom of the Founding Fathers in their sophisticated appreciation of the needed balance between the individual and the society, as well as the long-term and the short-term, is truly remarkable. Equally notable is the fact that, after the most intensive and partisan debates between Federalists and Antifederalists, the latter, as losers, acquiesced in the decision. The absence of obstruction played a large role in the successful implementation of the Constitution.

In conclusion: it is apparent that *the mutually supportive relation between the individual and the society has badly eroded in recent decades.* We shall see, in chapter 11, that *all the linkages among T, O, and P must be addressed in revitalizing it.*

CHAPTER 10

Linking Technical
and Personal Perspectives

In the next century, from out of the dust of the information explosion will emerge Homo optimisans se ipse, Autocreator, Self-Maker, who will laugh at our Cassandras (assuming he has with what to laugh).

Stanislaw Lem, *A Perfect Vacuum*

The main fuel to speed our progress is our stock of knowledge, and the brake is our lack of imagination. The ultimate resource is people—skilled, spirited, and hopeful people who will exert their wills and imaginations for their own benefit and so, inevitably, for the benefit of us all.

Julian Simon, *The Ultimate Resource*

In chapter 8 we addressed the linkages between technical and organizational perspectives, in chapter 9 those between organizational and personal perspectives. In this chapter we turn to the interactions between technical and personal perspectives. We begin with a sampling of the differences in the perception of probabilities between the technologist and the intuitive nontechnologist. Next, the role of the technical-personal perspective link is explored in connection with creativity. Finally, the concerns in relying on the technical perspective to enhance our humanity are examined.

THE TECHNOLOGIST VERSUS THE PEOPLE

The twin forces with which we are concerned, population and technology, raise the importance of this linkage to a new level. The tremendous population growth creates both crowding and the need for more energy and materials. The technology responds, and we have more massive processing, movement, and disposal of

263

potentially hazardous substances. The system complexity, the unparalleled quantity levels, and the innovations all lead to greater uncertainty of impacts. We have already noted the distinction between explanation and predictability (chap. 6). It is the latter that is usually of more practical interest in the real world and occupies us here.

Dror[1] divides the span between certainty and complete ignorance into four stages:

1. there is a simple causal model—the present is known, the future outcome is precisely predictable (as with Newton's laws of motion)

2. the reality is stochastic, that is, there are fixed probabilities associated with each of the possible, known outcomes (as with Mendel's laws in genetics)

3. the reality is complex, that is, there is both unpredictable randomness (chaos) and an underlying order or mechanism that permits limited predictability, typically a nonlinear dynamic system highly sensitive to initial conditions

4. there is complete ignorance and total surprise; the future is undetermined by the past and there is no predictability whatsoever

All four levels come into play. For the present we shall focus on level 2; in chapter 12, we shall come back to levels 3 and 4.*

The best predictive model is one based on simple causality (level 1). The next best is one based on fixed probabilities (level 2). But this already lands us in trouble. It assumes complete scientific rationality, but human beings do not conform to *this* notion of rationality—often for very good reasons. Consequently, theory often seems to diverge from intuition. Thus we see in the perception of probability a striking illustration of the difference in perspectives among scientists, engineers, and systems analysts on one hand and the general population on the other. Table 6.3 clearly indicates the divergence, and we saw examples in chapter 4 in connection with the *Exxon Valdez* case.

We have observed that individuals tend to misunderstand very low probabilities. They assume that an oil spill of a magnitude

---

*For another taxonomy of ignorance and uncertainty, see Smithson.[2]

calculated to have a probability of occurring only once in a very long time period will not occur in the near future. It is simply counterintuitive that an event with such probability may occur tomorrow.* We ignore the possibility of an event with very low likelihood, such as an earthquake, even though everyone can recount personal experiences of very low-likelihood events that he or she has had. People become careless about vaccinations such as polio and measles, because for years no significant outbreaks have occurred.

In fact, we have difficulty conceptualizing very high and very low numbers. This applies to astronomic distances at one end of the spectrum and to subatomic particles and very low probabilities at the other. While we can clearly distinguish between probabilities of 0.50 and 0.10, it is almost impossible for us mentally to distinguish 0.004 from 0.00004, despite the fact that the former is 100 times greater than the latter.

We generalize from an inadequately small sample. The commonly held conviction that, after five consecutive coin tosses yielding heads, the chances for tails on the sixth toss is higher than 50–50, is well known. The human mind integrates past information in an unmathematical way: recent input is weighted more heavily than more remote input.[3] This is, of course, simply a restatement of the discounting tendency inherent in the P perspective (chap. 9).

People tend to be risk averse to choices involving gains, whereas they are risk takers when the same choices are framed as involving corresponding losses. In other words, they decide differently even though the problems are actually identical. As Harvard mathematics professor Persi Diaconis observes, "our brains are just not wired to do probability problems very well."[4]

The media also tend to influence perceptions of probability strongly. Our estimates of the risk of widely covered events are exaggerated, while those of poorly covered events are underestimated. For example, New Yorkers overestimate the likelihood of being killed by a mugger and underestimate the likelihood of dying of a stroke or asthma. This reflects a major asymmetry: bad news sells better than good news. The media, as well as special interest groups, feed on vivid catastrophes while ignoring dull

---

*Mathematically, over an unbounded time period, a probability of event occurrence of zero does not negate the possibility that the event will occur.

safety records (chap. 8). The trend is exacerbated by the lawyer-ization of American society (chap. 9).

THE TECHNOLOGICAL FRONTIER
AND THE CREATIVE INDIVIDUAL

Previous societies used technology to extend human physical capabilities. Cranes and fork lifts multiplied the power of our arms; automobiles and airplanes multiplied the power of our legs; radio amplified the power of the our voice. Now information technology is extending the power of our brains. The astounding new abilities to obtain, process, store, and transmit information raise fascinating possibilities.

Information technology in all its aspects, from communication to computers, from neural networks to biochips, suggests innovative and fluid organizations. An unprecedented variety of linkages can be used in creative ways to develop fascinating new coordination-intensive structures. We are on the verge of opening up new means of processing information. Consider an example: We know now that the human brain does not operate on the same principles as today's digital computer. Neural networks have an architecture that is inspired by the structure of biological neural systems, that is, they try to mimic the modus operandi of our central nervous system and some of the sensory organs connected to it. They are trained rather than programmed, and they are self-organizing. They have several layers, with the processing elements, or neurodes, in each layer connected to a series of neurodes in the preceding layer. Each neurode receives a large number of input signals that, together, constitute a pattern. The pattern causes the neurode to reach some level of activity. If strong enough, this action generates a single output signal that is transmitted over the neurode's interconnections to other neurodes.

Parallel processing of signals, data storage throughout the networks, and a robustness that enables them to work with gross or garbled input distinguish them from conventional computers. The networks operate from an intuitive understanding of the structure of the task, from a learned internal model of the process they are involved in, rather than from a cognitive level. Hybrid systems, combining digital systems, artificial intelligence, and neural networks, may be developed in the coming decades.⁵ However, biological information systems lie beyond the horizon.

How can we use such "naturally intelligent systems" to augment our brains in managing our society, in making our institutions more responsive and flexible? We are in no way suggesting that human beings will approach obsolescence, as some purveyors of science fiction fantasize. Cranes and fork lifts have not made our arms obsolete; automobiles and aircraft have not made our legs obsolete. Indeed, vehicles now enable us to walk and trek in breathtakingly beautiful regions of the earth that, in earlier times, were never accessible to most of us. Neither will advanced information technology make human brains obsolete.

Consider computer-assisted product design and manufacturing in industry. The computer excels in analysis, rather than in design. The difference between design and analysis is analogous to the difference between writing an essay and checking it for grammar, spelling, and logic. Computers facilitate composition, but in a mechanistic rather than creativity-enhancing way. Not surprisingly, automation of product design is lagging far behind automation of analysis in manufacturing.⁶ The subtleties inherent in the capabilities—and limits—of the new computer-assisted operations should not be underestimated. The role of the human being in this evolving man-machine system remains critically important.

If innovation is the key to maintaining leadership in the economic competition of the knowledge society, creativity is the key to innovation. And the human being is the source of creativity. The three essential cognitive elements are intelligence, expertise in relevant areas, and a distinctive thinking style. We refer to a style that engages in "deviant" thinking guided by uncommon analogies and heuristics that include intelligent guesses. Mental wanderings may seem chaotic at times, but they arrive at new linkages and novel solutions. Personality characteristics that facilitate creativity are perseverance, high energy, curiosity, independent judgment, intrinsic motivation, and willingness to take risks. The key role of intuition has already been discussed in connection with the P perspective (chap. 6).

Hadamard, in *The Psychology of Invention in the Mathematical Field*, quotes the German physicist Helmholtz, who observed that "happy ideas" never came to him when his mind was fatigued or when he was seated at his work table. "After the fatigue . . . has passed away, there must come an hour of complete physical freshness before the good ideas arrive."[7] Poincaré, another well-known French mathematician, distinguished (1) fully conscious work, (2) illumination ("happy ideas") preceded by incubation, and (3) the quite peculiar process of the first sleepless night. The unconscious appears to consist of several levels: "It is quite natural to speak of a more intuitive mind if the zone where ideas are combined is deeper, and of a logical one if that zone is rather superficial. This manner of facing the distinction is the one I should believe to be the most important."[8]

In the case of exceptionally intuitive minds, even important links of deduction remain unknown to the thinker who has found them. Cardan's invention of imaginary numbers ($i = \sqrt{-1}$) is a beautiful example of the use of the nontraditional to leap from one rational to another rational domain.

In today's world we have the brilliant, young "computer nerds," Steve Jobs, Bill Gates, Steve Wozniak, and John Warnock, who create personal computers and software. Consider this description of Steve Jobs:

> He spoke slowly . . . You could almost hear his brain thinking out an idea. Then he'd leap from his seat, pick up a marker, and begin sketching diagrams and arrows on a whiteboard to explain a notion visually. His whole body would speak. His hands would come together as if he were holding a product in them. He would make you see what didn't yet exist.
>
> Nothing consumed Steve's interest more and nothing seemed more central to that dream [we shared for Apple Computer] than the doings of a team of young, dedicated fanatics who toiled under a pirates' flag in the Macintosh Building. Steve's "pirates" were a handpicked band of the most brilliant mavericks inside and outside Apple. Their mission, as one would boldly describe it, was to blow people's minds and overturn standards . . . Steve dreamed up the pirate metaphor . . . "It's more fun to be a pirate than to join the Navy," Steve would say. This group shunned corporate orthodoxy and the conventions of society . . . Even at midnight it was a place that burst alive with activity . . . One person, one computer was the route by which they planned to change the world.[9]

The role of the P perspective simply cannot be overestimated. A strong T-P link is clearly crucial to a society's place at the frontier of technology *and it is this strength that is the basis for what is likely to be America's cultural legacy to future civilizations.*

How long can a society stay at the leading edge? If we think of the analogy of a surfer riding the wave, we must recognize that even an expert surfer can only ride a wave for a very limited time and that there is a rhythm to the waves that lies outside his or her control.

Let us look for a moment at one indicator of a society's research strength, its Nobel prize recipients. We use the same kind of long-term trend analysis discussed in chapter 7. When the awards were first given at the beginning of this century, they went to Europeans. In 1907 the Americans appeared and won an award every few years. The Americans' steady ascent proceeded while the number of awards to Europeans declined. By World War II, America dominated, and, since 1958, there have been American awards each year. In 1988 Americans garnered five awards, in 1989 six. A graph of the prize "shares" (fig. 10.1) shows the distinctive trends for the U.S., Europe, and "other," which denotes the remainder of the world, including Japan, Australia, and Canada.[10]

Innovation is based on research, and another analysis shows that the time between invention and innovation appears to be shrinking.[11] The implication of the two analyses is that *the task of maintaining America's technological leadership may become more difficult in the 21st century even if all the supportive institutional steps are taken. Another society may take up the baton of leadership or there may be a worldwide slowdown some time during the century. We may find that both population and technology growth drives have spent themselves before the end of the 21st century.*

Recalling our discussion of cycles in chapter 7, *we certainly should expect the era of information technology to give way to another one, presently undefined, during the next century.* In this book we are confining ourselves to the beginning of the century and even this period—within the lifetime of most readers—we confront more *terra incognita* than ever before.

The T perspective on the creative process sweeps in both human and material resources. The human resources draw in knowledge, exposure to colleagues' thinking, and staff support. Most recent Nobel laureates have been honored for work done collaboratively. Their material resources range from communications networks (such as electronic mail, teleconferencing, and information retrieval systems) to major research facilities (such as

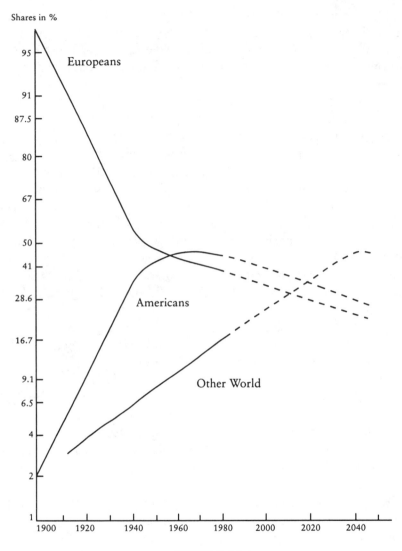

FIGURE 10.1
The Competition for Nobel Prizes

Source: Reprinted by permission of the publisher from "Competition and
Forecasts for Nobel Prize Awards," by T. Modis *Technological Forecasting
and Social Change,* 34, p. 98. Copyright 1988 by Elsevier Science Publishing
Co., Inc.

the Bell and Lincoln Laboratories). It is also significant that the
computer has in recent years assumed the role of a laboratory
tool, as demonstrated by the advances in chaos theory.

THE TECHNOLOGIST AND THE HUMANIST

In introducing part 4, we observed that pessimists have repeatedly been proven wrong because they lacked the imagination to see how looming limits could be overcome. Although the earth is finite, and we have emphasized that matter and energy use on a finite earth cannot grow indefinitely, the genius of human beings has consistently overcome perceived bounds. They have done so by using their intelligence. Thus human brains circumvent limits in the continuing evolution of life on this planet—and technology is a primary helpmate in this process. We have noted, for example, that we will have to invest heavily in the information infrastructure to remain competitive; that is, to stay at the leading edge we must exploit information technology to its utmost.

Our technological brilliance may well have blinded us to O and P perspectives. There is intellectual arrogance in the T-oriented claim that, if we can produce the technology that sends a man to the moon and back, we should be able to use technology to solve more earthly problems of society.

We do know that science and technology have presented us with the godlike ability to destroy all life on spaceship Earth, but they have not given us the wisdom to manage ourselves on this small planet. A clue that should give us pause is the poor educational achievement level observed in spite of the advances in information technology to date (table 7.1). *The world leader in information technology is producing trained incompetents and functional illiterates.* Too often the technological "cure" seems to have iatrogenic effects, that is, it exacerbates more problems than it resolves. Higher order negative impacts cancel out first order benefits.

Too often the engineer's world of models and analyses ignores the realities of human behavior. In part 2, we examined one case of human error: running aground in a well-mapped Prince William Sound in clear, calm weather. Other examples include: a sophisticated F-114 Navy fighter rolling off a carrier deck into the North Sea because a sailor failed to put blocks under its wheels; a DC-10 crashing because a cargo door was not properly closed; an air traffic controller in Los Angeles guiding an incoming airliner to land on top of a taxiing commuter plane on the runway; a Kansas City hotel construction engineer making a small, but fatal, change in the architect's design plan that resulted in a crowded walkway's collapse; an emergency cooling system shut off at the Chernobyl nuclear power plant for twenty-four hours in violation of regulations; failure to recognize that water being used to clean a chemical line at the Union Carbide Bhopal plant was seeping into a chemical storage tank—all stem from human frailties.

Too often we ignore the human limits to cope with change. Life in a big city moves much more rapidly than does life in a village. Similarly, life in the global megalopolis of 2020 will move more rapidly than did life in the city of a generation ago. In particular, inundation with information presents serious problems. There will often be far less time for making decisions if economic competitiveness or military security is to be maintained in a crowded global megalopolis. A response with unprecedented speed may be forced by a competitor or an enemy. Timely introduction of a new product in response to unanticipated market changes or an effective counter to a surprise missile strike offer cases in point. In each situation the decision maker may contend with information overload, confusion, misunderstanding, distortion, disinformation, and almost certainly inadequate time to analyze and think. Surprise cannot be eliminated, and automating the decision-making process is dangerous. Systems analysts point to the available decision support systems that should enhance the speed of decision making—management information systems, expert systems, and other computer-based means. "What-if" scenarios can be played out by computer. But these techniques all have critical limitations (see chaps. 6 and 12).

Furthermore, recent work indicates that, as the information flow rate increases, nonlinear systems speed up and take less time to reach chaotic or unstable states. Therefore, it may be dangerous to speed up the information flow in complex systems. If the system appears to oscillate or behave chaotically, it may be advisable to slow down information flow to reduce feedback, make the system more stable, and ease the task of the decision maker.[12]

Can we train people to operate effectively in this high-speed environment? Institutions can be restructured to facilitate rapid responsiveness with smaller staffs and flattened hierarchies. The high-reliability organizations discussed in chapter 5 are illustrative, but they constitute at best a partial answer as the questions assume increasing significance in the coming decades. We need a balance between the speed of information flow and the decision process, one that is difficult to determine and maintain.[13]

The complexity of the system and the fact that our knowledge will always remain incomplete should give technologists a measure of humility and wean them from an excessive reliance on their technical perspective. They may inadequately control the increasingly powerful technology and realize too late that this spaceship is a closed system that cannot withstand certain levels of abuse resulting from misuse of the technology. We do see accumulating damage to the protective ozone layer, to the oceans, and to the forests. Evolution strongly suggests that we are safe-fail systems (chap. 2), but we have no proof. Even in a local case such as the oil transportation system discussed in part 2, we concluded that it is a challenge to bring the system close to a safe-fail state (chap. 5). The possibility of a catastrophic failure of the megasystem cannot be ignored.

While we are not masters of the Earth, we are masters of ourselves. As Hannah Arendt notes, "Each new generation—indeed, every new human being, as he [she] becomes conscious of being inserted between an infinite past and an infinite future—must discover and ploddingly pave anew the path of thought."[14] Or, in the words of Ovid, a Latin poet, *teloque animus praestantior omni* (a mind is more excellent than every weapon).

Our strength as human beings derives from our intelligence: the ability to learn, invent, design, and organize. Indeed, self-organization is one of the most remarkable human properties. This "antichaos" aspect of complexity is central to the process of human evolution. Self-organization begins with the fertilized egg and shapes the complex human being, who, in turn, forms the family and communities ranging from the clan to global institutions.

As social animals we have evolved a sense of right and wrong, of morality and immorality, of justice and injustice. Our weak points include closedmindedness, extremism, greed, fear of change, egocentrism and myopia (the "me-now" of fig. 9.3).

Fear of change is related to the need for stability and for anchors or "fixed points" in our life. Change can be stressful. Some stress is vital to human health, but excessive stress is harmful. A major task in preparing for the 21st century—with the dramatic pace of change due to information technology—is to cushion the changes experienced by the individual.

In the past one anchor was the extended family. Its composition and location was largely fixed, except for births, marriages, and deaths. Another anchor was one's work. Once training was completed, the individual's work or career path was fixed. Today we have largely lost these anchors as the pace of technological change has accelerated. Instead, we have the mobile, nuclear family and increasing job obsolescence. The high divorce rate often removes even the nuclear family as an anchor. It is difficult for an individual to set personal goals and avoid high stress in such an unstable environment.

New anchors must be developed for the individual's health and effective functioning. In their absence, drugs, violence, family dysfunction, and personal uncommittedness will proliferate. We noted in chapter 8 the growing appeal of fundamentalist religious and nationalistic movements in a world perceived to be intractably complex. Many ministers hit a responsive chord with their thundering denunciations of moral decadence, democracy, and technology. They plead for personal behavior strictly regulated by the Bible, Koran, or ideologically pure nationalistic doctrine. At the same time, they cleverly exploit the new technology to shape the beliefs of their susceptible audiences into a rigid mental mold, stripping them of all human individuality. The anchor becomes a means to lock the human mind.

An antitechnology movement is hardly a new idea. Plato, in the *Phaedrus*, warned us long ago that the invention of writing would mean the decline and fall of culture. After the introduction of Gutenberg's printing press, scholars were concerned about information overload. The Industrial Revolution prompted Carlyle and Thoreau to articulate fears for humanity's future. The introduction of machinery in the weaving industry in England was fought by the Luddites, who smashed shearing frames and power looms in factories. Today there is talk of the "technological steamroller" and a call to dismantle the genetic engineering, nuclear, and computer technologies. Postman decries the onset of "technopoly," a society in which all forms of cultural life are subordinate to technology. He sees us forsaking humane values—reason, freedom, and individuality—for infatuation with technology and enchainment in a computerized world.[15]

Some strident environmentalists insist that all technology should be assumed guilty until proven innocent. Our response would be that, had this prescription been followed, human beings would still be living in caves. After all, the earliest human tools, such as bones and arrowheads, were used for both good and evil, to survive and build, as well as to kill fellow human beings. They could not be proven "innocent." All technologies involve the risk of misuse by human beings. Indeed, we have no reason to assume that evolution is, or can be, a riskless process.

Uncontrolled technology does have a dehumanizing effect. It can

- impair and endanger physical health, for example, by environmental pollution, harmful foods and drugs, genetic mutations, and information overload
- isolate human beings from a healthy social environment, for example, by substituting sterile artificial environments
- alter human consciousness
- replace humans with robots and computers while failing to give them challenging new work
- extend the life span of human beings without providing a constructive societal role for them in the years they gain
- create a new underclass of illiterate quasi-slaves

The challenge we face is to control technology intelligently, so that it enhances our humanity. The potential of the computer and information technology generally to extend the capability of the human brain is promising. But each individual must take responsibility for himself or herself; neither society nor technology can be a substitute. That inner strength presents the real test of our humanity.

Some intriguing questions need attention in the coming decade:

- How can the technology help create societal anchors and help the individual to engage with them?
- How can networking be used to improve the linkage of the individual to the community for their mutual benefit?
- How can telecommunications upgrade education and training?
- How can virtual realities, such as participative games and simulations, make the individual more comfortable with the pace of change?

- How can technology help reduce excessive human stress?
- How can technology provide more opportunities for individual self-fulfillment?

The last question is one of high priority. In the preceding section we were concerned with innovation. A primary aim must be to continuously create new jobs. With an increasing population pool and increased automation, the challenge is a daunting one. Available forecasts give no assurance that it can be met. For example, bioengineering can manufacture its products at a high level without using huge industrial plants and personnel. The new technologies are simply not creating skilled production jobs at the rate required in the U.S.

Suppose the rate of job creation cannot meet the human needs. What then? Part-time work, work sharing, and temporary employment arrangements are already on the rise. Protectionism is the common politicians' solution to save jobs, but it has a devastating long-term impact on economic competitiveness. Another possibility is the creation of artificial work. Cynics would insist that bloated bureaucracies in the public and private sectors, such as the armed forces, the civil service, General Motors, and IBM, already are masterful in making work for themselves.

The "cybernetic scenario" developed by Linstone and Turoff in 1975 no longer seems quite so farfetched. It postulated the creation of a top secret government agency to develop and offer to the public interesting work projects. The reason for the secrecy was that the problems involved had already been solved and it was considered important to maintain high motivation on the part of the workers.[16] It might be noted parenthetically that this path has also already been taken, albeit inadvertently. The Department of Defense sponsored studies on the effectiveness of strategic bombing following World War II, Korea, and Vietnam—in each case with the same disappointing lessons learned. In a bureaucracy it always seems easier to authorize a new study than to dig out relevant old studies.

To a child that is given a hammer, everything begins to look like a nail. The T perspective, if dominant, has the same effect: every problem must be addressed by analysis and resolved by technology. It is ultimately a sterile, even dangerous, prescription.

Coates reminds us that William James, a 19th-century psychologist, identified two important categories of people, the tough-minded and the tender-minded. Their perspectives on the future are strikingly different and there is little effective communication between them. It is the classic T-P gap, and Coates stresses the mutually beneficial effect of bridging it.[17]

For example, both tough-minded and tender-minded deal with the theme of transformation. The former focus on issues such as population and technology, the latter on introspection and self-awareness. The implicit idea of self-reform is that changes in personal attitudes and behavior will bring about desirable changes in the external world (see chap. 12). Both types deal with the concept of 'holism', one with whole systems in which all elements interact, the other with harmony of self and nature. Both deal with the global village, one with its unity in the physical sense, the other with its unity in the spiritual sense.

Synergistic contact between the two types should be encouraged. Indeed, *only if we consider the technical perspective jointly with the personal perspective will we realize its potential to enrich our humanity.*

CHAPTER 11

The Inseparability of T, O, and P

We have to abandon the arrogant belief that the world is merely a puzzle to be solved, a machine with instructions for use waiting to be discovered, a body of information to be fed into a computer in the hope that, sooner or later, it will spit out a universal solution . . .

The way forward is not in the mere construction of universal systemic solutions, to be applied to reality from the outside; it is also in seeking to get at the heart of reality through personal experience. Such an approach promotes an atmosphere of tolerant solidarity and unity in diversity based on mutual respect, genuine pluralism and parallelism. In a word, human uniqueness, human action and the human spirit must be rehabilitated.

Vaclav Havel

In chapters 8, 9, and 10 we considered pairwise linkages: T and O, O and P, T and P, respectively. But as the discussion in chapter 6 implies, and the triad schematic of figure 6.2 emphasizes, T, O, and P are inextricably linked and interdependent.

The recent Russian Revolution presented its reformers with the same hard lesson. As Aleksandr Yakovlev, Gorbachev's friend and advisor, observed:

The conservatism that has eaten into everyone and into society, like rust, has turned out to be more durable than we expected. Resistance to change has turned out to be very strong . . .

There was a clear-cut understanding that what existed had to be overturned—the authoritarianism, the command-bureaucratic economic system . . . But how could that be done, in what stages? . . . we had to find the instruments . . . the process of transformation: in politics, the economy, culture, the law, *glasnost*, and democracy. And it turns out that those instruments are very important—no less important than the overall concept.[1]

The first statement shows the link binding the individual and the bureaucracy (P and O), the second the link between technical and organizational perspectives (T and O).

It is our firm conviction that any realistic planning for the future of the United States must pay full attention to the three perspective types and their linkages. The discussion in this chapter builds on the linkages examined in the preceding three chapters. We focus first on two issues raised in chapter 9: *(a)* the short-term versus the long-term view and *(b)* democracy versus freedom. Next we take up, from chapter 10, technology and the creative individual. Finally, we return to the issue raised in chapter 8, an innovative U.S. in the competitive world of the new century.

THE SHORT-TERM VERSUS THE LONG-TERM VIEW

How can we improve the balance between short-term and long-term concerns raised in chapter 9? Referring to figure 9.3, we can see that, it is reasonable to consider ways we can bring distant event points (crises or opportunities) like x, y, or z within the heavily shaded area, that is, within our current field of perception or planning horizon. There are seven techniques suggested here, cued to the perspectives generating them:[2]

1.   O: A minicrisis can be created in the near term, effectively moving x to a. If a potential future catastrophe can be averted by such a strategy, we may be justified in considering it. Labor uses the concept in precipitating strikes and the military use it in creating "threats", such as the non-existent "missile gap" in 1960.

2.   T: We can use today's information/communication technology to bring space-distant events close to us, moving y to a. Television has dramatically foreshortened the space dimension and, with forceful impact, brought distant crises and triumphs instantaneously into virtually every American living room. The Kennedy assassination, the Apollo manned lunar landing, and the Gulf War come to mind.

3.   T: Telecommunications has occasionally been effective in similarly foreshortening the time dimension, that is, moving from x to a. Orson Welles's radio version of H. G. Wells' *War of the Worlds* offers an early example. A more recent one is *The China Syndrome*, a successful film presentation of nuclear industry dangers.

4.   T: An interesting approach suggested by the success of shrinking the space dimension technologically is to transform x,

say, an undervalued future problem for us, by substituting a current comparable problem elsewhere, *y*, and then bringing that *y* into our view at *a* via technology. If we see environmental pollution as a future crisis for us, we can bring images of similar current environmental destruction in East Europe and the Third World vividly into our homes, raising awareness of what is in store for our grandchildren.

Another example: Scandinavia is usually about fifteen years ahead of the U.S. in social legislation, such as allowing paternity leave following the birth of a child.[3] The significance of such precursor events can be communicated by insightful television coverage of descriptions and impacts that are already occurring.

5. T/P: A very different alternative is to extend the individual's field of perception or horizon, that is, expanding the heavily shaded area in the direction of point *z*. This can be done through education and is a slow process. Children today are becoming conscious of the significance of environmental degradation. In class they can develop good and bad scenarios as well as games depicting the evolution of the community in the lifetime of the students. Computer games are already available that familiarize the players with the long-term impacts of environmental changes (e.g., SimEarth) or urban growth (e.g., SimCity) and permit them to test various policy options. Virtual reality games will soon be coming to the market and offer a vehicle for intensifying the learning process.

6. O: The medieval Christians were taught that the reward for bearing the travails of their baneful existence will come in heaven. We need not go to such extremes. Corporations can provide incentives (e.g., bonuses) that are based on long-term performance rather than one-year results. Tax rules can be modified to more meaningfully reward long-term stock and bond ownership and penalize short-term profit seekers.

7. P: Most important of all is leadership that has a long-term vision and is able to articulate and communicate it, so that it galvanizes many others and becomes a shared vision of the future. This is equally true for corporations and governments.[4] This vision is what distinguishes the statesman from the politician and the Thomas Watson of IBM from the Ross Johnson of RJR Nabisco. National leaders, recognizing long-term problems, have at times effectively used scare tactics to bring them into near-term consciousness and gain wider support for controversial actions.

Franklin Roosevelt, unlike many Americans, saw Nazi Germany as a long-term threat to the U.S. in 1940. He implied that Germany might soon attack the U.S. even though it was at the time unable to cross the English Channel. Paul Volker, as chairman of the Federal Reserve Board, provided strong long-term leadership in reining in inflation. In our time we desperately need articulate leaders who will rouse the American people to extend their horizon. Such personal characteristic should become a strategic factor in filling key positions.

There are inevitably some dangers lurking in these seven approaches. For example, repeated creation of minicrises can dull their effectiveness (the "cry wolf" syndrome) and their misuse by vested interests can cause serious harm. However, these possibilities pale by comparison with the potential benefits. A longer-term horizon is crucial to alter the "me-now" attitude and veer the U.S. from a path that is pointed toward economic crisis and long-term decline.

DEMOCRACY AND FREEDOM

In the words of Robert Bellah and associates, "democracy means paying attention." They use this phrase to stress the need for the individual to care about his or her institutions, be it the family, the city, or the nation. They see this as the most critical prerequisite to revitalize our institutions for the common good[5] and focus on the connection between the individual and the society, O-P. Our contention is that *in the knowledge society, we must "pay attention" to the more complex dynamics of the connections of the triad T-O-P.*
The prospects of democracy and freedom are affected by:

- the *technical-organizational* perspective linkage (chap. 8): *(a)* the forecast of its institutional spread based on a 200-year trend (fig. 8.1), *(b)* the current globalization/localization trend, and *(c)* the power to enhance participation through information technology;
- the *organizational-personal* perspective linkage (chap. 9): *(a)* mutual accountability (i.e., the responsibility of the individual to the society and of the society to the individual)

and *(b)* the freedom of the individual and its endangerment by the global population explosion, tribalization pressures,* and the concentration of economic power resulting from "out-of-control" capitalism (scenario B_2 in our earlier discussion);

- the *technical-personal* perspective linkage (chap. 10): *(a)* education is a basis for individual empowerment; *(b)* information technology serves as the eyes and ears, the instructor and messenger, connecting the individual directly to the governance process.

The interactions are not invariably mutually reinforcing and beneficial in enhancing democracy. We now sketch five possibilities opened up by the technology.

First, *information technology, while weakening the nation-state community, can substantially facilitate "paying attention."* It can provide networks and "virtual communities," although they may not offer the personal contact that fully engages human beings and gives a sense of belonging. The individual may join many networks and participate in many virtual communities but have minimal identification with, or loyalty to, each. He or she may also feel little sense of responsibility for the success of each such community. On the other hand, the individual may see this kind of personal involvement as simpler, more direct, and productively engaging than the frustrating interaction with traditional private and public bureaucracies.

The new technology cannot replace human communities, but it may well make local communities with new linkages more effective for the individual than larger conventional ones. For example, it can allow ethnic communities to maintain their uniqueness, but give them full and equal access to all services and "virtual integration."

Second, *technology can promote a secure environment for the citizen in the global village, a prerequisite for a healthy democracy.*

---

*Freedom for an ethnic community does not equate with freedom for its individual members: "Where new democratic experiments have been conducted in retribalizing societies, in both Europe and the Third World, the result has often been anarchy, repression, persecution, and the coming of new, noncommunist forms of very old kinds of despotism . . . the antipolitics of tribalization has been explicitly antidemocratic: one-party dictatorship, government by military junta, theocratic fundamentalism."[6]

Germans and Japanese see no objection to registering their place of residence with the local police. Swedes have their personal health records on computers from cradle to grave. Thus, in an emergency, family members can be instantly notified and a hospital has available crucial medical data for an accident victim in seconds. Fathers avoiding their child support payments can be located and welfare system defrauders detected. All aspects of crime control become much easier. But in America, individualism is still idealized by many in terms of the Old West—free to roam incognito and tote a gun for protection. Mandatory registration and identification cards are seen, at best, as an unconstitutional invasion of privacy and, at worst, as a dreadful totalitarian menace.

It is ironic that information technology can now provide on a wallet-size plastic card all the information useful for personal identification, including driver's license, passport, voter registration, genetic profile, blood type, medical problems, and so forth. Each type of information can have its own security and accessibility level. The potential reductions achievable in crime and in government bureaucracy are truly mind-boggling. But most Americans equate this use of technology with Big Brother cybernetic surveillance. The fear of misuse is, unfortunately, not unreasonable. The technology already is being used in marketing to profile supermarket customers without their knowledge, while "secure" systems are proving vulnerable to tampering.

Third, *technology can serve as a powerful catalyst to affect the problem of individual freedom and its endangerment by population pressures.* In discussing the population and technology explosions in chapter 1, we observed that information technology is uniquely compatible with a crowded world because it does not have the characteristic of a zero-sum game, that is, spreading knowledge does not take it away from the donor. Knowledge is not limited as material resources are. More people can (not necessarily will) mean more total intelligence and knowledge.

Technology can redefine freedom. The urbanite's curtailed freedom of movement may be restored by redesigning cities to function with public transit systems rather than with cars. The ability of information technology to connect the individual with the entire world, with sources of knowledge, and with common interest groups, as well as the ability to create virtual realities, vastly increases the freedom of the individual even if physical movement becomes more constrained. Limits in available resources, such as energy, need not constrict the individual's reach. And the cost of information is constantly declining.

Fourth, *technology can enhance physical resources enormously to ease the pressures created by population growth.* "Necessity is the mother of invention" may be said to describe the relation between resources and technology. Looming shortages and bottlenecks provide strong incentives for intensified search for resources and development of new, and often superior, alternatives. Vast regions of the earth have never been explored for minerals and technological substitution has consistently solved resource shortage problems. The earth has potentially enough natural resources to supply the needs of populations much beyond the present one, provided cheap energy is available (see chap. 7).[7]

Fifth, *technology can enhance our human resources through education and training and thus help disarm the population explosion as a "threat."* The most severe challenge for technology is to create more jobs for the growing population, a task compounded by the labor-reducing characteristics of aspects of information technology, such as automation. Through innovation it must create new industries at a rapid pace. Consider how many of today's industries were nonexistent at the beginning of this century, for example, aerospace, computer, motion picture, television, and waste management. Advanced training and education technology are needed to bring the new industries to early realization. Another powerful assist from technology is the development of long-term reversible birth control devices for both men and women. Further improvements are likely and, as the cost declines, they can be made widely available.* In such ways technology can serve as the engine to realize economic health. History suggests that a rising tide of economic well-being may be the most effective device to reduce population growth: the affluent have far fewer children than the poor, and they have them later in life.

The challenge of the T-O-P balancing act leads us to search for specific cases that may provide insights. A relevant one for the democracy-freedom problem is that of the city-state of Singapore. It has been transformed from backward to well educated and

---

*That does not mean, of course, that they will be used. For example, in Nigeria it is estimated that only 6 percent of the men have ever used a condom. In this country, a recent survey reports that 54 percent of high school students say they have had sex and the majority uses no birth control.[8]

properous in one generation. Its per capita income has grown from $500 in 1965 to $12,000 in 1990; unemployment has declined from 14 percent to less than 2 percent. Inflation is low (2 percent to 3 percent annually) and foreign investment, particularly the high-tech variety, is high (more than $1 billion in 1990). More than 800 U.S. companies have invested $9 billion in Singapore. Its harbor is the world's busiest, with about 700 ships in port at any one time. In 1990 it handled 188 million tons of cargo. The city-state produces more than half the world's computer disk drives and is a major data processing center for Asia. Government efficiency is exemplified by the fact that a road pothole is filled within forty-eight hours of being reported to the Public Works Department. There are five primary reasons for this success story:

1. T: Small size (240 sq. miles and 3.1 million inhabitants, but not culturally homogeneous)

2. P: A strong and brilliant leader, Lee Kuan Yew (a graduate in law from Cambridge with highest honors, he has insisted on firm ethical standards: the incorruptibility of the government is illustrated by the fact that a bribe is a ticket to jail)

3. T/P: High level of competence throughout government (civil servants are well-trained and get salaries comparable to private industry)

4. T/O: Willingness to experiment (pragmatism is the rule—if it works, keep the policy, if it does not, change it)

5. O/P: Discipline (achieved by a myriad of enforced rules and regulations—encompassing activities such as gum chewing, littering, and toilet flushing—is accepted by the individual to a degree that would be inconceivable in the U.S.; Lee, who routinely suppresses human rights, states, "I'm not sure human rights are a traditional value, even in Christian societies")[9]

Singapore has achieved a highly successful free-market economy, but at a steep price: constraining democracy and limiting personal freedom, just as is the case under a fascist or communist regime. Nevertheless, this "soft authoritarianism" may serve as a model for other free-market societies, an alternative to American-style democracy.

Malaysia, with a Malay, rather than Chinese, cultural base, and a larger population (18 million), has exhibited a similarly spectacular growth pattern. Over the past twenty years, it has averaged a 6.5 percent annual economic growth rate, as an agricultural country has been rapidly industrialized. One in seven semiconductors imported to the U.S. is assembled in Malaysia. Foreign investments are welcomed; Matsushita, for example, has invested in eleven new manufacturing facilities since 1987, and its annual sales in Malaysia are estimated at $1.6 billion. As in Singapore, the primary force for the country's development has been one man, its premier since 1981, Mahathir Mohammed. As in Singapore, the approach is one of enlightened, stable, "soft authoritarianism," with an emphasis on an open economy fueled by foreign investments, rather than on democracy.[10] Malaysia is actually evolving into a serious competitor to Singapore, drawing enterprises from the city-state.

Los Angeles and California, discussed in chapter 7, should serve as living laboratories to try out new relationships between individual, society, and technology; that is, new structures of multicultural democracy. The present structure of the County of Los Angeles was established in 1913 when the population was a mere 8 percent of what it is today. Organizationally oriented proposals thus tend to dominate the planners' current debates. One of the proposals is to cut vehicle travel hours 34 percent by shifting 12 percent of new jobs to outlying counties while shifting 6 percent of new housing from outlying areas to the central city. Community-based projects, led by neighborhood residents, should provide housing and jobs. An Affordable Housing Production Trust Fund partly financed by corporations (with existing tax credit incentives) has been proposed to make loans to community-based groups for low-income housing and assist developers to create model affordable housing projects. Drop-in learning centers where families learn together can help increase literacy. Apprenticeship programs at the high school level are proposed to aid minority students to enter new fields of work.[11]

So far, surprisingly few of California's massive, talented human resources have been enlisted to design and implement imaginative experiments to exploit the technology to develop "unity with diversity" and to motivate individuals to come together and act as "extended neighborhood families" (e.g., coordination-intensive networking within and across enclaves).

California is, of course, not the only living laboratory. There are forty-nine other states where new concepts can be tested. For example, Oregon's innovative restructuring of high school education and Medicaid prioritization offer valuable learning experiences. Beyond U.S. shores, there are diverse "laboratories" that may prove instructive. Perhaps the most historic is Switzerland, where German, Italian, French, and Romansh cultures have lived peacefully and synergistically together for many centuries. The country's small size and geomorphology undoubtedly facilitated attainment of a balance between local autonomy and centralization long before the arrival of modern information and communication technology.

The key to this "living laboratory" approach is openmindedness and willingness to experiment, to risk failure and learn.

THE TECHNOLOGICAL FRONTIER
AND THE CREATIVE INDIVIDUAL

We next revisit the T-P linkage in technological creativity; we examined this in chapter 10. This linkage must not be interpreted as implying a minor role for the O perspective and its linkages T-O and O-P in the creative process (fig. 11.1). Important organizational needs are an open environment with freedom to pursue untried ideas, minimum external constraints, encouragement, recognition, challenge and stimulation. We can readily point out interactions among all these perspectives:

> P + O -> T: The creative individual provides input to the organization on the needed technical resources
>
> P + T -> O: The creative individuals and the technical resources determine the target areas for investigation and the scope of innovations to be pursued by the organization
>
> T + O -> P: The technical supporting staff often performs detailed, behind-the-scenes tasks and provides input to the organization that determines how best to proceed toward successful innovations

The O-P relationship with regard to creativity styles has given rise to an adaption-innovation theory.[12] Individuals who see

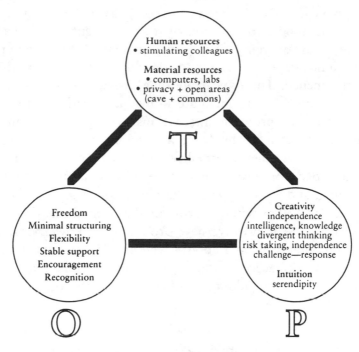

FIGURE 11.1
Perspectives on Creativity

change in terms of improvement in an existing setting, acceptance of established organizational assumptions, and concern for organizational stability are labeled "adaptors." Individuals who pursue radical or new approaches that do not fit the existing setting and assumptions, and that challenge the organization's stability, are termed "innovators." The former are more O-minded, the latter more P-minded. Recognition of these contrasting styles and their meshing can play a vital role in forming a smooth and constructive O-P bond.

"General management" skills are quite inadequate for managing innovation. Invoking multiple perspectives, we can develop insights about the creative process in an organizational setting that facilitate the unique task of managing creative individuals.[13]

Outstanding examples of creativity fostering organizations are The RAND Corporation in the 1950s, the Santa Fe Institute since the mid-1980s, and Apple Computer, Inc. All are characterized by an informal, exciting, highly stimulating setting that melds T, O, and P into a powerful innovative system.

Apple's John Sculley views his company as "a living laboratory for the model corporation of [the] new century . . . It's as if we already are a twenty-first century company."[14] A case in point is the innovative research and development complex that it recently opened. The typical building provides researchers with both private offices (to facilitate individual creativity) and common spaces (to promote interactions among team members). Labelled by architects the "cave and commons" concept, such a design combines quiet informality with intense commitment. The common spaces or "user-definable areas" serve as places for meeting, eating, and even sleeping. They have couches, stuffed chairs, and rugs, as well as white boards for sketching ideas. One of the company's most impressive new products, the Powerbook notebook computer, weighing about seven pounds, may itself change high-tech workplace operations. It makes portable and wireless computing practical, adding another dimension to organizational flexibility.[15]

THE U.S. IN A NEW WORLD OF COMPETITION

We now return to the central theme of part 4, the U.S. at the edge of the 21st century and, in particular, its role as an innovative society in a highly competitive world.

Using the schematic of figure 6.2, let us look at the U.S. in terms of its strengths and weaknesses in the global arena of the evolving economic competition. The U.S. has unquestionably been very strong in technology (T) and individualism (P), in particular, technological creativity. The strong T-P bond is reflected in the fact that *the U.S. has accounted for about half of all major innovations since 1730.*[16] Given the proper motivation, the U.S. has also excelled in collective (O) efforts:

- The Nazi and Japanese aggressions inspired America's all-out World War II effort, including the Manhattan Project
- The Soviet "threat" gave rise to a powerful military-industrial defense complex
- The Soviet Sputnik evoked a highly successful Apollo space program

In each case *a high degree of organizational cohesion, marked by innovation and rapid response, was achieved. It is significant that the motivation was a perceived foreign threat.*

Figure 11.2a suggests this situation schematically as a T-O-P profile. The same characteristics apply to organizations such as Apple, Microsoft, and the Lockheed skunk works in recent years, as well as to corporations such as IBM in earlier incarnations (fig. 11.2c). Unfortunately, aging giants such as General Motors and Chrysler Corporation lost much of their original T and P strength; bureaucratization (O) came to dominate, leading to organizational arteriosclerosis in the 1980's (fig. 11.2d).

One specific example suffices to illustrate this situation. When he became manager at Chrysler in 1985, Ron Zarowitz proposed a built-in fold-down child seat as an option for car buyers. The first such seats appeared in Chrysler cars only in 1992—and they are being sold as quickly as they can be made.[17] By contrast, the future innovation-oriented corporate leader (fig. 11.2e) will make full use of coordination-intensive structuring to maximize responsiveness and flexibility.

For comparison, the case of Japan is depicted in figure 11.2b. P is weaker than T and O in the Japanese culture, and this is reflected in the relative number of Nobel prizes in science: 151 for Americans, 4 for Japanese through 1989 (fig. 10.1). The Japanese culture plays a central role in linking T, O, and P. The pantheon of Shinto, Japan's state religion, includes the sun goddess, and all Japanese are considered to be her descendants, forming a cohesive nation that is "a family of families." The education system imbues the young with a sense of being an integral part of this Japanese "family" and serves as a powerful tool in molding Japanese society; it accentuates discipline, hard work, and technical training.

Kaizen describes a peculiarly Japanese philosophy relevant to economic competitiveness. It constitutes both a collective belief and a collaborative, total commitment to excellence in action. Unlike the result-oriented management that preoccupies Westerners, *kaizen* focuses on process-oriented thinking, facilitating the linkages between T, O, and P to attain improvement that is constant and steady, albeit less spectacular than the attention-getting Western innovation path. *Kaizen* covers total quality control, just-in-time, productivity improvement, workplace discipline, new product development, robotics use, and many other factors. It is concerned with building quality into its people and takes a long-term view.[18]

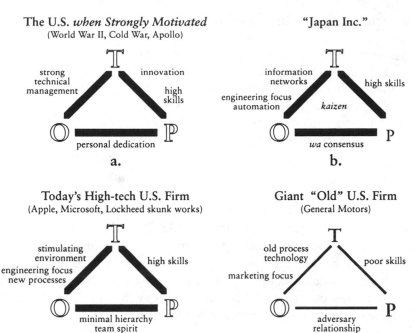

The U.S. *when Strongly Motivated*
(World War II, Cold War, Apollo)

strong technical management · innovation · high skills · personal dedication

a.

"Japan Inc."

information networks · high skills · engineering focus automation · *kaizen* · *wa* consensus

b.

Today's High-tech U.S. Firm
(Apple, Microsoft, Lockheed skunk works)

stimulating environment · engineering focus new processes · high skills · minimal hierarchy team spirit

c.

Giant "Old" U.S. Firm
(General Motors)

old process technology · poor skills · marketing focus · adversary relationship

d.

The Future-Oriented Firm

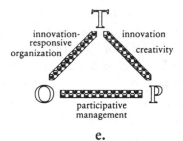

innovation-responsive organization · innovation creativity · participative management

e.

P   P   letter size and line thickness
━━━ ─── denote relative strengths
▨▨▨▨▨ = coordination intensive structure

FIGURE 11.2
T–O–P Profiles for Economic Competitiveness

One aspect of the "family" bond between O and P is the Japanese desire to avoid open conflict at all costs by seeking consensus. Harmonious teamwork is a societal objective. There may be wide-ranging discussion among individuals, but, once a decision is made, there is complete family unity. In the terms of our discussion, the agreed-upon O perspective displaces all P perspectives.*

The Japanese process to minimize the mismatch between T and O in the industrial system can be summarized in the form of five principles:

1. corporate self-regulation
2. interface management through networking
3. interdependence (differentiation and integration) among the constituents of T and O—the total is more than the sum of its parts
4. joint optimization of the technical and social systems
5. open system planning with rapid adjustment to change in the environment, constantly balancing competition and cooperation[19]

One major effect of this process is the ability to quickly and massively exploit new technological opportunities. It is evident that the Japanese culture has been instrumental in attaining a strong T-O-P profile. The U.S. achieves a very effective T-O-P profile when motivated by a perceived foreign military threat or in unusually dynamic corporate settings. It must be emphasized, however, that in a highly dynamic world, achievement and maintenance of a such an effective T-O-P profile is by no means assured. It will constitute a continuing challenge for both the U.S. and Japan.

Let us briefly look at some of the other actors on the forthcoming global economic stage, specifically those in Asia. A major reason for the difficulties experienced by developing countries in technology transfer and technology leapfrogging is the inadequate attention paid to the national technology climate or T-O-P profile,

---

*One area where the team approach is not proving effective is the software industry. The creative loner is encouraged in the U.S. and discouraged in Japan. Nor is the entrepreneurial atmosphere supportive of a Japanese Steve Jobs or Bill Gates. However, countries like India and Russia, where P is stronger than in Japan and mathematics has a notable tradition, are potential world-class competitors to the U.S. in this field. Their low wages for world-class programmers and the inevitable maturing of this industry favor their ascendancy.

which conditions the process of planning and implementation of decisions relating to technological change.

South Korea exhibits a climate that differs from that of Japan: strong T and P, but weaker O (fig. 11.3a). Korean decision makers are younger, more individualistic, and often trained in the West. Here the T-P bond facilitates technology transfer. The Korean electronics industry uses five steps to build its competitive competence: (1) random exploration, (2) technology introduction, (3) equity joint venture, (4) imitation and production, and (5) development of its own technology. There is a recognition that Korea cannot depend on the latest information from foreign firms and must aim for some improvements on received technology.

An example of step 5 is seen in the development of a printing mechanism and control unit for Korean alphabet office equipment.[20] Once again the P perspective comes to the fore. The guiding force behind Korea's technological development is the dynamic, American-trained engineer Dr. Hyung-Sup Choi, founder of the Korean Institute of Science and Technology. There is today a national push for innovation, reflected in government R&D-related loans and tax credits as well as direct funding of R&D. T-focused, analytic, top-down planning has recently been instituted and has identified product-oriented technologies (e.g., high-definition TV) and fundamental technologies (e.g., environmental engineering technology) for national R&D support. However, the T-P bond is most apparent in the massive private R&D thrust and the drive to succeed.[21]

In Thailand P and O are moderately balanced, but T is weak. In India, Pakistan, and Bangladesh, T and O are both weak and P is supreme. Due to the legacy of the colonial period, O works in a very peculiar way. Open peer review is not acceptable. Individuals often use knowledge and skill systematically to destroy peers. The limited T may be utilized to find ways and means to fulfill self-interest at the cost of collective development. Outspoken leaders are shunned, and mediocre ones, who do what is expected of them, are preferred. Consequently, the technology climate is poor.[22] Even so, there are islands of unquestioned technical excellence.

The cases of Singapore and Malaysia were noted earlier in this chapter. They are interesting in view of the rapid change in one generation. Singapore has acquired a very favorable technological transfer climate by the hard-driving (P) leadership of one person,

a. Four Asian Countries

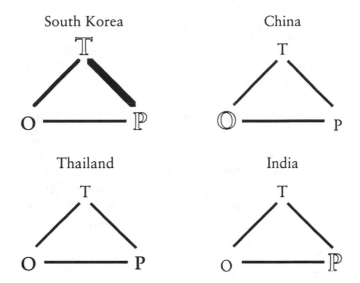

b. The Rise of Singapore

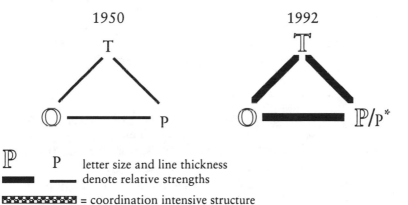

ℙ P	letter size and line thickness
▬▬ ──	denote relative strengths

▦▦▦▦▦▦ = coordination intensive structure

* Denotes strong leader, compliant followers

FIGURE 11.3
T–O–P Profiles for Technological Advance in Asia

prime minister Lee Kuan Yew (see fig. 11.3b). He has invigorated the T-O-P linkages using highly authoritarian means, making the 240 square mile city-state a very able world-class trading center. Malaysia has industrialized at a dizzying pace under the leadership of the nationalistic and charismatic prime minister, Dr. Mahatmir Mohammed.

Neither natural resources nor size is the key to the growth of Japan, South Korea, Taiwan, Hong Kong, Singapore, and Malaysia. The meshing of T, O, and P have proven decisive in creating value, specifically, well-educated workers, enterprise-friendly governance, and technology acquisition. For developing countries this technology flow is stimulated by a "make-some, buy-some" strategy.

In China the T-O-P linkages are poorly developed, and the dominant O perspective alone cannot achieve a good technology climate. We discussed some aspects of the Chinese problem earlier, in chapter 8. American systems analysts immersed in the Chinese setting provide some interesting insights about the weak T-P linkage:

> Our impression is that the three fundamental activities underlying decision analysis (generating alternatives, accounting for uncertainties, and eliciting decision makers' preferences) . . . are not being accepted nor are they likely to soon be accepted by the Chinese bureaucracy . . . [A]nalysis would proceed under the assumption of complete certainty about almost all aspects . . . In a formal, ritualistic culture that values the expected (certainty), assigning probabilities to events must seem strange indeed . . . The open questioning, cajoling, hypothesizing, and thinking out loud [in our role-playing exercise in utility assessment] probably seemed either preposterous or silly.[23]

> Management science/operations research [MS/OR] problem solving may suffer from the traditional Chinese tendency to categorical formalism reflected in the fondness for such constructs as "the Four Modernizations.".. Indeed the one weakness of my otherwise exceptional students was their tendency to force problems into a "type" like those studied in class. They displayed great ingenuity in solving formal problems, but when faced with less formal problem descriptions, it rarely occurred to them to think "outside" of the problem, to change it, or to wonder whether there was a problem at all. Even more tellingly, they seemed to lack skepticism about the tools, methods, and applicability of MS/OR.[24]

The language itself is an obstacle to modernization. It does not absorb new ideas readily and new characters are rarely invented. A geneticist at the Chinese Academy of Science described the situation thus: "We are being held prisoner by our language. There is an information explosion in the world that we cannot cope with in our language as it is now."[25]

Westerners have misperceived Chinese culture, and the Chinese Western culture, for at least two hundred years. The British mission led by Lord George Macartney in 1792 initiated a period of contacts that were notable for people from dissimilar mental worlds talking past each other. The British were fascinated with innovation, the Chinese could not grasp the idea of progress. The dynamic British went everywhere, the static Chinese went nowhere (being convinced that their society was already the best in the world).[26]

Today Westerners are often taken aback by the way the P perspective is culturally downplayed in China at the expense of O, individualism at the expense of conformity. *Danwei*, one's unit, is more important than one's name.[27] Without a *danwei* a person's existence was barely recognized until recently.[28] Privacy is not appreciated, and individuals are discouraged from expressing personal opinions. A set of guidelines for Americans doing business in China advises:

> The foreign businessman should not focus on the individual Chinese person, but rather on the group of individuals who are working for a particular goal. If a Chinese individual is singled out as possessing unique qualities, this could very well embarrass the person . . . In discussions with Chinese people, the foreigner should avoid "self-centered" conversation in which the "I" is excessively used. The Chinese view with contempt the individual who strives to display personal attributes, as Chinese are much more oriented to the group.[29]

It is noticeable that overseas Chinese, operating in a setting less organizationally dominated, unshackle formidable T-O-P strengths and often become successful scientists and technological entrepreneurs. Singapore (70 percent Chinese) and the U.S. offer many striking examples. Not surprisingly, Singapore has now become a model for China's leader Deng Xiaoping. In the words of a young Chinese government official, "Of course old Deng likes Singapore. It's run by Chinese, it's efficient, it's rich, and no one jabbers about human rights."[30]

The traditional dominance of the O perspective in China is also apparent in other centrally planned societies, such as those of the former Soviet Union and many Third World countries. It places the emphasis on organizational perpetuation, stability, and control at the expense of technical problem solving. It minimizes local participation, leading to policy failures and economic disasters. As a consequence, decentralization efforts are now in progress in China (see chap. 8), as well as areas such as the former Soviet Union and Mexico. In the Mayan (Yucatan) region of Mexico, for example, local resources have been brought together successfully for sustainable development projects. The T, O, and P perspectives are swept in by better horizontal communication and participation of diverse groups, creating constructive interaction and balance among them. Computer-mediated communications may prove particularly helpful in linking the various informal groups.[31]

Having glimpsed the characteristics of some competitors, let us return once more to the United States and the new era. Our discussion in chapter 8 suggested three alternative scenarios for the global economy:

- A, a struggle among three economic superpowers, North America, Europe, and the Far East
- B_1, smart capitalism in a global megalopolis, with less focus on national entities and more on local, regional, and global ones
- B_2, a global megalopolis becoming a global asphalt jungle, with rampaging, dog-eat-dog, quasi-anarchic, out-of-control capitalism

All three represent change, but only A and B_1 represent progress and innovation. Ipso facto, they are bound to evoke strong resistance and thus constitute severe challenges. We focus on them here.

Scenario A

In this scenario, the U.S. is back in a race, a commercial version of the military/space confrontation it undertook against the Nazis and Japanese in World War II and against the Russians in the subsequent cold war. In theory, it should be possible to reinstate the strong T-O-P balance of figure 11.2a. The question is, Will there be a trigger to motivate and unite the American public, one comparable to that presented by Pearl Harbor, the image of a demonic Hitler, the Soviet atomic bomb, or Sputnik?

We recall, for example, the constant concern after 1950 to be technologically ahead of the Soviet Union. The planning horizon was ten to twenty years. Intelligence was monitored assiduously to avoid unpleasant surprises. The Department of Defense and NASA were organizationally attuned to the rapid pace (see table 8.1). In the words of General Colin Powell, Chairman of the Joint Chiefs of Staff, "For forty years we have chased Soviet technology and have felt the hot breath of Soviet technology on our back. We were always trying to stay ahead of it . . . Our whole industrial base, and their whole industrial base, kept churning out. That is gone."[32]

During the Reagan-Bush years, there was no sense of urgency about an economic "threat," either at the corporate level or at the national level. At the corporate level, most eyes were focused myopically on the next quarterly report. The old American business adage that "a long-range planner is a person who brings a lunch to work" was only a mild exaggeration. At the national level there was no imaginative leadership (P), no equivalent of the National Security Council to develop an overall strategy to run the race (O), no political equivalent of the Department of State to coordinate with our allies (O), and no economic equivalent of the Department of Defense to organize and coordinate the human and material resources for the struggle (T-O-P). A serious effort would include elements such as:

- intelligence, with a focus shift from military to economic intelligence gathering and environmental scanning
- needs analyses, analogous to those done by the military to determine new system requirements based on future "threats"
- risk sharing between government and industry on new technologies
- joint government-industry research and development projects
- development of coordination-intensive structures to cut lead time for new products and maximize responsiveness to customer needs
- protection of new products from exploitation by members of other economic blocs for a stipulated period of years

- education and training to develop a highly skilled, world-class work force, supplemented by programs for lifelong learning and retraining

Leadership (P) is the first step in gearing up for the new competition. The absence of a trigger and the subtle nature of the evolution of this scenario present an enormous challenge to leadership. Franklin Roosevelt had a difficult time rallying the nation before Pearl Harbor. The "arsenal of democracy" and the military draft were difficult to sell to a country that did not feel directly threatened. John F. Kennedy sold the Apollo program only because Sputnik raised the fear in the public consciousness of a Soviet landing on the moon: that possibility was construed as a military threat. Even Dwight D. Eisenhower, a charismatic president (and military expert), was unable to convince the American public that this widely accepted interpretation was false.

The combination of an extended decline in the American economy, apparent growth of one or both competing economic blocs, and a restive, angry American public may galvanize the emergence of new leadership that articulates the new challenge and engages the nation in a collective response with the reestablishment of a strong T-O-P balance as in figure 11.2a. There is an inevitable gamble. We recall that in 1933 one depression-plagued democracy (the U.S.) chose Roosevelt, while at the same time another depression-plagued democracy (Germany) voted for Hitler.

Scenario B_1

We stressed in chapters 7 and 8 that information technology makes possible simultaneous localization and globalization in both public and private sectors. The impact is to weaken the nation-state, eliminate the national identities of corporations, and create many new networks and linkages.

In this information technology world, which invites, even demands, coordination-intensive restructuring, terms such as *U.S.* and *trading powers* may not mean what they did throughout most of the twentieth century. Traditional national borders will fade in significance as globalization and localization proceed. Traditional corporate boundaries will also change as restructuring paves the way for operations in a global village. We should remind ourselves that the Greek city-states, the Mediterranean republics (such as Venice and Florence), and the Hanseatic League of German and Scandinavian towns prospered economically without nation-state structures. Similarly, it is useful to step back and remember that corporate organizations have transformed themselves periodically for two hundred years (fig. 7.3). The "global tribes" described in chapter 8, that is, the Anglo-Saxons, Japanese, Chinese, Jews, and Indians, provide valuable global linkages today in commerce and technology through their networking. They thus help to advance the concept of "smart" capitalism in the global village.

Simultaneous localization and globalization reduces the central role of the federal government in commerce and gives rise to regional intergovernmental as well as nongovernmental confederations and compacts. A global marketplace, a global commons, with imaginative new coordination structures will prevail over nationalistic power politics. In Scenario B_1, "smart" capitalism, the focus is on a constructive blend of peaceful cooperation and competition. In *The Long View*, Schwartz comments on this:

> This world is entrepreneurial, multicultural, full of hope and harshness . . . Economic intelligence is the organizing principle . . . The flood of complexity is kept coherent by a new way of organizing society, built around communications links. It favors those who can understand the unpredictable dynamics of trade; it encourages cooperation, productivity, efficiency, and organizational learning. People work to achieve, not to control; decentralization and diversity are paramount, but do not block the flow of ideas between regions. Vested interests still exist, but they get found out and removed—not by government regulators, but by eager competitors and investors . . .
>
> Education [is seen] as the path to prosperity . . . Learning faster than your competitors comes to be seen as the only sustainable competitive advantage in an environment of rapid innovation and change.
>
> In this "market world," companies continually restructure, rise, fall, and intermesh . . . Boom-bust cycles are common and expected . . . This is a world learning to live with innovation as a constant.[33]

A good T-O-P balance is likely to be characteristic of the successful enterprises in this setting (fig. 11.2c, 11.2e). Procedures will be sought, copied, and adopted from many sources. Table

11.1 reviews some of the potentially valuable lessons drawn from the discussions on T-O-P linkages in this book. There will be successes and failures as survival of the fittest becomes a constant challenge in this evolutionary process. Mistakes will be made as the transfer of procedures from one society to another confronts differing cultural settings.

The global linkages will facilitate a continuous learning and self-reorganization process. For example, the Japanese cultural weakness in personal creativity and the American weakness in group discipline are not likely to persist in this scenario. The global village with its networking and mixing of multicultural people should, in the long term, move societal development toward a more complex level, following the process that has characterized the evolution of life from the primeval soup to the present.

RIDING THE INNOVATION WAVE

A common feature of scenarios A and B_1 is the concern with innovation. In the new competitive world, it is crucial for Americans to exploit the leading edge of technology, that is, to ride the wave like a surfer. It is a strategy Don Kash calls "perpetual innovation."[34] No events illustrate this strategy better than Sputnik and the first Soviet orbital flight. John F. Kennedy and Lyndon Johnson saw the opportunity to exercise leadership and gain wide public support. The U.S. space effort shifted into high gear, and the country successfully beat the Russians to the moon. Similarly, the Department of Defense for decades displayed a constant concern to maintain a technological lead in weapon systems, focusing its attention on an array of critical areas. The future "threat" served to propel the defense industry ahead technologically at a breakneck pace for decades.

The Japanese, drawing on American creativity, have been racing to market innovations at a similar pace in other sectors. Consider the case of the robot. The first programmable robot was created by George Devol in the U.S. in 1954. The first industrial "pick and place" robot followed in 1959, and General Motors installed the first robot, built by Unimation Corp., in 1962. The first sale to Japan was made by AMF Corporation in 1967. By 1970 the first Japanese-built robots were sold. At the end of 1982, there were about 32,000 robots in operation in Japan, 7,000 in the U.S. One reason for Japan's rapid deployment is the fact that the Japanese robot-producing companies, such as Mitsubishi, Hitachi, and Seiko, are also robot users themselves. This tactic has yielded exceedingly valuable hands-on experience with direct feedback to development. A *process* innovation, such as robots, takes more time than a *product* innovation. The pace of introducing new products such as videocassette recorders is appreciably faster. Not surprisingly, since 1983 Japan has gained, and the U.S. has lost, share in total patents in thirty-eight of forty-eight product categories, including computers and transportation equipment.[35]

A recent, fascinating example is the one of fuzzy logic. Fuzzy set theory is a mathematical concept created in the 1960s by Lotfi A. Zadeh of the University of California, Berkeley. It is a calculus that quantifies fuzzy or imprecise terms such as *heavy* and *far*, thus permitting their incorporation into a computer program. Its use makes it possible for a machine to make rough, commonsense judgments, thereby exercising a primitive form of intelligence.

While the American engineering community dismissed the concept as a product of "fuzzy thinking," the Japanese saw the potential for its use in product improvement. Today its subway trains use the technology for controlling brakes by feeding into the train's computer the approximate speed of the train and its distance to the next station. Japanese washing machines, cameras, vacuum cleaners, portable computers, and air conditioners incorporating fuzzy logic are already on the market. The Matsushita washing machine adjusts its operation as a function of the amount of dirt, type of clothing material, and load size it senses. Its vacuum cleaner shows by colored lights when it is picking up dirt. The Matsushita camcorder compensates for shaky hands when filming; the Sanyo electric razor adjusts its blade speed by measuring the thickness of the beard. Combining fuzzy logic with expert systems, Komatsu has cut the maintenance time for some of its heavy construction equipment by 24 percent.

The next advance is expected to be the combination of neural networks with fuzzy logic and expert systems to make the hardware still smarter. Not surprisingly, the Ministry of International Trade and Industry (MITI) has established two laboratories jointly with corporations such as Toshiba and Hitachi, as well as with universities, to explore the next stage of fuzzy logic.[36] As the competitive pressures to innovate intensify, the Japanese are moving inexorably toward more basic research. Other examples that

TABLE 11.1

Reviewing Lessons on Strengthening T-O-P Linkages

Country	T-O Linkage	O-P Linkage	T-P Linkage
U.S.	strong military-industrial high-tech complex team projects: Manhattan Project; Apollo Space Project; skunk works research on coordination-intensive structures	encouragement of technical entrepreneurs (e.g., Land, Jobs, Gates) minimal hierarchy in high-tech organizations	individualism promotes creativity (150+ Nobel science prizes) superior research facilities
Japan	"Japan Inc." spirit, MITI vertical and horizontal information networks intelligence, environment scanning/dissemination risk sharing (govt.-ind.) fast reaction—concept to marketable product government-industry R&D ties	*wa* (consensus concept) cooperative management-labor relations job security corporate stability long-term view "kaizen"	strong engineering training national skill certification job rotation quality circles

Country	T-O Linkage	O-P Linkage	T-P Linkage
Germany	culture values discipline good customer servicing	management-labor codetermination good social services	apprenticeship concept
S. Korea	five-step program to build technical competence		good technical training parallel advancement ladder for engineers, craftsmen strong leadership (Choi)
Singapore	favorable capital investment climate	imposed discipline capable government bureaucracy strong leadership (Lee)	good technical training

underscore the pattern of U.S. innovation followed by rapid and effective Japanese exploitation include active matrix liquid crystal displays and micromachines.[37]

The Japanese strategy calls for a long planning horizon, rapid change and acceptance of considerable risk and uncertainty. Akio Morita, the chairman of SONY, observes that American companies struggle to create a vision for the next quarter, while Japanese companies have a vision for the next decade. A Hitachi slogan is, "Though we cannot live one hundred years, we should be concerned about one thousand years hence." Thus the organization attempts to shape the individual's perspective. Furthermore, the individual is placed in an organizational setting that makes the individual comfortable with this level of change and uncertainty. (However, it is not nearly as effective in promoting individual creativity.) Peter Drucker has found that Japanese companies work simultaneously at three levels of innovation: incremental improvements, significant jumps, and true innovations.[38] By contrast, in the U.S. there are far too many cases where the organizational setting does not support innovation. There used to be a significant difference between the public and private sectors. In connection with a multiple perspective case study we did fifteen years ago, a city manager told us:

> Innovation does not occur within a governmental agency unless someone internal is interested in it or unless some fairly steady pressure is applied with a certain amount of directness . . . There are two reasons for that: you have a multitude of priorities and the inability to answer them all, and therefore when somebody suggests something new and wonderful, your tendency is to throw it in the wastepaper basket because you don't need one more thing. The second one is a very peculiar thing that applies only to government as far as I know and that is the tremendous price that anyone in government pays for failure.
>
> An example in the private sector: The Edsel was a tremendous failure of the Ford Motor Company. The guy who headed that program later went on to become the president of Ford Motor Company. Because all Ford asked after the catastrophe was, "Did you know what the hell went wrong?" And the project manager said, "Sure we do, we know that our pre-World War II marketing methods are totally unreliable in today's marketing. We've analyzed where we went wrong . . . and so we completely revamped our marketing techniques . . ." The Ford Motor Company went on with the Falcon and Mustang and

> just one success after the other from a marketing standpoint.
> And that's all they asked—do you know what you did wrong?
> The private sector accepts failure and that's part of the price you
> pay for progress. You've got to try the stuff out and if it doesn't
> work, you've got to cut your losses . . .
>
> Government can't accept failure. If somebody fails in the
> public sector, the political people are all over him because
> they've got to protect their ass; the editorial writers are all over
> him . . . you'd expect the resignation of whoever was connected
> with it, or the demotion, or certainly the end of their career . . .
> and it's this inability to accept failure that has made government
> officials very rigid about trying something new.[39]

Unhappily, the difference between the public and private sectors
has now diminished. Like their public sector cousins, private sector risk-takers are now not rewarded if they succeed, but are
penalized if they fail. The twenty-year struggle of Doug Keller and
Clay Smith and their Otisca Industries presents a case in point.
Having developed an innovative process for precleaning coal so
that it would burn much cooler and more cleanly than untreated
coal, they have languished in a morass of utility-industry aversion
to risk-taking.[40]

The enormous debt load now resting on the backs of many
corporations is a legacy of the leveraged buy-out and junk bond
era. This burden has resulted in substantial cuts in industrial
research and development. Overall, American industry's spending
on industrial research grew annually at an average of 7.5 percent
(in constant dollars) from 1980 to 1985; the figure dropped to 0.4
percent from 1985 to 1991. America's government-funded
research has remained focused on defense; in 1989 this area
received 65.5 percent of all research funds. By contrast, the German government spent only 19 percent of its research budget on
defense, and the Japanese a mere 9 percent in the same year. In the
1988–91 period, Japan, Germany, France, Italy, Sweden, and
Great Britain significantly increased their support of civilian
research, so that the cumulative level rose from 25 percent above
the U.S. level to 34 percent.[41]

According to a recent Arthur D. Little study of corporate
innovation practices, *U.S. executives are most concerned about
the dearth of managers who know how to drive the creative
process.* There are, fortunately, some admirable exceptions that
inspire imitation, among them Silicon Graphics, Amgen, and

Rubbermaid.[42] The National Science Board recommends that the National Science Foundation broaden its mission to include "the education of future corporate leaders of both high-technology and traditional industries."[43]

Don Kash perceives Americans as being rooted in the "secular trinity" of the individual, the free market, and efficiency: "In the United States the secular trinity is the lens through which we see and understand reality and is the standard that guides our decision making and management. It is also an illusion. Reality is changing, but the secular trinity is not. Increasingly, it is providing a distorted view of reality and is prescribing actions that are inappropriate to the nation's problems and needs."[44] The implication is that fundamental institutional changes must be made. But are they really so basic? We observed earlier that what the defense and space sectors accomplished for many decades was indeed a form of perpetual innovation. Highly effective linkages were established between government, industries, and universities. In effect, America has been operating in these sectors to some extent the way Japan operates in the commercial sector (see table 8.1). There is the longer-range thinking, the public-private sector collaboration, and the support of basic and applied research. One possibility is conversion of the Defense Advanced Research Projects Agency (DARPA), which has effectively supported advanced high-tech ventures in the defense area for decades.

Historically, agriculture has been a sector of very effective technological innovation and transfer in America. As already noted, America's work force engaged in farming has dropped from 50 percent at the beginning of this century to about 2 percent. Besides the motorized farm machinery, a most impressive factor has been the land grant college system established by the Morrill Act in 1862. This provided close linkage between the research community and the farmer through county agents (and was opposed by academe almost to a man).

There have been modest government efforts to encourage partnerships between universities, nonprofit institutes, and industry during the last decade. Engineering Research Center grants have been given by the National Science Foundation. These have fostered the transfer of technology from university or institute to industry and, less anticipated, from industry to university and institute. An example of the former flow is Sematech Corporation in Texas, a Pentagon-industry consortium that is helping to restore U.S. dominance in the semiconductor industry through advances in chip manufacturing techniques and standard testing guidelines. An example of the latter is MIT's Project Athena for campus-wide integration of interactive graphics and computers; this had a heavy industry knowledge component flowing into the university. Unquestionably this T-focused approach has proven beneficial; however, the T-O issue remains: *a common observed weakness is the lack of concern with the management of technology transfer.*[45] This is precisely the issue highlighted by the U.S. executives in the 1992 Arthur D. Little report noted earlier.

In 1992 the federal government took another step: establishment of the Critical Technologies Institute (CTI) at the RAND Corporation. Its task is to provide a T perspective on science and technology policies by which the government can assist business to develop its products. In particular, new areas (see chap. 7) that will keep the U.S. at the leading edge of technology are to be examined.

In academe we run into the obsolete compartmentalized organization, which is often unsuited to interdisciplinary technological innovation needs. Faculty see their tenure and promotion tied to old-line departments, and the awarding of advanced degrees raises jurisdictional problems typical of bureaucracies. According to J. Williams, head of engineering at Carnegie-Mellon University, "Over the next twenty years, more and more exciting problems will be interdisciplinary—in contrast to the pure science and narrow focus as at Yale or Princeton."[46] Consortia may help to drag academe, even if screaming, into the 21st century world of unbounded systemic thinking.

The P perspective is as important a factor in managing the innovation process as it is in promoting technical creativity. Consider, for example, the role of leadership for success in developing consortium arrangements:

- Erich Bloch of IBM was instrumental in inducing twenty semiconductor and computer corporations to pool research budgets into a superfund to create Sematech Corporation
- George Cowan, the head of research at Los Alamos, saw the need for a new, holistic approach to complex systems and became the driving force in creating the scientifically high-powered Santa Fe Institute

- John Young of Hewlett-Packard was the key fundraiser behind Stanford's Center for Integrated Systems, which brings together scientists from many major electronics companies

- James B. Hunt, Jr., governor of North Carolina, provided leadership in obtaining a fund to create a Microelectronics Center in the Research Triangle Park in his state.[47]

We recall from the discussion of long-term trends in chapter 7 that innovations tend to cluster during periods of decline in the long wave pattern. They lead to a spurt in the creation of new industries and substitution of technologically new products for old during the next growth phase. Figure 7.3 indicates that we can expect this phase in the period 1996 to 2024. We are rapidly approaching that time.

When a technology is in its infancy, the competitive strategy surrounding it should be one of maximizing performance. At the next stage of adolescence, maximization of market share is the key. When mature, cost minimization is in order, and in old age, offshore turnkey plants and disinvestment are appropriate. To stay at the leading edge, therefore, competition is based more on product performance than on price. The fact that this country is running a trade deficit *in high-tech goods* means that major changes are in order to pursue this strategy successfully.[48]

Ideas that are worthy of consideration include:

- changing the innovation process: more time devoted to thorough product planning and little change once the specifications are determined

- deferring corporate bonuses to base them on long-term performance rather than short-term results[49]

- creating a National Finance Committee to provide debt capital to sunrise industries (a concept attributed to economist Lester Therow)

- controlling the export of new technology (not products) for a limited time, say five years[50]

- making the tax credit for research and development permanent

- placing a moratorium on the U.S. Treasury regulation that induces American corporations to move research overseas

- enacting legislation to increase work force mobility by transfer of seniority, to stimulate training and retraining programs by incentives to industry (like the French sales tax rebate to companies that train apprentices), and to create incentives to individuals for continuing education

- accelerating product innovation by increasing worker participation in management

- raising productivity in manufacturing by stimulating process control, software, materials, and semiconductor research and development[51]

- providing for incentives to develop technological substitutes and reduce the "lifeline" obsession, justifying huge defense expenditures and support of Third World dictators to protect resource supplies

Finally, there is now specific historical evidence that corporate innovative capacity is strongly correlated to investment in the building of the public education infrastructure. Peaks in corporate innovative capacity follow peaks of investment in educational infrastructure (K–12), and troughs are similarly related.[52]

As we hurtle into the new century, staying on the leading edge will be the best, perhaps the only, strategy for creating enough new jobs to satisfy the need. This will be an exceedingly tough challenge, and will require constant attention to T, O, and P. As any surfer knows, riding a wave can be very tricky—and very tiring. Technological leadership cannot be maintained without tremendous drive and risk propensity. The balance between technology, individualism, and collective commitment is crucial. Policies that have been in place for a long time must be changed. Unless there is inspired leadership, it may take a severe domestic crisis to shake up the mindset of enough individual citizens to generate sufficient pressure and energy in public and private institutions to set the process in motion.

The most powerful and exhilarating way to stay at the leading edge combines scenario B_1, "smart" capitalism, with coordination-intensive restructuring in governance and corporations (chap. 8). Simultaneous localization/ globalization in the public sector and decentralization/ centralization in the private sector offer the organizational means to unlock the new era. Technology must provide the engine for economic success; it will make strong demands on creativity (chap. 10). And little can be accomplished without adopting a longer space/time horizon (chap. 9).

PART 5

Complexities
and Imperatives

CHAPTER 12

Accepting the Challenge

What a piece of work is a man! How noble in reason! how infinite
in faculties! in form and moving, how express and admirable! in
action, how like an angel! in apprehension, how like a god!

William Shakespeare, *Hamlet*

I may not be much, but I'm all I've got.

Anonymous

Parts 2 and 4 focused on substantive "case studies," part 3 on our
approach in viewing such complex "systems." In part 5 we draw
overarching implications from our explorations in the context of
managing ourselves and our technology in the years ahead. But
we begin by revisiting the idea of 'complexity'.

BACK TO BASICS: COMPLEX SYSTEMS
AND HUMAN PERCEPTIONS

The more our knowledge grows, the more complex we perceive
the world to be. Suppose we drop a rock into the center of a
placid pond that is so large that we cannot see the shore. The
expanding wave front emanating from the drop offers an analogy
to our predicament. Let the expanding wave represent the limits
of our knowledge at any point in time, separating the known
from the unknown. Ever since the dawn of humankind, the
domain of understanding has been expanding steadily (fig. 12.1).
However, the unknown region beyond the known hardly seems to
have shrunk; it is also expanding. Each age, which can see farther
as the wave front moves out, does so only to become conscious of
a still larger unknown expanse.

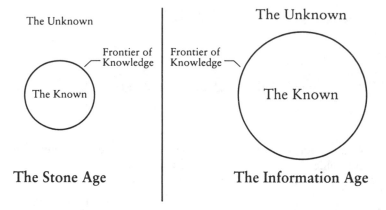

FIGURE 12.1
The Moving Frontier

Let us look at a case in point. For thousands of years we knew little about weather forecasting other than the regularity of seasons and the existence of a relationship between clouds and rain, sun and heat. With the development of science in Greece came meteorology. Aristotle wrote *Meteorologica*, a pseudoscientific treatise on the weather, in the fourth century B.C., and Andronicus of Cyrrhus designed the horologion, the first weather station, in the first century B.C.. We developed an understanding of weather fronts and temperature inversions but still could not forecast hurricanes and typhoons. The balloon and airplane overcame this difficulty.

Today, weather satellites and computers permit still better forecasts. For example, a computer model can now forecast El Niño conditions—warm waters and prevailing trade wind disruptions resulting in heavy rainfall—in the Southern Pacific Ocean two years in advance. Even so, current weather models yield forecasts for moderate latitudes that become misleading beyond day 5 in at least half the cases. The outer limit for weather forecasts is about fourteen days.[1]

In the last two decades we have gained considerable insight into the *inherent* limits in long-term weather forecasting. Extremely minute and totally unpredictable changes locally can produce radical changes in distant weather patterns. A butterfly flapping its wings in Hawaii can have an effect on the weather in Chicago (hence the term *the butterfly effect*). This is the result of the discovery of previously unsuspected chaotic behavior of many normal-appearing systems. Thus, we have expanded our range of knowledge *and* an awareness of our range of ignorance at the same time. As we learned in chapter 6, *better understanding does not automatically and necessarily lead to better prediction.*

Human beings have always exhibited a drive to know, to understand. As the unknown becomes the known, as we can explain it and learn how to deal with it (even when we cannot predict future behavior), we also tend to shift it in our minds from the realm of the complex to the simple. Humankind created gods and religions to help it understand the world in terms that were psychologically satisfying. Thunder was seen as a signal from the gods; prayers were said and dances performed to entice the gods to bring rain. In modern times, science and technology serve to provide explanation and amelioration (for example, hurricane alerts, air conditioning, cloud seeding). They have become our new religion.

Among the many concepts that undoubtedly seemed puzzling and complex to our ancestors but are taken for granted today are farming, writing, electricity, and money. In the seventeenth century, Newtonian physics was considered advanced; today, compared to modern physics, it is considered simple. Some seventy years ago, the model of the atom with a nucleus surrounded by orbiting electron rings was at the leading edge of knowledge. Now this concept is considered naive. The continuing discoveries of even more elementary particles, such as mesons, quarks, and antiquarks, suggests a much more complex subatomic system. Similarly, our universe seems to become more complex the farther we penetrate it. We develop new techniques only to find that they lay bare new complexities, leading to a kind of compounding effect.

We turn to either dogmatic ideology and religion (e.g., fundamentalism) or science/technology (e.g., models based on abstraction and idealization) to find answers—or we bear the unsettling burden of new uncertainties.

If the term *complex* seems subjective, so is the label *system*. We talk about a telephone system and an ecological system, a digestive system and a health care system. Living systems range from simple cells to supranational systems, such as those listed in table 8.2.[2] The systems scientist's definition of a system as "an integrated set of components with relationships among them and among their properties, with the whole functioning as more than the sum of its parts" constitutes a useful organizing principle. Its generality has inherent advantages and limitations. We gain insights applicable to many kinds of systems, but the system's structuring approach is both constraining and subjective. Only certain elements are brought into coherent relation; some elements and properties are invariably left out.

Systems scientists love to organize, to categorize, and to order. Consider the sequence "1, 2, 3, 4, 5, 6, 7, 8, 9"; we can readily describe the basis for its ordering. Now study the sequence "8, 5, 4, 9, 1, 7, 6, 3, 2"; it seems random. However, it is also simply ordered—alphabetically (beginning with *e* for eight). The recognition of order is, like complexity, a property of the observer, not of the observed system.

The preceding definition of a system made reference to relationships among its components. A simple example makes an important statement about such connections and gives us a clue concerning the difficulties of dealing with systems that have many parts. Suppose a system consists of just three elements, A, B, and C. How many interactions of the kind "A has an impact on B" (or "A -> B") are possible? High school students usually answer that there are 6: A -> B, A -> C, B -> C, B -> A, C -> A, and C -> B. A better answer is 49.* This includes such combinations of subsets as AB -> C, AB -> A, C -> ABC, and A -> A. If we add one element, the same formula tells us that the number of possible interactions rises to 225. A system of just ten elements has over 1 million possible interactions among its subsets. Actually the formula used still only gives us a lower limit, because it assumes there is only one relation between any two subsets. Is it any wonder that idealized models of complex systems often fail to capture significant relationships? Indeed, this reasoning goes far toward explaining the "normal accidents," in which an unlikely combination of interactions causes a catastrophe; that is, the occurrence of low likelihood, severe consequence events like the Alaska oil spill, Three Mile Island, Chernobyl, and Bhopal (chaps. 2 and 5). The enormous number of possible, but improbable, combinations of human-technical subsystem element malfunctions cannot possibly be encompassed by any preplanning effort.

The connections among system subsets often imply "feedback

---

*The formula for the number of interactions is $(2^n-1)^2$, where *n* is the number of elements in the set. Thus $(2^3-1)^2 = 49$.

loops." An example is A -> B -> C -> A. Such a closed loop may be damping (stabilizing) or deviation amplifying (destabilizing). An example of a stabilizing loop is a home thermostat: if the temperature in a room declines below the set comfort level, the furnace comes on and the temperature rises; when it exceeds the desired level, the furnace turns off.* An example of a deviation-amplifying feedback loop is a bank in crisis: when customers hear of the crisis, they withdraw their money; this action depletes the bank's reserves, and, in turn, this exacerbates the crisis, causing a further rash of withdrawals. The process can continue until the bank collapses.

Numerous, often unrecognized, interactions imply that there will be many unnoticed feedback loops. Such loops act to counter or to amplify the effect of our decisions. This is the reason we often are surprised or frustrated by the unexpected impact, or lack of impact, of our actions. It also explains why we find ourselves confronted with seeming paradoxes (fig. 8.2). More of something initially thought to be good leads to less, and vice versa. Building freeways to ease traffic congestion soon leads to more congestion as people realize that, using freeways, they can work and shop farther away from their homes.

The systems of interest to us involve technology and human beings in settings where virtually "everything interacts with everything" and "you can never merely do one thing." This is sometimes known as "Garrett Hardin's Law of Complexity."

It is understandable that we feel a desire to idealize, but we pay a heavy price: idealization denies the richness of complexity. The tendency is to reduce the number of elements in the system as well as the number of interactions and feedback loops considered. But *this step may lead to disillusionment at best and human-induced catastrophes at worst.* The ignored interactions may prove decisive, nullifying the desired effect or producing a devastating unanticipated effect.

Simplification is often attempted by a process that cyberneti-

---

*Feedback loops may conceivably even operate at the planetary level, with life on the earth modulating its physical environment (e.g., through interactions of gases and organic compounds) to maintain itself. In this Gaia Theory, the entire thin biosphere layer around the earth constitutes a living system. This "earth system" view might explain why the earth's mean temperature has remained constant and favorable for life for 3.6 billion years, in spite of the 25 percent rise in the sun's luminosity, and why the life-sustaining atmospheric oxygen level has remained nearly constant for 200 million years. (Gaia is the Greek goddess of the earth.)

cist Heinz Von Foerster calls his First Law: "The more complex the problem which is being ignored, the greater are the chances for fame and success."[3] If we cannot solve a problem, we reduce it to a simpler one. If we still cannot solve it, we keep reducing it. In this way we finally arrive at a problem we can solve—and thereby achieve "success." The final version may, of course, have very little to do with the original problem.

The famous Limits to Growth model projected the state of the world to the year 2100 in terms of five variables—population, industrial production, agricultural production, natural resources, and pollution—and relations among them.[4] Economists produce theorems for the behavior of a stable economic system, but rarely for a rapidly changing, unstable system. Thus there is little expertise relevant to the current transition in the former Soviet Union from a centrally planned command economy to a free-market economy. One examination of President Carter's Energy Plan identified forty-six variables with sixty-seven connections but consideration of only nine feedback loops. It should come as no surprise that many well-intentioned plans do not yield the expected results. They are negated by subtle feedback loops.

The savings and loan crisis offers a striking example. Under pressure in an increasingly competitive environment, savings and loan institutions were strengthened by loosened controls and the creation of a federal agency to insure deposits. Thus they attracted more money—and, with it, less scrupulous managers. With customers protected, they could make higher-risk investments and even reap huge personal profits. The result was grave losses and insolvency for many such banks. Instead of being strengthened, the industry was devastated—and the federal government had to supply more than $200 billion to cover the losses. Had the loop been recognized at the time the legislation establishing the federal deposit insurance program was under consideration, the savings and loan industry (as well as the taxpayer) might have been spared financial disaster.

Another example: NASA developed a bonus program to encourage employees to report defects, hoping thereby to lessen the chances of a space disaster. The number of defects promptly increased, leading management to suspect that some employees had tampered with critical components in order to collect the bonus. Thus a vicious circle may have been created: an action designed to reduce the number of defects results in an increase.

The problem is spread around, rather than solved.[5] These two examples show what happens when feedback loops are ignored.

In large organizations, simplification is often sought through compartmentalization. Problems are separated, and then the pieces are assigned to distinct units. Potential interactions are thus artificially cut. For example, How can a national energy policy be developed when more than thirty federal agencies have responsibility for energy? The Department of Energy was created to manage U.S. energy policy, but energy involves the Department of Transportation, the Department of Defense, the State Department, the Department of the Interior, and a host of other agencies.

Discounting (fig. 9.1) offers another way to deny complexity: it closes one's eyes and ears to all but the "here and now." As this is such a common human (O and P) tendency, it is hardly surprising to see it reflected in the media. Distant space-time consequences of current actions, feedback linkages, and interactions are ignored. They tax the abilities of the media producers and are assumed to turn off most readers and viewers.

We must distinguish the search for real insight from the misguided search for superficial simplification. Frequently a qualitative analysis can become an aid to understanding, even though it makes no attempt to simulate or predict the full system behavior.

An example of valuable qualitative insights that we can obtain from modeling is an understanding of a system's regions of stability and instability. We find that complex engineering systems and complex living systems exhibit a very different characteristic. The former are designed to be *fail-safe* and prevent failure. The latter are *safe-fail*; they overcome failure and learn from it through feedback (see chaps.2 and 5).[6]

In chapter 10, we defined four steps from certainty to complete uncertainty or ignorance. The first corresponds to the causal model, the second to a stochastic model with known probabilities. The third level deals with truly complex behavior, which is characteristic of nonlinear, dynamic systems, precisely the kind that are of most interest to us. We find that such systems can be stable, converging to an equilibrium; oscillate stably; diverge unstably; exhibit persistent chaotic behavior within predictable bounds; or move "on the edge" between stability and chaos. The impact of the exciting recent work on these systems is seen by some to represent a scientific shift in thinking as revolutionary as the theory of relativity and quantum mechanics earlier in this century.

In the past two decades, chaos theory has yielded fascinating insights about complex systems. For example, we now know that systems once thought to behave in an orderly fashion may suddenly shift to chaotic or unstable behavior. Apparent patterns may, in fact, be random, and the information that these systems appear to hold may be illusory. On the other hand, systems that appear disorderly may, nevertheless, have an underlying order that yields some predictability. Furthermore, disordered systems may spontaneously crystallize into a high degree of order. The linking of chaos and order is observed in the S-shaped growth pattern that occurs so commonly (fig. 7.2). Chaotic oscillations mark the beginning and end of the S-curve; in a sequence of such curves, they characterize the transition from one curve to the next.[7]

Another significant characteristic of such systems is their fractal behavior. By this we mean that their pattern is repeated at each scale. For example, the irregular outline of the land-ocean interface appears similar whether we look from an orbiting space vehicle, an aircraft, a standing position on the beach, or kneeling and looking through a magnifying glass.[8]

Chaos theory is now beginning to prove valuable in a wide range of system applications, from cardiology to marketing. It may soon lead to exceedingly valuable insights about organizing complex social systems. For example, such systems may at times perform in oscillatory or apparently random ways that are not caused by external factors but are inherent. In interpersonal relations, a very small input can produce disproportionate, that is, nonlinear effects. A glance, a grimace, a change in inflection can change the outcome of negotiations.[9]

"Complexity science," even newer than chaos theory, studies systems that exhibit aspects of both chaotic and predictable behavior. An example is a sand pile. Dropping sand grains on a plate builds an artificial heap until the sides grow too steep, at which point one additional grain, clearly an arbitrary and unpredictable one, causes a sudden avalanche collapsing the heap. Adding more sand causes a repetition of the buildup-collapse process. This self-organization implies a maximum slope, with quasi-stability and quasi-chaos phases.

Biological evolution, characterized by survivability, adaptability, and diversity, seems to proceed at this boundary region between order and chaos. Researchers in complexity science are extracting valuable clues on the evolution of living systems of increasing complexity from the study of the behavior of genetic material. Such dynamic systems are stable enough to store information and evanescent enough to transmit it.[10] The process of gaining feedback and learning from it is crucial to the survival of the fittest in particular, their adaptability to changes in their environment. Through trial and error, knowledge is accumulated in, and distributed throughout, the neural network. While the system elements follow remarkably simple basic rules, the result of the interaction of many elements is very complicated and quite unpredictable. As one of the complexity researchers put it, "the mysterious 'something' that makes life possible is a certain kind of balance between the forces of order and the forces of disorder."[11] Sociotechnical systems (including the stock market) similarly suggest a combination of order and chaos.[12]

One way to depict the evolution of complex systems is as alternating between processes of separation and combination, fragmentation and integration, or decentralization and centralization (fig. 12.2). A simple hierarchical system, say, a tribe or a small company, grows until it can no longer be effectively controlled centrally. Then it separates into smaller units with considerable autonomy. When there is too much decentralization and the system is no longer effective, reunification occurs, usually at a higher level of complexity than existed previously. In other words, successful evolution proceeds to increasing system complexity by periodic restructuring involving swings between differentiation and integration. We observe this behavior at scales ranging from cosmological schemes to business organization plans.[13] One can construct a spiral like Fig. 12.2 for chemical elements, with hydrogen at the center and the heaviest elements at the rim, or for life forms, with DNA at the center and human societies at the rim.[14] This pattern, repetitive as we change scale, also suggests the behavior we have defined as fractal.

The alternating phases suggested in the sand experiment also appear in the sequence of S-curves in figure 7.2b: order (homeostatic growth through development of system components in an internally compatible way with overall stability), followed by disorder (instability or disequilibrium as a total system change occurs), followed by order (stable growth again), and so on.[15]

A most intriguing question is this: How does the *simultaneous* centralization-decentralization capability of 21st-century infor-

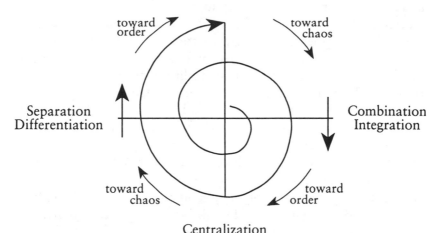

FIGURE 12.2
The Evolution of Complex Systems

mation technology affect this evolutionary pattern of *alternating* phases? Another fascinating question arises with the recognition of the nature of biological systems: they are learning-programmed, not goal-programmed. We recall that, in table 6.2, "learning" was associated with P, "process" with O and "product" with T. P corresponds most closely to the *biological* system; T reflects an *artificial*, goal-oriented human creation. O is an intermediate stage, the *social* system. This helps to account for the difference in discounting as we move from P, most discounting, to O, less discounting, to T, least discounting (see table 6.2). Does the multiple-perspective approach facilitate balance and synergism between learning-programmed and goal-programmed systems?

It is sobering to realize that *the more we learn, the more we realize how little we know.* Although another adage reminds us that a little knowledge can be a dangerous thing, the growing complexity of life on this planet has made us, at least until now, a more survivable "safe-fail" system. Chaos and order, feedback and self-organization have worked remarkably well in advancing the evolution of living systems. How long this pattern can be sustained is far beyond our understanding.

This brief look at "the basics," together with our earlier discussion on forecasting (chap. 7), tells us that we are learning a great deal about complexity and have derived highly useful practical knowledge. Among the insights we gain are ones that address regularity and surprise, order and disorder; robustness; and the need for expanding our inquiring systems.

Regularity and Surprise, Order and Disorder

In chapter 7 we noted the universality of invariants, S-shaped curves, and cycles. They suggest the predictability of many macrotrends, for example, in energy, transportation, and economic prosperity. But we now recognize that, for complex systems, disorder is as normal as order and everything interacts with everything. Thus the prediction capability of science and technology is constrained, and so, too, is our imagination. Surprises will therefore be certain in our future. A century ago, our predecessors could envision some surprises and could not envision others, that is, "surprising surprises." There are surprises that cast recognizable shadows before them and surprises that do not. Major climate changes give ample warning. At the beginning of this century, science-trained novelist H. G. Wells foresaw television (or "home theater"), world wars, lunar exploration, and the military use of aircraft and nuclear energy. Into the category of total surprises in 1900 would fall genetic engineering and wireless communications. For example, a 1900-vintage forecast was accompanied by a drawing showing aerial reconnaissance of the battlefield by dirigibles with wires hanging down to the ground for communication.

In the case of technology, the impacts of a centrally controlled technology are easier to assess than those of a pervasive one. Nuclear weapon development was concentrated at one location, Los Alamos, centrally controlled by one user, the War Department, and focused on one function, the atomic bomb. That is not true for information technology, which spans the spectrum from infotainment to robotics to surveillance to genetic manipulation. Thus the impacts of nuclear technology are easier to assess than those of information technology, the overarching technology that is creating a totally new era—the knowledge society—which reaches into every American's life in ways that constantly create new surprises.

The emerging quality of living, learning, self-organizing systems and the spiral of growing complexity suggest that the 21st century will witness the restructuring and reorganizing of corporations, governance, and education systems in entirely new ways. We look to complexity science to give us further clues in the coming years.

In chapter 8, we pointed out that concepts of 'growth' and 'sustainability' have changed dramatically from the hunting society to the industrial society. We also recognized that we do not yet see very clearly what growth in a knowledge society will mean. Will it be qualitative rather than quantitative growth? Our values will undoubtedly change as they have done repeatedly over recorded history. A California yuppie of today would describe a "desirable quality of life" in terms very different from those his or her gold rush ancestor would have used 150 years ago.

The inevitability of surprises leads us directly to a second insight.

Robustness

Robustness describes a strategy that can be effective in an environment inherently subject to all kinds of surprises and shocks. It should maximize future options, not constrain them. For example, it recognizes that enterprises will not be able to plan and control complex systems in the ways that businesses have traditionally sought to do. We have a good clue: there is a crucial difference between business planning and military planning in coping with uncertainty. The business planners typically look for the "most likely," that is, highest probability, scenario in determining a strategy, ignoring the fact that such probabilities are illusory. Given "high," "medium," and "low" forecasts, they will choose the "medium" forecast for planning purposes.

The training of officers using numerous war games and the development of contingency plans reflects a different strategy. Monitoring an unpredictable environment, maintaining strategic flexibility, and developing crisis management capability become essential.* With regard to the scenarios of chapter 11, this means that an organization should try to be reasonably prepared for any

---

*An example of robust corporate planning comes from Southern California Edison Co., described in chapter 8. In their own words: "How much confidence do we have that the future will be as depicted under [the single] "reference" scenario"? *Very little* . . . How much confidence do we have that we will be prepared to deal with whatever happens? *A great deal.*"[16]

one of the three scenarios discussed (and modifications thereof) and closely monitor the signals that would help identify the actual direction.

In the case of a technological system that involves a "very low likelihood/severe consequence" hazard, a robust strategy would be of a different nature. Once again, probabilities are misleading. Instead the worst-case scenario should be considered and the question asked, Is the consequence tolerable to the community or society? The concept of 'safe-fail' becomes central in a robust strategy. Survivability in the event of failure is the key rather than minimization of likelihood of failure.

The Need for Expanding Our Inquiring Systems

A third insight about complex systems is on the limited ability today of the T perspective, that is, the inadequacy of traditional inquiring systems (i.e., data-based and model-based ways of gaining knowledge. System complexity implies interconnectedness, ambiguity, unboundedness, and uncertainty as well as the inability of any one perspective or worldview to explain and forecast the system's behavior.

The concept of categorizing modes of gaining knowledge dates back to Spinoza and has been developed for our times by Churchman.[17] Widely used modes of inquiry or inquiring systems (IS) include the following:

- empirical—agreement based on observation or data; truth is experiential
- theoretical models—using formal models; truth is analytic
- multiple models—empirical and theoretical complement each other; truth is synthetic
- dialectic—confrontation between opposing models; truth is conflictual

None of these suffices for complex living systems. Churchman sees the need for a higher level system that "sweeps in" as diverse an array of inquiring systems as is useful.* It is termed "higher level" because it recognizes the enriching value of different ways

---

*This IS is often labelled "Singerian," after philosopher Edgar A. Singer, Jr., or "unbounded systems thinking" (UST).[18]

of "seeing" or "knowing" and opens up the knowledge-gathering process. This kind of "new thinking" allows us to include ethical bases other than scientific logic or rationality, goals other than problem solving (see table 6.2). It bridges the gap between the idealized world of the analyst or modeler and the real world of the human and technological explosions.

The challenge presented by this "new thinking" should not be underestimated. The willingness to "see" in different ways, often conflicting, makes many humans uncomfortable and is frowned upon by many of their institutions. Most religions insist on exclusive acceptance of a single worldview. One can be a Catholic or Protestant, Christian or Jew, but not both simultaneously.* Inquisitions and tribunals have dealt harshly with "heretics" who have questioned the single "correct" dogma, be it the Ptolemaic view of the earth as the center of the universe or the infallibility of Stalin. Even in science those who have offered multiple theories to replace a single one** have found themselves hounded. The "new thinking" calls not only for openness and wisdom in the "sweeping-in" process but also for a willingness to accept the nonterminating, never-completed nature of such a knowledge-gathering process.

As the sciences of the 21st century develop, the T perspective may encompass more and more of the abilities of O and P. In other words, T will become a source of insights for which we now must rely on O and P. We already noted (chap. 6) that recent T-based work on complex adaptive systems recognizes that unique variations in the sub-sub-systems provide an individuality to such systems, while still maintaining common overall behavior. Through T we may strengthen the ability to play games that involve self-organization and the "complexity of life." In this way, policy games may yield a realistic grasp of the impact of policy decisions. One may even be tempted to speculate that, in some applications, the multiple perspective approach may ultimately be represented as alternative virtual realities, just as different perspectives of a physical object are now producible in computer-aided design. But, for now, we must do the best we can with the approach we have. Decisions on complex systems cannot wait.

---

*The insistence is not universal, however. In Japan, for example, individuals may practice both Buddhist and Shinto rites.

**Examples are the wave and particle theories of light in physics, Euclidean and non-Euclidean geometries in mathematics.

T-O-P AS AN EXPRESSION OF WESTERN CULTURE

In parts 2, 3, and 4 we used multiple perspectives—*technical (T), organizational (O), and personal (P)*—to offer one concrete way to stimulate such new thinking. The three categories emphasize the bond among *technology, institution, and individual* or among *product, process, and self.*

The scope of each of the two specific cases discussed, the Alaska oil spill (part 2) and the United States in a changing world (part 4), was very different: one case was local and the other national, one specific and the other broad, one retrospective and the other prospective. At totally different levels, each illustrates the intimate and subtle linkages among technology, society, and the individual.

We unabashedly admit that the reason we find T, O, and P appropriate lies in their distinct Western cultural basis. Indeed it should by now come as no surprise that the representation we used in chapters 6 and 11 (figs. 6.2, 11.1, 11.2, and 11.3) also serves to depict the ideal of secular Western culture itself. This is shown in figure 12.3.

If, as Sigmund Freud claimed, Western culture is enslaved by its own image, that criticism certainly applies to our representation.

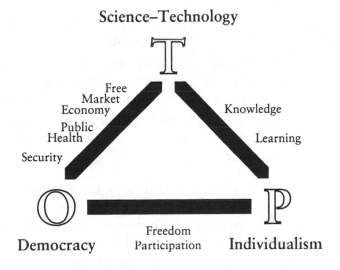

FIGURE 12.3
The Basis of Secular Western Culture

All our perspectives inherently reflect the bias of this anthropocentric culture. After all, O and P deal with human beings, and T is the product of the human mind. Our starting point in chapter 1, the view of population and technology as an explosive combination, reflects the same Western values. It is also the West that pioneered

- late marriage and small families (hence population control)
- private property and capitalism as generators of wealth, and the use of technology to create more wealth

By contrast, Eastern philosophy has traditionally focused more strongly on spirituality rather than on material wealth, so that population control and technology management have not become such central concerns. This was already implied by our discussions of Asia in chapters 8 and 11 (see fig. 11.3). We also need to remind ourselves that democracy is a product of Western values and is by no means viewed as an ideal form of government in Islamic and other cultures that compose much of the Third World.* In other words, the perspective types we are using sweep in only some of the many that could be used.

Finally, the action orientation implied by "managing" our society and our technology is itself a Western cultural conceit. An alternative would be to do nothing, in the belief that

it is presumptuous and self-deluding for human beings to think they can "manage" truly complex systems, and/or

emergent societal system evolution and self-organization, balancing chaos and order, has long proceeded, and will proceed, without human attempts to superimpose control, and/or

societal cycles (fig. 7.3), point to a natural rhythm, a kind of "cosmic heartbeat," outside our conscious control that determines social changes

In fact, the management of science and technology does justify serious concerns. First, there is the hubris that is an inevitable outgrowth of two centuries of continually accelerating success. Physician-novelist Michael Chrichton expresses the concern over a common science-technology mindset as follows:

---

*In Islam the law of God, rather than the law of man, is supreme.

You know what's wrong with scientific power? It's a form of inherited wealth . . . Most kinds of power require a substantial sacrifice by whoever wants the power. There is an apprenticeship, a discipline lasting many years . . . You must give up a lot to get it. It has to be very important to you . . . It can't be given away: it resides in you. It is literally the result of your discipline. The discipline of getting the power changes you so that you won't abuse it.

But scientific power is like inherited wealth: attained without discipline. You read what others have done and you take the next step. You can do it very young. You can make progress very fast. There is no discipline lasting many decades. There is no mastery: old scientists are ignored. There is no humility before nature . . . you can stand on the shoulders of giants, you can accomplish something quickly. You don't even know exactly what you have done, but already you have reported it, patented it, and sold it. And the buyer will have even less discipline than you. The buyer simply purchases the power, like any commodity. The buyer doesn't even conceive that any discipline might be necessary.[19]

Second, the irrelevance of much ongoing research to the challenges of the real world is indisputable. Bureaucratic and personal perspectives tend to dominate, resulting in a flood of academic papers ("publish or perish") that satisfy the organization and the individual. For the organization, compartmentalization and self-perpetuation are important, for the individual, prestige and power. For both, short-term concerns overwhelm long-term ones. The endless production of abstract findings and theoretical propositions has become the main business of research institutions and an end in itself. In our terms, T may be brilliant in Western culture, but T, O, and P are rarely joining forces to adapt to the new realities of the knowledge society in the small world of the 21st century.

THE SEVEN IMPERATIVES FOR COPING

Our discussion of the two cases in parts 2 and 4, as well as the methodological discussion in part 3 and here in part 5, leads us to propose seven basic requirements in managing ourselves and our technology in the decades ahead:

#1. balance the thinking more effectively between the near ("me-now") and the far in both space and time

#2. achieve a better balance between individual and collective rights and responsibilities, recognizing that an organization or society in which individuals do not strive for collective well being cannot prosper in the long-term

#3. constantly question assumptions and drop the obsolete ones, as they perpetuate the widespread bureaucratic arteriosclerosis in the corporate, governmental, educational, and scientific establishments

#4. recognize that, in the global village created by the population and technology explosions, it will often be found that "less is more" and that "more is less"

#5. design coordination-intensive structures that will facilitate success in an environment requiring unprecedented adaptability and flexibility—in particular, public and private networks that foster distribution of control (e.g., simultaneous globalization and localization)

#6. reinvigorate science and technology: in science, press for deeper, holistic understanding of complex systems, their self-organization, order and chaos; in technology, intensify the innovation process and focus it on societal needs

#7. strive to minimize the disparity between technological and organizational/societal change; in particular, manage the increasingly powerful technologies in a manner both ethical and enriching in human terms

Four points should be stressed in connection with these imperatives:

A. The imperatives demonstrate the need for deeply interweaving diverse technical, organizational, and personal perspectives. Strictly technical or organizational or individual approaches have little chance of success. Technological change without individual and collective attitude changes will not be effective. Every one of the aspects examined brought to the forefront the central role of individual and organizational actions. The first three imperatives begin with O and P and then must sweep in T; the last four begin with T and then must bring on O and P. For the information technology era, the technical perspective is obviously vital, but it is in no sense adequate.

The fundamental differences in the paradigms associated with the three perspective types (table 6.2) clearly impede mutual understanding and fruitful interaction. But even two perspectives of the same type, say two organizational perspectives, are likely to differ significantly. In particular, each perspective tends to make assumptions about the other perspectives that are erroneous (see Appendix A).

The widening gaps discussed in chapter 9 certainly exacerbate misunderstanding. As the affluent become more affluent, they inevitably lose touch with the rest of the society. The movers and shakers live in a world of power breakfasts, country clubs, fund raisers, executive conferences, retreats with legislators, *Wall Street Journals*, private schools, and suburban, gated communities. It becomes increasingly difficult for them to "see" the concerns of the factory worker, the school teacher, the farmer, and the small-store owner. Similarly, the middle-class suburbanite "sees" the inner-city dweller differently than the latter sees himself or herself. Consider one concrete example: for a long time "law and order" was interpreted by the enlightened college-educated as a synonym for a right-wing conservative position. But in the inner city, the term is taken very literally. Here the people feel themselves in constant physical danger and yearn for law and order to feel more secure. For them the term has nothing to do with conservative politics.

It should not be surprising then to hear George Bush confess that, when he was president, he did not correctly sense how pessimistic his countrymen were about the economy. Nor did he grasp the common voter's concern about escalating health care costs until he was stunned when Dick Thornburgh, his attorney general, ran for the Senate in Pennsylvania and was soundly defeated on this issue by a little-known opponent. The president lived in imperial isolation.

Differences in perspectives are at the root of racial tensions. In the words of Bill Bradley, a senator from New Jersey,

> Until whites reject seeing most blacks as Willie Horton and blacks reject seeing most whites as Archie Bunker, there is no hope. Unless we remind each other of our common humanity, too many of us will continue to think of ourselves as islands of privilege and comfort.

White Americans will have to understand that their children's standard of living is inextricably bound to the future of millions of nonwhite children. To allow them to self-destruct because of penny-pinching or timidity about straight talk will make America a second-rate power.[20]

In industry, CEOs are insensitive to the widening gap between their pay and that of the workers. Americans and Japanese make different assumptions about American workers' industriousness. Only 27 percent of the Japanese admire Americans' industriousness, a view reflected in Prime Minister Kiichi Miyazawa's statement to the Japanese parliament that American workers are lazy, greedy, and lack a work ethic. It is not a view shared by the Americans, nor is it supported by data; they show that the average U.S. employee puts in 163 more hours annually than he and she did in 1970 and 320 more hours than do German workers.[21]

Scientists see the importance of logically sound theory but are often dismissive of ethical collective action. With few exceptions, such as the "complexity scientists" at the Santa Fe Institute, they cherish their narrow disciplinary perspective and scorn a holistic view of systems. Engineers see the potential advances of technology but are often blind to the fears and concerns of ordinary citizens, who, in the words of Shakespeare's Hamlet, would "rather bear the ills we have / Than fly to others that we know not of." Youthful computer hackers may not only be socially maladjusted but also fail to see the difference between the virtual and actual reality. Their games and pranks may have devastating societal consequences unimaginable from their narrow perspective.

A lament often heard in industrial organizations is that there is incredible talent, but it is difficult to mesh it. With differences in background and education, the same words mean entirely different things to engineering, marketing, and production people. The differences become liabilities instead of assets.

Understanding of other—very different or alien—perspectives can be psychologically painful and disconcerting. The same is true of shifting to a longer-term view, that is, reducing one's discounting of time and space. It is far more comfortable to retain a single perspective and focus on the "me-now." It is the simplicity of a single perspective that makes the siren call of a religious or ideological fundamentalism particularly enticing. The disciple avoids having to wrestle with the conflicts and complexities that attend multiple perspectives. There is only one worldview, one truth. A corollary is the dislike and even hatred of those espousing multiple perspectives. In particular, the "global tribes" (chap. 8)—Anglo-Saxons, Japanese, Chinese, Indians, and Jews—frequently bear the brunt of much resentment in their diasporas, where they constitute minorities.

There is much to be said for deeper examination of the perspectives of the stakeholders and their embedded assumptions (see Appendix A on assumptional analysis). Navigating the assumptional minefield is but one of the means to galvanize the vital interaction and integration among perspectives.

B. The imperatives are all connected to each other.   For example, imperative #1, the need for better balance between the near ("me-now") and the far in both space and time ("out of sight, out of mind") ties in directly with imperatives #2–#7:

#2—Egocentrism goes hand-in-hand with a myopic planning horizon. It means the individual cares about himself or herself and not about either future generations or today's impoverished Third World (in the South Bronx or in Africa). Irresponsible procreation with no concern for the consequences condemns offpring to lives of misery. In a corporation it means that the management and workers see each other as adversaries, each intent on maximizing their immediate gains, rather than a single team intent on collective excellence and long-term development.

#3—A short-term view will never recognize impending changes in definitions and boundaries. A short-term view will not see the importance of continual learning. Thus organizational rethinking and restructuring will always be "too little, too late."

#4—If we focus in figure 8.2 only on the near term, that is, if we stay close to the origin (point A), all we see is that "more leads to more"; we miss the distinctly different long-term behavior entirely. Saving jobs now by protectionist measures is likely to lead to job losses later as retaliation by the affected countries cuts U.S. exports. Saving jobs now by keeping polluting plants open may lead to enormous costs later to repair the environmental and human damage. The manufacturer of an intrauterine birth control device (IUD), A. H. Robins, neglected the serious medical problems detected early during its product testing in order to maximize near-term sales and profits, only to be forced into bankruptcy years later by the numerous lawsuits instituted by IUD users. Similarly, larger immediate profits for Johns Manville asbestos sales led to the later bankruptcy of that firm.

#5—The structure of the successful organization in the information-technology age is totally different from that common today. We noted in chapter 8 how the introduction of computers into the office and robots onto the factory floor requires a deep rethinking of the operation. Basic, long-term changes are implied, ones that contrast with the myopic, superficial adjustments usually proposed. In the public sector, politicians appoint a commission to study the problem; corporate leaders effect a superficial reorganization that moves people around but leaves the old structure intact. Neither comes to grips with the fact that information technology demands a macroview rather than a microview, strategic change rather than tactical change.

#6—Complexity science is beginning to yield deeper understanding of the central role of process, rather than goal, focus for long-term system development. Reinvigoration of technological innovation calls for a long-term commitment to a stronger research and development process.

#7—Paying attention to potentially dangerous technologies means taking action long before crises erupt. For example, the impact of the depletion of ozone at high altitudes or of the greenhouse effect due to increasing carbon dioxide in the atmosphere may not be provable for many decades. The short-term view is therefore to simply monitor the situation and wait to learn more. Scientists can be relied upon to include uncertainties and caveats in their conclusions that appear to justify executive inaction. One favorite industry tactic, exemplified by the tobacco industry, is to label the available scientific information as inconclusive to prove the long-term harm. Another, illustrated by the automobile and utility industries, is to brand ameliorative action as so expensive that it would necessitate a shut-down, causing mass unemployment. A long-term perspective in an industry dealing with potentially dangerous technology invokes concepts such as a new high-reliability organization. A long-term perspective in a manufacturing organization is concerned as much with process innovations as with product innovations.

C. The imperatives propel us forward in meeting the 21st-century challenges and should inspire "new thinking."　In the public sector, consider the issue of the protection of our ecological environment for future generations.

#1—Only extension of the planning horizon in space and time can lead to adequate care of the planet briefly entrusted to our care. As noted above, many impacts of environmental degradation have a very long incubation period, and the same is true of ameliorative measures. If action is postponed on the warming of the atmosphere until an obvious crisis impends, achievement of a rapid reversal of the trend is highly unlikely. As stewards of the planet, we must strive for intergenerational equity, sharing the "global commons" with the generations to follow.[22]

Myopic focus on the near term ignores the consequences of our actions today; excessive focus on the long-term imposes a straitjacket that inhibits creative and innovative action today. The key word in imperative #1 is *balance*. In some quarters, environmentalism has become an irrational ideology opposed to scientific and industrial progress, and it thereby impedes economic and social development. In the words of Vaclav Havel,

> *The only solution . . . lies in the combination of economic growth and respect for the environment . . .* This is not just a technical, economic or ecological task. This tremendous challenge has a moral and spiritual dimension . . . The world we live in is made of an immensely complex and mysterious tissue about which we know very little and which we must treat with utmost humility. [emphasis ours][23]

#2—Balance between the individual and society is the basis for preservation and equitable sharing of the beauty of the natural environment. The use of wilderness areas, land development, pollution from agricultural and industrial sources for the benefit of individuals can exacerbate the ruination of the natural environment. It is the classic "Tragedy of the Commons" (chap. 6).

#3—The optimistic assumption that future generations will solve ecological problems that we seem unprepared to confront now, for example, stratospheric ozone depletion and ocean pollution, means we are playing dice with the earth. Many would consider this ethically unwarranted even if the chance that inaction now will result in a global disaster were low. The soothing assumption that problems can be compartmented and solved separately is invalid in a world that, through technology, is becoming a global village and must be treated as an integrated whole. The earth has never seemed smaller than it does now.

#4—"Less is more" describes the essence of the conservation ethic. Lower population growth eases the burden on the environment. It eases the strains on all forms of life on the planet.

#5—Complexity science symbolizes a "science of the 21st century" and is vital in learning to better manage ourselves in the changed environment. Coordination-intensive structures facilitate monitoring of the environment and permit earlier action to minimize deterioration.

#6—Reinvigoration of the technological innovation process facilitates the substitution of safe energy, materials, and processes for some that are hazardous to humans and their environment.

#7—The management of unavoidably dangerous technologies in an ethical manner gives the environment the greatest protection from technologically induced catastrophes.

In the private sector, consider the example of the corporation looking toward success in a changed world.

#1—There must be full awareness of the need to look globally at the market as well as suppliers and production. For example, General Motors and its global sourcing strategy PICOS (chap. 8). Similarly, there must be full appreciation that one of America's strongest long-term assets, its research and development capability, cannot be expected to show a short-term payoff (chap. 10).

#2—The adoption of more egalitarian work arrangements with minimal hierarchies and more individual worker (or "associate") responsibility has been pioneered in this country by firms such as Honda (Marysville, Ohio, plant) as well as the high-tech companies and adopted by General Motors in its Saturn plant in Tennessee. Autonomous, self-managed work teams are in place in organizations such as Federal Kemper Life Assurance Co. (chap. 8). Such worker empowerment changes the role of the manager.

#3—Companies must recognize that, in an information technology era, unique competence will be the basis for competitiveness. Strategic thinking and insight regarding complex interactions will be more important than traditional tactical thinking. In an era where knowledge is power, a mark of the effective organization will be its ability to learn quickly, to see future products before customers recognize their need. SONY's Walkman and Microsoft's DOS software illustrate the point. "Time is money" will take on a whole new meaning.

#4—The obsolescence of old tenets such as "bigger is better" has been recognized in organizations such as Quanex and Lockheed's "skunk works" (chap. 8).

#5—Coordination-intensive structures will alter organizations dramatically and mark the leading enterprises in all industries. An early example is Frito-Lay Inc. (chap. 8).

#6—Technological innovation, that is, riding the innovation wave, is a strategy that builds on America's greatest resource in a competitive global environment. Examples of the failure to exploit new, homegrown ideas such as the robot and fuzzy logic should alert business leaders (chap. 11). The 21st century sciences must be closely monitored, as they not only form the foundation of future products but also will transform economics and management.

#7—The contrast in the management of nuclear power by the US Navy and the nuclear power industry and the Exxon attitudes on the oil transportation system offer striking testimony to the need for management rethinking. The need for crisis management ability and high-reliability organizations will be a vital aspect of the ethical management of dangerous technologies in the future (chaps. 2–5).

D. The imperatives underscore the point that the renewal of our society will require decades and the revitalization of our corporations will take years. The full impact of emerging new sciences and technologies may not be felt for many years. Problems that have accumulated for decades cannot be "solved" by a change of administration or a new technology. Children that have grown to adulthood in an environment of family anarchy cannot be reclaimed by a single government program. We recall Hardin's Law of Complexity, "you can never merely do one thing."

The population explosion cannot be reversed even in a few decades, and personal attitudes and norms cannot be legislated. We see around the world how feuds have persisted for centuries, even though they appear to be no longer relevant to the needs of the society. *A period of personal sacrifice and societal austerity in the United States is as necessary now as it was to battle the Axis powers in World War II. The struggle was won then and it can be won now. But we must recognize that it is likely to be a much longer battle than the one our country faced during the 1940s. Figure 7.3 suggests that the nadir between the two boom peaks, roughly 1969 and 2024, is reached in the 1990s. From then on, the path should be upward.*

A corporation that has been stodgy for decades cannot be energized and reorganized by a three-day seminar with a management guru. However, it is fortunate that the private sector organization has inherently a shorter reaction time and stronger feedback system than the general society. A looming bankruptcy or decisive loss of sales concentrates the executive mind in a powerful way. In fact, such built-in responsiveness is one of the great advantages of the private sector and constitutes a powerful argument for downsizing to maximize flexibility. Thus, rapid implementation of the necessary changes in time to take advantage of the start of the next cyclic upturn is a realistic prospect for corporate management.

GETTING OUR ACT TOGETHER

Americans should look for a society that is in tune with the 21st century, in its thinking and its evolution as a living system. Figure 7.3 suggests inexorable trends at work. If we consider the next economic boom, must we also consider the next war? Furthermore, the next boom by no means implies that the U.S. will maintain its status as number one in the world. The past peaks around 1800 and 1856 occurred when this country was not in that position. The decline of the nation-state in the global community of the 21st century may well alter the meaning of "number one." In fact, the inexorability of the trends itself must remain open to challenge. The population and technology forces at work may, through their powerful impacts, override the cyclic patterns in the 21st century.

We have discussed in this chapter the basic constraints on the ability to forecast the behavior of complex systems. Although we have at best partial ability to control our future, we can do much to help ourselves. Failure to act would mean that American society reached its zenith with the Apollo landing on the moon; success would mean that the zenith still lies ahead. The longer Americans wait to get their act together, the more traumatic the task will be. The longer they let the problems fester, the more likely is also the rise and acceptance of extremists with simplistic, ephemeral solutions.

If giant General Motors and IBM had not been controlled by aged, ponderous bureaucracies and acted in a timely way on available signals to restructure themselves, their transformation would have been much smoother. It is the same for other large organizations and the society as a whole.

Above all, Americans must stop deluding themselves, blaming others, and refusing to face realities. There are no easy, painless "solutions." Making the frayed infrastructure and debt-ridden economy sound again carries a heavy cost that must not be dumped on future generations. For example, a lower personal monetary income is likely if full employment is to be achieved. Entitlements must be significantly reduced to eliminate the federal deficit, free money for infrastructure revitalization, and upgrade the work force. Personal sacrifice and societal austerity will be "a tough sell" even for the most inspired political leadership and administration.

One hundred years after de Tocqueville, Franklin Roosevelt reaffirmed his faith in America. Taking command of the ship of state in a period of deep national crisis, he effectively combined the T, O, and P perspectives through rational experimentation (T), constitutional process (O), and humanity (P) to reestablish confidence and recharge the American people.[24]

If a de Tocqueville were to revisit America today, the most puzzling question he would probably ask himself is this: Why does the most successful and advanced society on earth let so much of its capability dissipate instead of marshalling it to address its tasks and assure its success in the new century?

If America wants to be a pioneering society in the new era, the "me-now" focus is a sure path to failure. *Unity with diversity is desirable, but can be achieved only through intense communication and constructive interaction among the many conflicting perspectives in the society or corporation. The first step is understanding perspectives other than one's own. The second is searching for common ground.* The citizens must be open-minded, willing to learn, accept responsibility and sacrifice. They must experiment, take bold risks (and often fail), be creative, and protect the options of future generations. Riding the innovation wave requires great skill and energy. *Successful action at the societal level as well as the corporate level requires*

- P: *individual leadership to formulate visions and responsible citizen/worker participation to support them*

- T: *technology that makes the visions practically possible*
- O: *organizational restructuring to make the possible a reality*

IS ETHICS THE ULTIMATE CHALLENGE?

Accomplishment, both individual and collective, will be determined by our readiness to manage ourselves and our technology at a higher ethical level than has been necessary in the past. The three-pronged perspective approach is vital here also. It is useless to expect either technology or legislation to assure personal morality and commitment, to expect either individuals or analyses to assure justice, to expect either organizations or individuals to assure rational decision making. *Ethical management implies the ethical integration of T, O, and P. It means simultaneous, balanced action: individually, in a moral way; technically, in a rational way; and organizationally, in a just way.*

The linkages among these ethical dimensions are once again important (fig. 12.4). The norms of morality of the individual are inexorably linked to his or her society (O-P). Appreciation of the systemic nature of the complex system means that there is a connection between scientific "solutions" and institutional justice (T-O). For example, science and technology have greatly expanded the world's food supply, but millions are still starving because of organizational and institutional ethical failures. Medicine in the United States has superb technical capability, but health care is inadequate for many millions.

The linkage between the technical and personal ethical dimensions (T-P) is exemplified by the discounting dilemma (chap. 9). Personal morality and humanism must be recognized by the objective, rational technological mind. The vast areas of unknowns and unknowables must be acknowledged and confronted. With artificial life on the horizon, ethical problems along the T-P dimension will rise to a new level that has no precedent in history.

While scientific logic or rationality has evolved a structure and set of rules that are generally accepted by the scientific community, we are not in such a happy state with regard to the other two ethical dimensions. We recall that a dimension, or perspective *type*, may encompass many different perspectives. There are many social entities and individuals, each with unique values, and hence ethics (as suggested schematically in fig. 12.4).

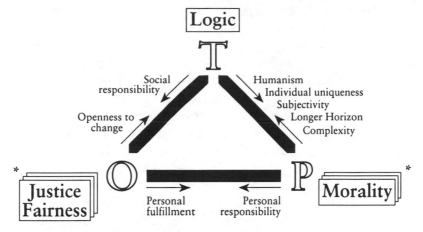

FIGURE 12.4
Ethics and Perspectives

Whose morality is "right" and whose concept of justice is "correct"? Religions and ideologies have tried to provide single, authoritative answers. Monotheism made it necessary to deny divinity to any god but one and galvanized aggressive religious intolerance. This heritage gravely compounds the task of ethical management and is the source of horrendous conflict even now. We are all familiar with ethnic and religious communities that have engaged in seemingly unending, brutal violence on grounds of moral standards and ethnic purification.

Even where the major religions appear to speak with one voice, the level of individual morality is of grave concern. We noted earlier that, in 1991, one murder was committed in the U.S. every twenty-one minutes and nearly 2,400 bank robberies occurred in just one major American city. This country holds the world record in the number of prison inmates, with nearly 900,000 confined at the end of 1992. In the corporate world, criminal greed motivates many executives at the top and workers at the bottom. The home stability of rich and poor is buffeted by drug and alcohol addiction, sexual promiscuity, child and spouse abuse. Many of the women murdered are the victims of family members.

Myopia or focus on the "me-now" at the expense of a longer space-time horizon has led societies to decline more inexorably than attack from without. *A superior ethical level must, in our view, evolve in the 21st century. It must include an unprecedented openness to other perspectives. It means more than tolerance of other views; it requires their synergistic interaction, leading to constructive coexistence. Ethical management is not possible without a deep appreciation of multiple perspectives.* In this connection, we must recognize and confront two sources of grave concern.

First, *the enemy of multiple perspectives is the unshakable belief in a single perspective and intolerance of all others, in other words, extremism.* It is reflected in naive solutions, reactionary rigidity, and blind dogmatism. It is seen in arrogant scientists, rabble-rousing politicians, callous tycoons, militant religious fanatics, and shrill environmentalists. The preferred future does not lie with "right wing," "left wing," or "more of the same." It does not lie in "economic growthism" or "environmentalism."

It is the fear of extremism that prompted the Heidelberg Appeal to the heads of state attending the 1992 Earth Summit in Rio de Janeiro (signed by 264 scientists and intellectuals, including 27 American Nobel laureates):

> We are . . . worried, at the dawn of the twenty-first century, at the emergence of an irrational ideology which is opposed to scientific and industrial progress and impedes economic and social development . . .
>
> We fully subscribe to the objectives of a scientific ecology for a universe whose resources must be taken stock of, monitored and preserved . . . We do, however, forewarn . . . against decisions which are supported by pseudo-scientific arguments or false and irrelevant data.
>
> The greatest evils which stalk our Earth are ignorance and oppression, and not Science, Technology, and Industry whose instruments, *when adequately managed*, are indispensable tools of a future shaped by Humanity, by itself and for itself, overcoming major problems like overpopulation, starvation, and worldwide diseases. (Emphasis ours)

Second, *the concept of multiple perspectives can be manipulated and misused to blur and question all values, encouraging moral relativism and ultimately lowering the ethical level. Openness to other views must not mean tolerance of evil or nihilistic deconstruction that leaves a moral void.* Ideologues and revisionist historians (chap. 1) have introduced perspectives that distort actual events and shift culpability for immoral acts to their victims or to third parties. Evolving information technology will compound the gravity of this concern. The ease with which virtual realities can be created plays into the hands of revisionists and fanatics. Telecommunications and networking magnify the influence of such perspectives, drawing them from the periphery into the mainstream of the society.

The forces of population and technology have the effect of shrinking time and space, thereby greatly confounding the challenge of ethical management (chap. 9). Prior to the creation of the global village, it was easier to ignore other societies and their concerns; they were distant and did not intrude in most of our lives. Before the massive and pervasive impacts of powerful new technology, it seemed feasible to focus on near-term, local consequences. If the environment was degraded, there were cleaner, unspoiled areas available. If one regional market was depressed, there were others that were unaffected. It was ethically more defensible to focus on the "me-now" (fig. 9.3). That time is gone.

Just as artificial life is coming closer to realization,* the possibility of a major species "extinction spasm," with a 20 percent reduction in global diversity in the next generation, is also a distinct possibility. Five such spasms have already occurred in the geological history of life on the planet. We have no way of grasping what impact the loss of many species will have and how long recovery of the biosystem will take.[26]

Let us consider three tough questions at the *societal level*:

Is it ethical to place our own generation's material well-being ahead of that of our descendants?

Our decisions and actions will have a strong bearing on the life of future generations that are a totally unrepresented constituency even in the most democratic decision processes. In one view, *a moral way for individuals, and a just way for institutions, is to act so as not to constrict the options of future generations.* This idea is closely related to the concept of 'sustainable growth', defined as satisfying our needs without diminishing the chances for future generations (chap. 7). Churchman suggests that "we should undertake to design our societies and their environments so that the people of the future will be able to design their lives in ways

---

*With artificial intelligence, computers model thought processes; with artificial life, computers model the basic biological mechanisms of evolution and life itself.[25]

that express their own humanity."[27] Reverting to our discussion of discounting (chap. 9) and assuming our descendants will be as materially focused as we are, this would be equivalent to refraining from discounting future dollars. Suppose that we want our grandchildren to have a better life. If their definition of a good standard of living remains similar to ours, it must therefore rise faster than the economy. The debt burden must not be placed on the backs of future generations, forcing the present generation onto a new and unaccustomed level of stringent discipline and sacrifice. In economic terms, it means the use of a negative real (after inflation) discount rate, implying that our grandchildren are more important than we are. To some, this defines a "strong culture." Just as a positive real discount rate acts as an incentive to deplete nonrenewable resources more rapidly, deemphasize R&D, and ignore long-term debts, so a negative real discount rate serves as an incentive to take actions more appropriate to a future generation.[28] We must remember, however, that this line of argument depends strongly on the debatable assumption that our descendants' values will not significantly differ from ours.

Is it ethical to place our affluent society's descendants' well-being ahead of that of the poor world that surrounds us today? Or, is it ethical to insist on the preservation of biodiversity when, according to a 1992 World Bank report, 34 million of our global village neighbors now die annually of poverty-related causes? Indeed, how can we live with the knowledge that the tender lives of 35,000 children are snuffed out every day?

Figure 12.5 depicts this latter ethical dilemma in the schematic form used in chapter 9 (fig. 9.3). It suggests the possibility of a balanced approach, moving toward point z rather than toward either x or y.

Is it more ethical to encourage all means to cut the birth rate and thereby help to reduce global starvation or to "be fruitful and multiply," therefore, to ban abortion and birth control? Is a man-centered (anthropocentric) or ecosystem (ecocentric) worldview more ethical for the future?

We should remind ourselves that the biblical moral code evolved when the planet was sparsely populated. At the time of Christ, the global population was only about 200 million, 2.4 percent of the anticipated 2025 population!

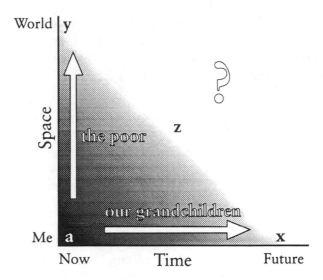

FIGURE 12.5
An Ethical Dilemma in Extending our Horizon

The explosive growth of population and technology confronts us with truly agonizing ethical choices that we would rather shun. Consider, for example, Garrett Hardin's grim lifeboat analogy.[29] He represents the rich 20 percent of the world by, say, fifty people filling a lifeboat, with the poor 80 percent represented by 100 people swimming in the ocean, desperately trying to get into the lifeboat. The lifeboat capacity is limited to, say sixty. If we take them all in, the load of 150 is likely to sink the boat. If we let ten struggling swimmers in, how do we pick them? If there are con-science-stricken boat passengers, perhaps they yield their place to swimmers. To add to the dilemma, we find that the number of swimmers is growing much faster than the boat passenger popula-tion. Every human born adds demands on the environment, at least in the near term. The inevitable "moral outrage" expressed at such stark choices ignores the reality that starvation and home-lessness are tolerated today as the affluent provide token help and assiduously guard their wealth. The boat passengers are throwing a few crumbs to the struggling swimmers, while righteously pro-claiming that "God helps those who help themselves." Those in the water, convinced that the passengers have no special, God-given rights to the boat, may try to overturn it, even if all then perish.

The Judeo-Christian moral code promulgates a strikingly anthropocentric vision of the "earth system." God may be a kind of chairman of the board, but man is the CEO. According to the story of the creation, a primeval history based on Mesopotamian sources,

> God said, "I will make man in my image, after my likeness; let him subject the fish of the sea and the birds of the sky, the cattle and all the wild [animals], and all the creatures that creep on earth" . . .
>
> God blessed [male and female], saying to them, "Be fertile and increase, fill the earth and subdue it; subject the fishes of the sea, the birds of the sky, and all the living things that move on earth."[30]

Thus, real-estate developers exploiting valuable wetlands, oil companies exploiting the Arctic National Wildlife Refuge (chap. 3), the timber industry cutting down forests, whalers, Arctic seal hunters, African elephant hunters, and, most important of all, opponents of contraception and abortion, appear to be following God's will. Humankind's supreme achievement, technology with its rationality ethic, has contributed heavily to the destabilizing disparity between birth and death rates and to the redefinition of "survival of the fittest."

Our recognition of true biosystem complexity inexorably moves us away from orthodox anthropocentrism. Rather, the human being is seen as one unique element of the biosphere system. Our approach recognizes the total system and its complexity. In this spirit, we have:

- rejected exclusive reliance on any single perspective, ranging from the traditional reductionist technical perspective, with its discipline compartmentalization (table 6.2), to the similarly reductionist political or religious ideology that claims sole possession of the "given truth"
- advocated the constructive interaction of multiple perspectives to deal synergistically with the ethical management of the total system
- focused on new concepts of growth (chaps. 7 and 8)
- stressed the need to reduce space and time discounting (chap. 9)

Some modification of the anthropocentric ethic is essential if the population and technology explosions are not to mercilessly degrade our own humanity and despoil this wondrous biosystem. Much damage has already been done, and moral outrage is justified. This small, delicate, and very complex biosphere will have to serve as home for all our descendants. Let them not say that it was once a fertile, garden nurturing life in all its forms and has become a slum where humans fight each other and devastate life around them to survive.

In the long run, of course, the natural system can be counted on to heal itself. Any wanton, explosive growth must be a temporary phenomenon on a finite earth.*

At the *corporate level*, we face ethical questions just as difficult as those at the societal level. Two examples suffice:

Should the imbalance in the corporation's concern for stockholders and its other stakeholders, such as its employees, community, and country be rectified? If so, how is the "return on investment" to be calculated for each stakeholder? Is a policy of caveat emptor (let the buyer beware) justified on the part of tobacco companies when non-buyers suffer harmful effects through passive smoking?

What is the societal or civic responsibility of a corporation that is global, transcending national boundaries?

Perhaps the deepest sources of difficulty for ethical management are at the *personal level*:

How can children at their most impressionable period of growth absorb high ethical standards and simultaneously develop free of the pressure to conform to a single perspective?

The accelerating pace of change is creating a chasm between generations. Parent-child communication is lessened when children learn much that is totally unfamiliar to their parents, such as computer-aided information search. Technology has contributed to

---

*In the arts, the point is made allegorically in Wagner's *Ring of the Nibelungen* with the *Götterdämmerung*, the twilight of the greedy, power-hungry gods who have ruthlessly desecrated the natural environment. Perhaps the ultimate optimist was George Bernard Shaw, who predicted that mankind will disappear and something better will take its place.

the creation of different cultural environments for young and old. Mobility has (1) facilitated the atomization of the family into very small units, often consisting only of a single adult, and (2) reduced the active contact time between children and their elders, making the teaching of ethics far more difficult. At the same time many parents respond defensively to the rapid change around them by forcing conformism on, and inculcating intolerance in, their offspring.[31] Openness to multiple perspectives is hardly encouraged in such a setting.

Some of the aspects on which ethical management must focus in the two cases we have examined in parts 2 and 4 are summarized in table 12.1, where they are classified by perspective type. In the words of Catholic theologian Hans Küng, "ethics must again become public instead of merely personal." He was the main author of the Declaration of a Global Ethic, which was signed in 1993 by twenty leaders of major faiths, including Catholics, Protestants, and Buddhists. It condemns killing in the name of religion, denounces environmental abuses and sexual discrimination, and calls for disarmament.

Paul Tillich, Reinhold Niebuhr, and Dietrich Bonhoeffer have also stressed the linkage of ethics to governance. Their theology offers a constructive answer to the two grave concerns we have expressed: (1) simplistic religious or ideological fanaticism, and (2) moral vacuum. Self-respect is linked to respect for others, personal behavior to community connectedness, rights to responsibilities, and "tough love" to tough standards of institutional ethics. These theologians espouse a "liberal" frame of mind, where the word denotes freedom from narrow prejudice and a mind open to diverse perspectives. It is the opposite not of "conservative," but of fanatical, bigoted, and intransigent.[32]

Cynics respond that resistance to multiple perspectives or worldviews and insistence on the myopic "me-now" or discounting of distant space and time are endemic to Homo sapiens. They see venal corruption, insatiable greed, and contempt for society's rules and their enforcement as innate human characteristics. They point to the record-breaking societal destructiveness of the 20th century, its two world wars and murderous holocaust, as evidence of unchangeable public ethics and scoff at ethical management as a quixotic quest to alter human nature. We do admit that we empathize with the futurist who acknowledged that he was optimistic, but really could not justify his optimism (part 4). However, human history also tells us that individuals have responded to daunting challenges, corporations have become outstanding contributors to the communal wellbeing, and societies have risen to greatness. Ingredients such as altruism, compassion, love, hope, creativity, drive, intelligence, and openness to new ideas are gifts generously spread among us.[33]

The foundation of ethical management must be an affirmation of humaneness, a love of man and the world. It may be religious, as with Dietrich Bonhoeffer's Christian *Menschlichkeit*, or nonreligious, as reflected in Albert Camus's writings, particularly *The Plague* and *The Rebel*.[34] Ethical management itself builds from the local level, the individual and the family. Role models for self-worth, morality, trust, love, and discipline are vital. Respect for the rule of law and for education must be demonstrated. The bottom-up approach to this ethical focus must then move beyond the family. Community and business leaders can set the tone by concretizing ethical management at a higher level. They must further extend the concept by developing the links to existing networks and establishing new ones. The linkages will expand and deepen progressively, encompassing more individuals, larger corporations, and more diverse communities.

The road will be strewn with obstacles, as the entrenched thinking senses threats to the old order and becomes increasingly entrapped in illusions. The recent trends of weakening of the family and widening of the gap between social strata pave the way for the multitudes of weak and poor to embrace fundamentalist faiths and hypnotic demagogues. There will also be strong pressure to maintain the status quo by the established elite in business, labor unions, education, medicine, armed forces, and civil service. Ming China, the Ottoman Empire, and Spain are exemplary reminders of the successful preservation of an orthodox status quo for centuries.

In chapter 9, we quoted poet William Butler Yeats in connection with the strain of the widening gaps wrenching our society: "the center cannot hold . . . / The best lack all conviction, while the worst / Are full of passionate intensity." Near the end of this century, we have plenty of extant examples of militant conservatism and virulent extremism, as well as an alarming lack of conviction in the mainstream.

Europeans from de Tocqueville to the World War II period were impressed by the brash, open-minded, energetic, and gener-

TABLE 12.1
Ethical Dimensions of Management*

Alaska Oil Spill	U.S. in the Knowledge Society
Individual Perspective	
Acceptance of personal responsibility by Exxon and Alyeska top executives	Stronger moral teaching in childhood
Less short-term and more long-term focus by corporate executives	Longer-term and less "me-now" focus
Sobriety on the part of ship's crews	More concern with responsibilities and less with entitlements
Responsibility of Alaska governors for the physical preservation of the state entrusted to their care	Recognition that the individual has a duty to serve as citizen of his/her community, the nation and the world
Recognition by technical staff that probabilities are subject to personal misinterpretation	More concern with lifelong education
Stoppage of the use of the ocean as a convenient garbage dump by ship's crews	Less resistance to, and fear of, change
	Acceptance of multiple perspectives on complex problems
	Avoidance of hubris, i.e., reliance on, or blind faith in, one perspective
Organizational Perspective	
Full recognition that Exxon has vital stakeholders besides its stockholders (communities, employees, etc.)	Openness to innovation, adaptability, and reorganization, e.g., power more distributed, less concentrated
Maintenance of high effectiveness standards promised by Alyeska in its contingency plans	Courage to discard obsolete assumptions, to experiment, try bold solutions (and possibly fail), learn from feedback

U.S. *in the Knowledge Society*

Modification of the anthropocentric ethic

Greater concern with future generations
(e.g., sustainable growth)

Focus on management of potentially dangerous
technologies, waste, and pollution

Science's first duty: to serve humanity

Use of coordination-intensive structures
for interaction among perspectives

Recognition of limits of discipline-based
"solutions" in complex systems

Care for, and protection of, the biosphere

Use of technology in a humanizing role

Stress on creativity and innovation

Openness to, and support for, sciences of the 21st
century (e.g., complexity)

Alaska Oil Spill

Organizational Perspective

Adoption of attitudes of high reliability organizations

Real, rather than token, environmental
protection actions at federal level

Technical Perspective

Rethinking of proper use of coordination-
intensive structures

Better grasp of system coupling and
interactions to make the system safety-
reinforcing, rather than error-inducing

Stronger concern with spill cleanup technology

*See also Valdez Principles, chapter 3.

ous Americans they encountered. With their drive, democratic nature, and lack of stultifying tradition, they radiated an unlimited confidence in the future. The year 1969 marked both the peak of the economic cycle (see fig. 7.3) and the crowning technological achievement of the manned landing on the moon. But change was already in the air as social unrest, triggered by the Vietnam War, made itself felt. Only recently has the societal decline of the last quarter century, particularly the erosion of the middle class, been widely perceived. Now, near the nadir of the economic cycle, the somber realization is dawning that societal reinvigoration is not assured, even if the projected cyclic upswing of the economy does occur.

Public and private sectors everywhere cannot escape the transition to the new realities of the 21st century. The technology for the knowledge society era is at hand, but our human systems are far from ready. Failure to manage both ourselves and our technology more effectively will spell grave danger. Continuation of the recent societal drift can breed highly ordered, but stifling, insularity or uncontrolled instability, either of which would doom our vibrant democratic system.

CODA: NEW THINKING FOR A NEW CENTURY

We have attempted in this book to help the reader in thinking about, and dealing with, the messy systems that constitute life in the decades ahead. There can be no single model, no comforting "solution to the problem," no reliance on "more of the same." New growth implies the willingness and drive to take risks and experiment. Wise self-management means learning to stay on the edge between order and chaos as the system evolves. The accelerated pace of change means learning must be faster. Everything interacts with everything. Tight coupling and intricate interactions abound in the system that is ours in the new century (see tables 2.1 and 2.2). This complicates the self-management task: decentralization is preferable with intricate interactions, while centralization is preferable in managing tight coupling. It is fortuitous that information technology permits us to do both simultaneously as never before.

Unbounded systems thinking is the ideal and multiple perspectives give us an immediate, practical way to approach such thinking in managing ourselves and our technology. The three kinds of perspective—technical, organizational, and personal—have distinct paradigms (table 6.2 and fig. 6.1) and there may be diverse perspectives within each of these categories.

Multiple, interacting perspectives are vital, and the schematic triangle used in figures 6.2, 12.3, and 12.4 tries to communicate this concern. They constitute a prerequisite for the continuing drive to find a better balance between short-term and long-term, between the welfare of our neighbors today and our grandchildren tomorrow, between the rights and obligations of the individual, corporation, and society, between technological advance and control, between democracy and freedom, and between local, national, and global management. The top-down, often T-focused, approach must meld with the bottom-up, usually O/P focused, approach.

Mismanagement in our past world of modest population and limited technological power could be tolerated far more readily than it can in the crowded, high-tech global megalopolis of tomorrow. As information flow accelerates in complex systems, there will be potent new ways to create system order and unexpected confrontations with system instabilities.

All purveyors of simple solutions as well as prophets of impending doom should evoke mistrust and suspicion. We know too much to believe in simple solutions and not enough to be pessimists. We have only partial control over our future, but that must not induce an attitude of passive resignation. The emergence of the global village may spawn a more complex society with superior public and private sectors that we cannot even envision today.

We have emphasized "unbounded thinking" and "ethical action," and the good news is that human beings have proven they have the capacity for it. The systems philosopher sees that combination as the basis for wisdom: "Wisdom may be thought which proceeds with deep ethical concerns and an unbounded approach."[35]

This, then, is the challenge we must grasp as we approach the newly interconnected world of the 21st century with its unprecedented risks and unique new growth opportunities.

APPENDIX A:
A SAMPLING OF WAYS TO
INTEGRATE PERSPECTIVES

The three techniques presented here suggest how T-based systems analysis approaches may prove helpful in dealing with multiple perspectives.

MAMP

The multiattribute/multiparty (MAMP) framework[1] creates a structure to analyze the decision-making process. First, the parties and their various concerns are identified and depicted in a matrix. Then the process is viewed as consisting of rounds, each having four facets:

1. problem formulation with stated assumptions and questions
2. initiation point, such as an application for a permit, a request for information, an agency approval, or a government intervention
3. interaction among the parties, represented by the positions and arguments of the parties
4. conclusion, which may be a decision, a stalemate, a change in the focus of the debate, or an exogenous event

To illustrate the framework, the first round in the liquified natural gas (LNG) facility siting process in West Germany is summarized in table A.1. In this siting process the parties include the applicant, government agencies (federal, state, local), and a local citizens' group. The concerns: national (need, import policy), regional (e.g., industrial development), local (e.g., economic benefits, risks to population), and applicant-specific (profit, image, control over sources).

TABLE A.1
Round A in the MAMP Framework for LNG Facility Siting
in the Federal Republic of Germany (1972–July 1976)

I. Problem Formulation

- Assumptions:
  1. Natural gas is an important source of energy and its benefits are generally accepted
  2. The possibility exists to import Algerian LNG
  3. The site at Wilhelmshaven is an area created to encourage industrial development

- Question: Given its feasibility, is the proposed LNG project suitable and desirable for Wilhelmshaven?

II. Initiation

- Two companies, Ruhrgas and Gelsenberg announce to Lower Saxony (the state in which Wilhelmshaven is located) their intention to build an LNG terminal (1972). The companies form a subsidiary, DFTG. Wilhelmshaven is considered to be the most appropriate harbor by DFTG and the Lower Saxony Ministry of Economic Affairs and Transportation (ME&T).

III. Interaction

Party	Position	Arguments
DFTG, gas companies	For site at Wilhelmshaven	Need for natural gas, fits regional development plans, technology is safe
Wilhelmshaven	For site at Wilhelmshaven, subject to environmental impact conditions	Contribution to industrial development, safety and high degree of environmental protection have to be ensured
ME&T	For site at Wilhelmshaven subject to conditions on business structure	Beneficial to regional economy, support of gas supply companies in Lower Saxony

IV. Conclusions (1976)
- Lower Saxony and Wilhelmshaven commit themselves to support the project at the selected site
- Gas companies and DFTG agree on certain conditions (settlement contract)

Note: There were four rounds in all, extending to July 1979.

Source: H. C. Kunreuther et al., *Risk Analysis and Decision Processes* (Berlin: Springer Verlag, 1983), p. 46.

The structuring of the decision-making process clearly helps to understand the perspectives of the actors and their interactions. But it has its limitations.

Any structure has the advantage of creating order and the disadvantage of forcing everything into one mold. As one reviewer put it,

> you seem to be imprisoned by the notion that all the various partisan participants in problem solving agree, during any one round, on a definition of the problem . . . But it is a common characteristic of interactive problem solving that many, perhaps most, of the participants carry a distinct version of what "the problem" is in their minds . . . They are not working on any one given problem, nor do they think they are.[2]

In our terms, MAMP is itself a T-type representation. The implication is that such a structure cannot help but be repressive; it buys structure at a price. One should really seek multiple MAMP versions that reflect the different perspective types.

INTERACTION MAPPING

It is also useful to adapt a T-oriented system structuring procedure, directed graphs or "digraphs," to exhibit the relationships among perspectives. The general rules for working with digraphs are as follows:

1. System elements are represented by points or nodes and the relations between them are shown by arrows:
   - If an increase in A causes an increase in B and a decrease in A causes a decrease in B, the arrow from A to B carries a + sign.

- If an increase in A causes a decrease in B and a decrease in A causes an increase in B, the arrow from A to B carries a – sign

2. A causal loop denotes a sequence of arrow relationships that ends at the starting node, that is, A -> B -> . . . -> A (see chap. 12):

   - A loop is positive or impact-amplifying if it has zero or an even number of minus signs

   - A loop is negative or impact-counteracting or damping if it has an odd number of minus signs (see chap. 12)

We now consider an application to the interaction among perspectives. An arrow marked + from T to O_1 means that the T perspective is supportive of O_1, that is, the T analysis presented to the organization favors implementation of the organization's preferred strategy. An arrow marked – from O_1 to T may signify that it is in the organization's interest to scuttle or discredit the T analysis. An illustration, drawn from the case of the evolution of the M-16 rifle in the U.S. Army, is shown in figure A.1.[3] The perspectives are Springfield Arsenal designs (T_1), outsider designs (T_2), Army Ordnance Corps (O_1), Army Caliber Board (O_2), the rifleman/sharpshooter myth (P_1), maverick inventor Spencer (P_2), and President Theodore Roosevelt (P_3).

Army Ordnance created the Springfield Arsenal and the mutual support relationship is shown by the two arrows connecting O_1 and T_1. The Western rifleman myth strongly influenced O_1 as shown by the + arrow P_1 to O_1. Outsider Spencer evolved a different technical approach (P_2 to T_2). However, Army Ordnance opposes the outsider (O_1 to T_2). Teddy Roosevelt and the Caliber Board subsequently also supported an outside design (P_2 to T_2, P_3 to T_2). The relation O_1 to O_2 signifies the Chief of Staff's disapproval of the Caliber Board recommendation.

We note that there are four loops in the directed graph:

1. $T_1 + O_1 + T_1$ (positive or reinforcing)
2. $T_2 + O_2 + T_2$ (positive or reinforcing)
3. $T_2 + O_1 - T_2$ (negative or countering)
4. $T_2 + O_1 - O_2 + T_2$ (negative or countering)

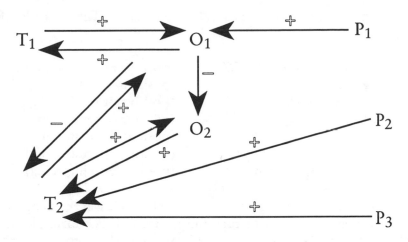

FIGURE A.1
Example of a Diagraph of Perspectives

Source: H. A. Linstone, *Multiple Perspectives for Decision Making: Bridging the Gap between Analysis and Action* ©1984, pp. 21, 322. Reprinted by permission of Prentice Hall, Englewood Cliffs, New Jersey.

As O_1 was far more powerful than O_2, the reinforcing loop $T_1O_1T_1$ was the dominant positive effect, and T_1, part of one strong positive loop, triumphed over T_2, part of two negative loops and one weak positive loop.[4]

Maruyama offers another application involving multiple perspectives, the role of bribery in socialist or postsocialist countries such as Poland, Russia, Romania, and China. Available goods and services are of poor quality and scarce, so one uses bribery to obtain basic necessities. Bribes may be not only in money but also in goods and services, such as apartments, telephones, university entry, a passport, or information. The system is one of mutual reciprocity and trust, and it serves as a stabilizing factor in the society.[5] It comprises the following elements:

- Technical: productivity (P); quality and quantity of products and services (Q); low value of money (L)
- organizational/institutional: planned economy (E); authority to control flow of materials and services (A); shortcut of cumbersome official procedure by bribe (S)
- personal/individual: willingness to work hard (W); bribing by giving things or services (B)

Figure A.2 presents the causal loop diagram. The causal loops are:

1. W + P + Q – A + B – W	(positive or reinforcing)
2. E – W – E	(positive or reinforcing)
3. E + A + B – W – E	(positive or reinforcing)
4. A + S – P + Q – A	(positive or reinforcing)
5. W + Q – L + B – W	(positive or reinforcing)
6. E – P + Q – L + B – W – E	(positive or reinforcing)
7. W + P + Q – L + B – W	(positive or reinforcing)

All loops are reinforcing or "vicious circles." However, changes in any element can begin to turn things around. For example, the shift from a planned economy (E) to a market economy (E*) changes E – P to E* + P after some time has elapsed. The time lag is due to the need to replace obsolete machines and outmoded management systems.

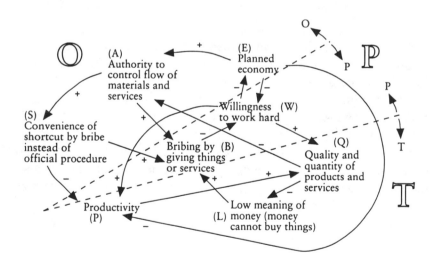

FIGURE A.2
Interactive Causal Loop Diagram of Bribery in Poland, Russia, Romania, and China

Source: Reprinted by permission of the publisher from "Overcoming the Socialist Aftereffects in East Europe," by M. Maruyama, *Technological Forecasting and Social Change, 40,* pp. 297–302. Copyright 1991 by Elsevier Science Publishing Co., Inc.

Similarly, the effect of E* on A and W will face delay. The market economy provides incentives to work hard (W), that is, E* + W, but many workers will initially be unwilling to change their ways and therefore oppose E*, that is, W − E.* Thus, for some time, the new loop 2 will be damping, E* + W − E.* Later, we will see E* + W + E* as the workers see the benefit to them of working hard (more pay) and the reinforcing effect takes hold. It becomes evident that the transition to a free market system must proceed with an appreciation of the dynamics of relationships that derive from different perspectives. Thus, the unique presocialist cultural traditions (O) prove as important as economic theory (T) in effecting a successful transition.

The digraph technique itself reflects the advantages and limitations of the T perspective. The structuring helps to identify, and often uncover neglected, relationships. The situation is usually quite complex and the digraph can serve as a roadmap to the interactions. It can test the effect of policy changes by tracing through the linkages of the roadmap. The disadvantage is that it abstracts and simplifies the interactions. The T-oriented procedure may therefore give a false sense of comprehensiveness and fail to represent the way O and P perspectives "see" the problem.

ASSUMPTIONAL ANALYSIS

A major source of difficulty in integrating perspectives can be traced to the fact that the different parties make different assumptions about the system or "the real nature of the problem." Then each collects input that confirms its view of the problem and accentuates the differences in the assumptions. We stress that assumptions are the property of the stakeholder, not the property of the problem.

A most helpful step in the integration process is the clarification of the assumptions made by all the parties, in particular, the degree of certainty and the importance attributed to them. We shall use an actual case study undertaken by I. I. Mitroff and J. Emshoff for a drug company.[6]

The problem is that McNeil, a drug manufacturer and a subsidiary of Johnson and Johnson, produces a well known, narcotic-based prescription pain killer under its brand label. It has suddenly learned that a generic substitute version has appeared on the market at a much lower selling price.* The situation portends potential catastrophe for McNeil, as much of its profit comes from this brand-label drug. Twelve key McNeil executives have been brought together to advise the CEO. The twelve have coalesced into three groups, each favoring a different alternative in pricing the brand-label drug: group A—lower the price; group B—raise the price; group C—no price change.

Group A wants to out-generic the generic version. Group B wants to communicate to the marketplace that it is of far superior quality. It assumes that many people equate price with quality. Groups A and B are oriented to the marketplace, but group C presents a different argument: While keeping the price unchanged, the company can raise profits by cutting internal costs, specifically, by eliminating the research and development arm of the company.

Each group collects data, reinterpreting data it has in common with the others to back up its case and procuring new data to further strengthen it. The process unconsciously becomes a circular one: each alternative is used to direct its believers to gather data that confirms the validity of that alternative.

Here is where assumptional analysis comes in. There are at least ten parties that have an interest in this problem: Johnson and Johnson, McNeil's management, McNeil's sales force, suppliers, stockholders, customers or patients, the competition, the federal government, pharmacists, and physicians. For example, the government is involved through the Federal Drug Administration and the regulations concerning imports of opiates from foreign countries.

The assumptions with regard to the physician prove to be the most critical. Consider groups A and B. Group A assumes that skyrocketing health care costs are making physicians price sensitive. They will prescribe the generic drug if it is significantly cheaper. Group B is assuming that physicians are motivated by the traditional model of health care, that is, the well-being of the patient irrespective of cost. Thus they are assumed to be price-insensitive. Another assumption by group B is that the patient will rely on the doctor and ignore the pharmacist's alternative suggestion.

We now determine how the three groups view the stakeholder assumptions along two axes: their importance or criticality (x),

---

*In some states the pharmacist is required to tell the patient coming in to fill a brand-label prescription that a generic version is available.

and the confidence in their validity *(y)*. Figure A.3 maps these views and tells us that the assumptions about the physician are both the most important and most uncertain. For group B this means the assumption that the physician is insensitive to price is not only critical but also highly questionable. It now becomes apparent that the data available to the group does not provide any enlightenment about this assumption.

Another party regarding which assumptions are important but uncertain is Johnson and Johnson. Is it more interested in maximizing profits or market share, more profit per drug sold or more volume? Why has McNeil not tried to find out the parent company's goal if it is important to the decision? Possibly the McNeil corporate culture has so far precluded such a step. Figure A.3 also shows that there is little uncertainty about the patient: a low-cost, quality product is desired.

The procedure actually helped the McNeil executives to agree as a whole to try alternative B in selected areas. The price of the drug was raised in test locations and the results were monitored. It could then be quickly determined whether the market would tolerate the price increase.

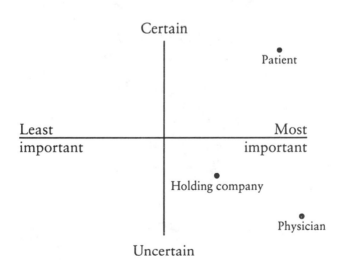

FIGURE A.3
Assumptions Associated with Stakeholders

APPENDIX B:
GUIDELINES FOR MULTIPLE
PERSPECTIVE USERS

The following summary should help those who want to use multiple perspectives in addressing their specific problems.[1]

GENERAL GUIDELINES

1. Each perspective offers insights not obtainable with the others. Together they provide a far more meaningful basis for decision than reliance on any one perspective. Shifting perspective types, that is, from T to O and P, shifts paradigms and minimizes retention of the same assumptions and biases. Conversely, the use of multiple perspectives shows with great clarity the severe limits in relying on a single perspective. Thus it has become clear that a computer model, such as the world dynamics model (chap. 6), cannot possibly be refined and enlarged by adding more variables and relationships to overcome its inherent weaknesses. One quickly reaches a point of diminishing returns, one that still leaves a wide chasm between model and reality.

2. The choice of perspectives requires judgment. There is no "correct" or "complete" set of perspectives. One can never include all possible perspectives. We must remember that making a decision involving complex systems almost always means doing so with inadequate information. It is typical of an analyst to conclude a study with the admonition that additional study will have to be done, that there are unresolved problems, that this study is only a first step, that uncertainties remain. The decision maker, individual and organizational, never has this luxury. The time is severely limited; there is an agenda and one must move on to the next important item. Thus the key to good decision making is good judgment in selecting and integrating perspectives.

3.   *The O and P perspectives are usually case-specific.* As the organizational and individual actors associated with any system are unique to that system, we cannot deal with generalized O and P perspectives. The dynamism of complex systems means that a given O or P perspective may change over time and that some organizations may disappear during the time of concern while new ones will emerge. A P perspective may become an O perspective as an organization accepts an individual perspective as its collective view. At times it may be difficult to distinguish between an O and a P perspective. Perspectives may reinforce each other or conflict with each other.

4.   *The means of obtaining input for T perspectives differ significantly from those for O and P perspectives.* Technical reports are typical means for communicating a T perspective. However, O and P perspectives may be politically sensitive. A company annual report will state the corporate objectives but omit mention of the parochial objectives of its divisions. For example, one division may wish to take over the functions of another division. The military publicizes its needs as if there were no internal conflicts between the services, and even within a service, such as infantry and armored units, or surface and submarine navy. Thus it cannot be expected that input on O and P perspectives can be obtained by reading technical reports. We have found that person-to-person interviews are often the best means of gathering O and P perspective inputs. (However, first-hand sources may not always be accessible.)

Interviews require a talent not usually found in T-oriented persons, such as academics. They often have a pronounced T bias that makes them poor interviewers for O and P. The good interviewer must be a good listener, open to unexpected comments and clues. What is not said may be as important as what is said. Therefore, a highly structured interview may be self-defeating.

Being an insider can be both an advantage and a disadvantage. It greatly facilitates obtaining information and conducting interviews. But the insider may also be preconditioned and biased by long immersion in the organization.

It is critical that language and cultural differences be understood and overcome in this process. Our experience in China showed that the Chinese well understood what was being probed with O-type questions. The Chinese culture is bureaucratic and hierarchic, so that O games and strategies are known to all.

Power relationships are enshrined in all kinds of slogans such as "two down, one up" for the planning process and "the pyramid of power." P perspectives present more of a hurdle. Often the answers to questions were too spare and general; one must strive for concrete examples and anecdotes to flush out the insights. All translation becomes interpretation and this requires a sophisticated knowledge of the local culture. Since simple word-for-word translations are not possible and the Chinese culture contains many untranslatable nuances, metaphors, similes, and allusions, very well trained and sophisticated interpreters are essential. Natives or insiders are of particular value.

5. *A balance of effort among T, O, and P is desirable.* If a T-oriented analyst were to apply multiple perspectives, it would not be surprising to find the bulk of his effort devoted to T, with O and P included almost as asides or embarrassing appendices to a legitimate T analysis.* Similarly, an O-oriented person is likely to pay lip service to T and P. It is therefore important to make a conscious effort to balance the attention devoted to each of the three types.

We observed in chapter 6 that an aspect of the uniqueness of the individual is the balance, or more likely, the imbalance, between T, O, and P in his or her own mind. If individuals with a good balance are unavailable, an effort must be made to create a team with a good collective balance. We stress that this does not mean a taxonomic approach, say, an engineer, an economist, and a psychologist. They are all likely to favor T. A mix of, say, an engineer, a lawyer, and an entrepreneur is more likely to provide the desired balance as they would be expected to have developed with an emphasis on distinctly different paradigms. For example, an engineer uses data- and model-based modes of inquiry, whereas a lawyer is at home with dialectic modes of inquiry.

6. *Interaction or cross-cuing of perspectives is vital at all stages of the decision or problem-solving process.* Whether the

---

*Misunderstandings should also not be surprising. The analysts' theories of subjective probability or utilities may use subjective input, but such input still comprises "data," and the theories still involve abstract "models." It is hard to free oneself from a deeply entrenched mindset. Thus, one excellent T-oriented analyst saw a "parallel" between decision-making theories and T, O, and P (cost-benefit analysis as T, social choice theory as O, and decision theory as P). But he also recognized that all three theories use economics and two use engineering as "intellectual roots," a clue to their true T nature.[2]

perspectives are developed in parallel or sequentially, there must be ample opportunity for cross-cuing or interaction. If the process is sequential, it is crucial that continuous interaction or feedback between rounds take place. Changing actors between rounds, for example, can easily lead to missed feedback and opportunities for cross-cuing, to the detriment of the entire task. An example is the controversial 1980 California Medfly Eradication Program. The participants viewed the problem from three perspectives sequentially. The procedure amounted to the use of three different filters with the only information used being that which passed through all three filters. As pressure mounted for action, managers actively tried to inhibit cross-cuing and critical information was lost.[3]

7.   *There is no "correct" weighting formula for the integration of perspectives.* The decision maker must integrate the perspectives in his or her own way. However, a prototype integration may, together with the perspectives, be provided to the decision maker. It is helpful to consider the American courtroom where the jury is the decision maker. It is presented with varying testimonies, that is, perspectives. Then the prosecutor integrates the perspectives, weighting them to suggest the guilt of the defendant. Next, the defense attorney offers a different integration of the testimonies to suggest the inocence of the defendant. The jury can accept either integration or undertake its own integration by examining the original testimonies.

Now consider a CEO in a corporation. Let us say that a decision has to be made whether to initiate a new line of business. The CEO has a staff study, a T analysis, showing the costs and benefits of several alternative actions. He or she now seeks other perspectives. Meetings with division managers will indicate which units will be enthusiastic about the new line and which ones will be against it (O perspectives). The reasons for opposition may have nothing to do with the merits of the line, but with parochial factors, such as losing development budget in coming years if the new line is implemented (it will go to another unit). The CEO may bounce the idea off an old friend who runs another company and whose judgment he respects (P perspective). When a reasonable number of perspectives have been obtained, the CEO integrates the input in a unique way and makes the decision.

8.   *Perspectives are dynamic and change over time.* Complex systems involving human beings are rarely static. Their dynamic nature means that organizational and individual actors change

and that the perspective of a given actor may change over time. The interaction of perspectives in conferences yields compromises and these alter the perpectives of the participants. According to RAND's highly respected Charles Hitch, one of the vital early lessons learned there was that every significant problem they tackled had to be redefined after the analysis was under way.

The multiattribute-multiparty (MAMP) framework, discussed in Appendix A, clearly shows how actors and problem definitions change from round to round. An understanding of the sequential nature of the process proves most valuable in developing insights.

In its ideal form, the multiple perspective process should be nonterminating. Although this is impractical in the real world of decision deadlines, it may at times be possible to revisit the process and modify the implementation accordingly.

9. *The T perspective tends to dominate in the early stages prior to a decision; the O and P perspectives dominate in the action or implementation phases.* In the early stages of a decision process, there are often few individuals and organizations involved. The potential gainers and losers are not yet identified. Often the work is confined to academic institutions or planning staffs with little personal stake. As the decision time approaches, the parties at interest become clear and stands are taken, that is, O and P perspectives are formed. Frequently the T-oriented analysts lose interest, feeling their completed report fulfills their assigned task. Once a decision is reached, implementation becomes paramount and O and P usually dominate T.

10. *The P perspective is particularly effective in communicating about complex systems.* Scenario writing presents a good case in point. The effective scenario writer is rarely a T-oriented person. An exception is Peter Schwartz, who has had many years of experience at SRI International and Royal Dutch/Shell and is a good communicator. His advice:

> Scenarios have to be simple, dramatic, and bold—to cut through complexity and aim directly at the heart of an individual decision . . . In the end every scenario is personal . . . Being a scenario-planner . . . means becoming aware of one's filter and continually readjusting it to let in more data about the world, but without becoming overwhelmed.

> Stories open people to multiple perspectives, because they allow them to describe how different characters see in events the

meaning of those events. Moreover, stories help people cope with complexity.[4]

Focusing on the business world, Peters and Waterman note that "we are more influenced by stories (vignettes that are whole and make sense in themselves) than by data (which are, by definition, utterly abstract)."[5]

Novelist Leo Tolstoy provided perspectives on the Napoleonic invasion of Russia that conventional historians' accounts and chronologies never could. Science-trained novelist H. G. Wells was a far more effective scenario writer than professional scientists and engineers. Plays such as Shaw's *Pygmalion* and Ibsen's *The Wild Duck*, as well as television's *Upstairs, Downstairs* successfully express social issues through very individualized perspectives. The enormity of the Nazi holocaust cannot be grasped by statistics such as six million murders, but it is communicated at least in part by the TV production *Holocaust* and the film *Schindler's List*. Ronald Reagan was known as "the great communicator" because he would use very personal anecdotes, "grabbers," that resonated with his audience. Recognition that, to a degree, the medium really is the message is the first step to skillful communication of perspectives.

A WORK PLAN OUTLINE

The following plan should be clearly understood as an illustrative example, not as a prescription. It must in every case be adapted to specific needs.

1. *An initial problem statement:* Begin with a preliminary problem definition. This will usually be based on a decision that needs to be made or a problem discerned that requires action. The definition will also include preliminary problem boundaries.

2. *Perspective identification:* Identify what appear to be important T, O, and P perspectives. This requires determination of the stakeholder organizations and key individuals who appear to be affected by, or who will affect, the decision or problem.

3. *Initial assessment:* Examine the problem from the point of view of each T, O, and P perspective. This requires techni-

cal expertise for T and in-depth interviewing for O and P. Each perspective may see, and thus define, the problem or system differently, may use different assumptions (including assumptions about the other parties at interest), and system boundaries (important factors to include and exclude). Each perspective may favor a means of resolving the problem or favor a certain decision. Each may have significant input concerning implementation.

4. *Interactions:* Probe the interactions among the perspectives. The procedures outlined in Appendix A may prove useful. It will become apparent what commonalities and conflicts there are in definitions, boundaries, assumptions, preferred decisions, and means of implementation. Search for potential impacts.

5. *Iteration of steps 1–4:* Retrace the sequence of steps 1 to 4 on the basis of the insights drawn so far. For example, there may be additions or subtractions of stakeholders, changes in definitions, boundaries, and assumptions. It will usually not be possible to resolve major conflicts at this stage. Therefore, this should not be an attempt to develop "final" versions of disputed aspects.

6. *Prototype integration:* Perform a prototype integration and indicate the expected consequences. This is the user's subjective summation and proposed decision or problem resolution and implementation plan. It is analogous to a prosecutor's or defense attorney's summation of a trial to the jury. Also explain the uncertainties and possible subsequent actions to deal with anticipated and unanticipated impacts.

7. *Conference—phase A:* Bring together in a conference representatives of each perspective and review the perpectives and prototype integration to initiate the discussion. If possible, the decision maker should participate in this session. Aim for a new integration based on the discussion.

8. *Conference—phase B:* Develop a base scenario using the new integrated version and reasonable assumptions for its evolution or implementation. Impose possible shocks or surprises on the base scenario and discuss their impacts.

9. *Action proposal:* Construct an action proposal based on the conference evaluations. The proposal should include:

what is to be done, when, and by whom. Should there be a test or experimental program? How should progress be monitored? What are milestones for subsequent decisions? How should unanticipated crises be managed?

10. *Communication:* Communicate the output of steps 7–9 in a format designed specifically for the decision maker or other client. Keep in mind guideline 10 on communication in this appendix.

NOTES

In the case of newspaper references, only the date is given, as the context makes the citation unambiguous.

PREFACE

1. E. T. Morison, *Men, Machines, and Modern Times* (Cambridge: MIT Press, 1966), p. 114.
2. Morison, *Men, Machines, and Modern Times*, pp. 116–22.
3. *Wall Street Journal*, Apr. 7, 1980.

CHAPTER 1

1. *New York Times* Apr. 18, 1992.
2. *New York Times*, Sept. 18, 1990.
3. "Anatomy of an Oil Spill," written and reported by John Tuttle, produced by Oregon Public Broadcasting System, *Frontline*, Mar. 20, 1990; J.-M. Cousteau, Interview on *CBS This Morning* (television program), Mar. 23, 1990.
4. Cousteau, Interview; *Los Angeles Times*, Mar. 18, 1990; *New York Times* Sept. 18, 1990.
5. *Los Angeles Times*, Feb. 22, 1990, Mar. 18, 1990.
6. *New York Times*, Sept. 18, 1990.
7. *New York Times*, Mar. 14, 1991, Apr. 30, 1993.
8. *Oregonian*, Sept. 18, 1990; *New York Times*, Apr. 30, 1993.
9. *Los Angeles Times*, Feb. 20, 1992; *New York Times*, Apr. 30, 1993.
10. *Los Angeles Times*, Jan. 12, 1993; Mar. 26, 1993.
11. World Health Organization, *Reproductive Health: A Key to a Brighter Future*, 1992.
12. *Los Angeles Times*, Jan. 6, 1993.
13. *New York Times*, May 31, 1992.
14. *New York Times*, Apr. 30, 1992.
15. *Los Angeles Times*, Feb. 6, 1990.
16. *New York Times*, July 25, 1992.

17. M. Cetron and O. Davies, *American Renaissance: Our Life at the Turn of the 21st Century* (New York: St. Martin's Press, 1989), chap. 2.

18. *New York Times*, July 20, 1992.

19. *Time*, Jan. 2, 1989.

20. *New York Times*, Dec. 16, 1984.

21. E. W. Lawless, *Technology and Social Shock* (New Brunswick: Rutgers University Press, 1977), pp. 208–16.

22. F. Braudel, *The Mediterranean and the Mediterranean World in the Age of Philip II* vol. 1, trans. Sian Reynolds (New York: Harper & Row, 1972), pp. 20–21.

23. C. P. Snow, *The Two Cultures: And a Second Look* (Cambridge: Cambridge University Press, 1964), pt. 1, in particular, pp. 11, 76, and 77.

24. G. T. Allison, "Conceptual Models and the Cuban Missile Crisis," *The American Political Science Review*, vol. 63, no. 3 (1969), pp. 689–718; G. T. Allison, *Essence of Decision: Explaining the Cuban Missile Crisis* (Boston: Little, Brown & Co., 1971).

25. J. W. Davidson and M. H. Lytle, *After the Fact: the Art of Historical Detection* (New York, Alfred A. Knopf, 1982).

26. I. Berlin, *The Hedgehog and the Fox* (London: Weidenfeld and Nicholson, 1967), pp. 1–2, 34.

CHAPTER 2

1. E. W. Lawless, *Technology and Social Shock* (New Brunswick: Rutgers University Press, 1977), p. 229.

2. Alaska Oil Spill Commission, *Spill: The Wreck of the Exxon Valdez*, Final Report, Feb. 1990, p. 169.

3. "The Future of Big Oil," *Fortune*, May 8, 1989, pp. 46–54.

4. M. H. O'Leary, Testimony for Cordova District Fishermen United to the Alaska Oil Spill Commission, 1989.

5. Department of Transportation, National Response Team *The EXXON VALDEZ Oil Spill: A Report to the President,* May 1989.

6. Alaska Oil Spill Commission, *Spill*, appendix K, pp. 6, 43.

7. Department of Transportation, National Response Team *The EXXON VALDEZ Oil Spill*, pp. 17, 22.

8. O. Harrison, Statement to the Alaska Oil Spill Commission, Aug. 3, 1989.

9. *Los Angeles Times*, Feb. 18, 1990

10. *Anchorage Daily News*, Aug. 3, 1989.

11. *Los Angeles Times*, Feb. 18, 1990.

12. Alaska Oil Spill Commission, *Spill*, p. 122.

13. National Research Council, *Tanker Spills: Prevention by Design* (Washington, D.C.: National Academy Press, 1991).

14. R. D. Woithe, "Emergy Analyses of the *T/V Exxon Valdez* Oil Spill and Alternatives for Oil Spill Prevention" (Department of Environmental Engineering Sciences, University of Florida, May 1992).

15. C. E. Perrow, *Normal Accidents* (New York: Basic Books, 1984), pp. 170–231; *Time,* Feb. 8, 1993; *Wall Street Journal,* Feb. 12, 1993.

16. "The Future of Big Oil," p. 50.

17. C. E. Perrow, *Normal Accidents.*

18. Perrow, *Normal Accidents,* p. 172.

19. *Los Angeles Times,* December 29, 1991 and January 9, 1994.

20. *Los Angeles Times,* Mar. 26, 1993.

21. Lawless, *Technology and Social Shock,* p. 231.

22. A. F. Dickson, "Navigation Problems (Tankers)" (*International Tanker Safety Conference,* International Chamber of Shipping, London, 1971, p.2) (quoted by Perrow, *Normal Accidents*).

23. Perrow, *Normal Accidents,* pp. 170–231; *Time,* Feb. 8, 1993; *Wall Street Journal,* Feb. 12, 1993.

24. American Petroleum Institute, *The American Petroleum Institute Task Force Report on Oil Spills,* June 14, 1989, p. ii.

25. *New York Times,* Apr. 11, 1992.

CHAPTER 3

1. "Anatomy of an Oil Spill," written and reported by John Tuttle, produced by Oregon Public Broadcasting System, *Frontline,* Mar. 20, 1990.

2. *Wall Street Journal* Mar. 16, 1990; *Los Angeles Times,* Mar. 17, 1990; *New York Times,* Mar. 13, 1991.

3. *New York Times,* Mar. 14, 1991.

4. *New York Times,* Nov. 19, 1992.

5. Exxon Corporation, information accompanying testimony at the Alaska Oil Spill Commission Hearing, Anchorage, Alaska, Aug. 2, 1989.

6. "The Future of Big Oil," *Fortune,* May 8, 1989, p. 50.

7. *Anchorage Daily News,* Mar. 25, 1989; *Time,* Mar. 26, 1990.

8. "The Future of Big Oil," p. 52.

9. "Future of Big Oil," p. 52.

10. *New York Times,* Nov. 19, 1992.

11. *Anchorage Daily News,* Apr. 20, 1989.

12. *Anchorage Daily News,* Apr. 29, 1989.

13. *Anchorage Daily News,* Aug. 4, 1989.

14. *Time,* Jan. 22, 1990, p. 51.

15. *Wall Street Journal* Mar. 6, 1992.

16. "The Ultimate P. R. Man," *Connoisseur*, Mar. 1990, pp. 74–79, 130–32.

17. *Time*, Mar. 26, 1990.

18. *New York Times*, Mar. 14, 1991; Aug. 23, 1993.

19. J. Kemeny et al., *Report of the President's Commission on the Accident at Three Mile Island* (New York: Pergamon Press, 1979).

20. *Wall Street Journal*, Mar. 16, 1990.

21. "Anatomy of an Oil Spill."

22. Alaska Oil Spill Commission, *Spill: The Wreck of the Exxon Valdez*, Final Report, Feb. 1990, p. 12.

23. Alaska Oil Spill Commission, *Spill:*, p. 27.

24. "Anatomy of an Oil Spill."

25. C. E. Perrow, *Normal Accidents* (New York: Basic Books, 1984), p. 183.

26. S. Wiborg, *Wo er steht, ist Hamburg* (Hamburg: Christians Verlag, 1992), p. 68.

27. "Anatomy of an Oil Spill."

28. *Time*, July 24, 1989.

29. *Time*, Mar. 26, 1990.

30. *Wall Street Journal*, Mar. 16, 1990.

31. *Wall Street Journal*, Mar. 16, 1990.

32. *Wall Street Journal*, Mar. 16, 1990.

33. *Time*, Mar. 26, 1990.

34. *Mobil World*, vol. 56, no. 8 (Oct. 1990).

35. *New York Times*, Apr. 25, 1991.

36. *Wall Street Journal*, Mar. 16, 1990.

37. *Los Angeles Times*, Mar. 23, 1990.

38. *New York Times*, Mar. 21, 1991.

39. *Wall Street Journal*, Oct. 2, 1991.

40. J. M. Blair, *The Control of Oil* (New York: Vintage Books, 1978), p. 145; *Los Angeles Times*, Jan. 18, 1993, Feb. 11, 1993.

41. *Anchorage Daily News*, Aug. 3, 1989.

42. *The Oregonian*, Aug. 16, 1989.

43. *Anchorage Daily News*, Apr. 27, 1989.

44. Alaska Oil Spill Commission, *Spill,* p. 35.

45. *Anchorage Daily News*, Apr. 21, 1989.

46. Alaska Oil Spill Commission, *Spill,* p. 39.

47. *Wall Street Journal*, July 6, 1989.

48. *New York Times*, Aug. 5, 1991.

49. *Los Angeles Times*, Feb. 4, 1990.

50. *New York Times*, Oct. 27, 1991.

51. *Los Angeles Times*, Mar. 20, 1992.

52. *Wall Street Journal*, July 6, 1989.

53. *Anchorage Daily News*, Aug. 3, 1989.

54. *Anchorage Daily News*, May 14, 1989.

55. *Wall Street Journal*, July 6, 1989.

56. *Wall Street Journal*, July 6, 1989.

57. *New York Times*, Aug. 5, 1991.

58. *Los Angeles Times*, Mar. 17, 1992.

59. *Wall Street Journal*, July 6, 1989.

60. "The Future of Big Oil," p. 52.

61. American Petroleum Institute, *The American Petroleum Institute Task Force Report on Oil Spills*, June 14, 1989, p. i.

62. T. LaPorte, "High Reliability Organization Project" (Unpublished memorandum, Department of Political Science, University of California, Berkeley, 1989). For information, contact the author at the Department of Political Science, University of California, Berkeley. See also table 5.1 in this book.

63. *Los Angeles Times*, Mar. 16, 1990.

64. *Wall Street Journal*, Feb. 9, 1990.

65. *Oregonian*, Sept. 7, 1990.

66. *The New Yorker*, Aug. 7, 1989, p. 67.

67. Alaska Oil Spill Commission, *Spill*, p. 156.

68. *Anchorage Daily News*, Aug. 4, 1989.

69. H. A. Linstone, *Multiple Perspectives for Decision Making* (New York: North-Holland, 1984), p. 166.

70. Alaska Oil Spill Commission, *Spill*, p. 150.

71. *Los Angeles Times*, Mar. 18, 1990.

72. *Los Angeles Times*, Feb. 28, 1990.

73. *Los Angeles Times*, Feb. 28, 1990.

74. *Los Angeles Times*, Mar. 17, 1990.

75. *Wall Street Journal*, Mar. 21, 1990.

76. *New York Times*, Mar. 21, 1991.

77. *New York Times Magazine*, Aug. 4, 1991.

78. *New York Times*, May 4, 1991.

79. *New York Times*, Mar. 21, 1991.

80. *New York Times*, Mar. 13, 1991.

81. *New York Times*, Apr. 25, 1991.

82. *New York Times*, Apr. 25, 1991.

83. *New York Times*, Apr. 30, 1991.

84. *New York Times*, Aug. 5, 1991.

85. *Anchorage Daily News*, Apr. 20, 1989.

86. *Anchorage Daily News*, Mar. 30, 1989.

87. *Los Angeles Times*, Mar. 18, 1990.

88. *Los Angeles Times*, Feb. 23, 1990.

89. *Los Angeles Times*, Mar. 16, 1992.

90. M. H. O'Leary, Testimony for Cordova District Fishermen United to the Alaska Oil Spill Commission, 1989.

91. *Wall Street Journal*, Mar. 30, 1990.

92. *New York Times*, Mar. 24, 1991.

93. Alaska Oil Spill Commission, *Spill*, pp. 66–84.

94. *New York Times*, Mar. 24, 1991.

95. *Los Angeles Times*, Mar. 16, 1990.

96. *Los Angeles Times*, Mar. 29, 1990.

97. P. A. A. Berle, "The Valdez Principles," *Audubon*, vol. 91, no. 6, November 1989, p. 6.

98. *Wall Street Journal*, Feb. 12, 1993.

99. *The Oregonian*, Aug. 20, 1993; *New York Times*, Aug. 24, 1993.

CHAPTER 4

1. "Anatomy of an Oil Spill," written and reported by John Tuttle, produced by Oregon Public Broadcasting System, *Frontline*, Mar. 20, 1990; *Wall Street Journal*, July 27, 1989.

2. "Anatomy of an Oil Spill."

3. *Time*, Mar. 26, 1990.

4. *Time*, Mar. 26, 1990.

5. *Time*, Mar. 26, 1990.

6. *New York Times*, Mar. 21, 1991.

7. *New York Times*, Apr. 25, 1991.

8. *Wall Street Journal*, July 6, 1989.

9. *Wall Street Journal*, Mar. 16, 1990.

10. *Los Angeles Times*, Mar. 24, 1990.

11. *Los Angeles Times*, Mar. 25, 1990.

12. *Los Angeles Times*, Mar. 24, 1990.

13. *New York Times*, July 11, 1992.

14. C. Perrow, *Normal Accidents* (New York: Basic Books, 1984). See also chap. 5 of this book.

15. "Anatomy of an Oil Spill."

16. *New York Times*, Mar. 21, 1991.

17. *New York Times*, Mar. 21, 1991.

18. *New York Times*, Apr. 25, 1991.

19. "Anatomy of an Oil Spill."

20. *Los Angeles Times*, Feb. 7, 1992.

21. *Los Angeles Times*, Feb. 7, 1992.

22. M. H. O'Leary, Testimony for Cordova District Fishermen United to the Alaska Oil Spill Commission, 1989.

23. *Money*, July 1989, pp. 77–83.

24. *Los Angeles Times*, Mar. 28, 1990.

25. *New York Times*, Sept. 18, 1990, Mar. 14, 1991.

26. *Anchorage Daily News*, Apr. 30, 1989.

27. *Los Angeles Times*, Mar. 18, 1990.

28. J.-M. Cousteau, Interview, *CBS This Morning* (television program), Mar. 23, 1990.

29. *Business Week*, Apr. 2, 1990.

CHAPTER 5

1. E. W. Lawless, *Technology and Social Shock* (New Brunswick: Rutgers University Press, 1977), p. 208.

2. *Los Angeles Times*, Jan. 12, 1993.

3. *Observer* (Regional Citizens' Advisory Council of Prince William Sound), 3, no. 1 (Jan. 1993).

4. *New York Times*, Aug. 23, 1992.

5. *Los Angeles Times*, Feb. 16, 1993.

6. I. I. Mitroff and T. Pauchant, *We're So Big and Powerful That Nothing Can Happen to Us: An Investigation of America's Crisis Prone Organizations* (New York: Birch Lane Press, 1991); T. LaPorte, "High Reliability Organizations Project" (Unpublished memorandum, Department of Political Science, University of California, Berkeley, 1989).

7. LaPorte, "High Reliability Organizations Project."

8. Alaska Oil Spill Commission, *Spill: The Wreck of the Exxon Valdez*, Final Report, Feb. 1990.

9. J. Kemeny et al., *Report of the President's Commission on the Accident at Three Mile Island* (New York: Pergamon Press, 1979).

10. B. Bowonder and H. A. Linstone, "Notes on the Bhopal Accident: Risk Analysis and Multiple Perspectives," *Technological Forecasting and Social Change*, vol. 32 (1987), pp. 183–202.

11. V. Legasov (first deputy director of the principal Soviet atomic research institute), quoted in *Washington Post*, Aug. 22, 1986.

12. J. H. Ausubel, "Political Fallout: What Fate Awaits Chernobyl in the New World Order?" *Sciences*, vol. 31 (Nov.–Dec. 1991), pp. 16–21.

13. B. Wahlström, "Avoiding Technological Risks," *Technological Forecasting and Social Change*, vol. 42 (1992), pp. 351–66.

14. B. Bowonder, "Integrating Perspectives in Environmental Management," *Environmental Management*, vol. 11 (1987), pp. 305–15.

CHAPTER 6

1. J. G. Miller, *Living Systems* (New York: McGraw-Hill, 1978).

2. J. Kane, "A Primer for a New Cross-Impact Language—KSIM," *Technological Forecasting and Social Change*, vol. 4 (1972), p. 141.

3. R. Tomlinson and I. Kiss, *Rethinking the Process of Operational Research and Systems Analysis* (Oxford: Pergamon Press, 1984), p. xi.

4. J. Forrester, *World Dynamics* (Cambridge, Mass.: Wright-Allen Press, 1971), p. 18.

5. H. A. Simon, "Prediction and Prescription in Systems Modeling," Speech at the IIASA Conference on Perspectives and Futures, Laxenburg, Austria, June 14–15, 1988. Edited version in *Operations Research,* vol. 38 (1990), pp. 7–14.

6. D. B. Lee, Jr., "Requiem for Large-Scale Models," *Journal of the American Institute of Planners*, vol. 39, no. 3 (May 1973), p. 174.

7. J. Fallows, *National Defense* (New York: Random House, 1981), pp. 140, 170.

8. J. Casti, *Searching for Certainty* (New York: William Morrow & Co., Inc., 1990), p. 407.

9. Simon, "Prediction and Prescription in Systems Modeling," Speech at the IIASA Conference.

10. Casti, *Searching for Certainty*, p. 407.

11. Johann Wolfgang von Goethe, *Faust,* pt. 1 and sections from pt. 2, trans. Walter Kaufmann (New York: Doubleday, Anchor Books, 1961), p. 93.

12. O. Young, "International Environmental Governance: Building Institutions in an Anarchical Society," Paper delivered at IIASA Conference, Laxenburg, Austria, May 12, 1992.

13. N. Machiavelli, *The Prince*, chap. 22, in *Machiavelli*, ed. J. Plamenatz (London: Fontana, 1972), p. 125.

14. N. Machiavelli, *The Discourses*, chap. 13, in *Machiavelli*, ed. J. Plamenatz, (London: Fontana, 1972), p. 231

15. N. Machiavelli, *Discourses*, chap. 55, pp. 204–5.

16. A. Jay, *Management and Machiavelli* (New York: Holt, Rinehart & Winston, 1968), pp. xi, 4, 7.

17. G. T. Allison, *Essence of Decision: Explaining the Cuban Missile Crisis* (Boston: Little, Brown & Co., 1971), pp. 110–11.

18. C. W. Churchman, personal communication, Oct. 31, 1991.

19. M. Thompson and M. Warburton, "Decision Making under Contradictory Certainties: How to Save the Himalayas When You Can't Find Out What's Wrong with Them," *Journal of Applied Systems Analysis*, vol. 12 (1985), pp. 17, 33.

20. Jay, *Management and Machiavelli,* p. 142.

21. Thompson and Warburton, "Decision Making," p. 9.

22. A. Wildavsky and E. Tenenbaum, *The Politics of Mistrust* (Beverly Hills: Sage Publications, 1981), p. 300.

23. H. W. Schneider, ed. *Adam Smith's Moral and Political Philosophy* (New York: Hafner Publishing Co., 1948), p. 247.

24. L. Tolstoy, *War and Peace*, epilogue 2, 1869, translated by L. Maude and A. Maude (New York: Simon & Schuster, 1958).

25. H. Smith, *The New Russians* (New York: Random House, 1990), p. 6.

26. C. W. Churchman, "A Philosophy for Complexity," in *Futures Research: New Directions*, ed. H. A. Linstone and W. H. C. Simmonds (Reading, Mass.: Addison-Wesley Publishing Co., 1977), pp. 88–90.

27. C. W. Churchman, *Thought and Wisdom* (Seaside, Calif.: Intersystems Publications, 1982), pp. 78–79.

28. C. W. Churchman, *Challenge to Reason* (New York: McGraw Hill, 1968), p. 86.

29. A. Koestler, *The Call Girls* (New York: Random House, 1973), p. 94.

30. H. Von Foerster, "The Curious Behavior of Complex Systems: Lessons from Biology," in *Futures Research: New Directions*, ed. H. A. Linstone and W. H. C. Simmonds (Reading, Mass.: Addison-Wesley Publishing Co., 1977), p. 106.

31. A. Toynbee, *A Study of History*, revised and abridged by A. Toynbee and J. Caplan (Fairlawn, N.J.: Oxford University Press, 1972), p. 97.

32. J. Hadamard, *The Psychology of Invention in the Mathematical Field* (Princeton: Princeton University Press, 1945), p. 21.

33. J. Larkin et al., "Expert and Novice Performance in Solving Physics Problems," *Science*, vol. 208 (4450) (June 20, 1980), p. 1342.

34. J. Salk, *Anatomy of Reality: Merging of Intuition and Reason* (New York: Columbia University Press, 1983), p. 79.

35. Two CEOs quoted by R. Rowan, "Those Business Hunches are More Than Blind Faith," *Fortune*, Apr. 23, 1979, p. 112.

36. H. Gardner, *Frames of Mind: The Theory of Multiple Intelligences* (New York: Basic Books, 1983).

37. D. Shapley, *Promise and Power: The Life and Times of Robert McNamara* (Boston: Little, Brown & Co., 1993).

38. *Wall Street Journal,* June 25, 1993, p. A12.

39. The Futures Group, *The Futures Group Reports* (Glastonbury, Conn., Oct. 1992).

40. K. Kelly, "A Distributed Santa Fe System," *The Bulletin of the Santa Fe Institute*, vol. 7, no. 1 (Spring–Summer 1992), pp. 4–6.

41. W. Ascher and W. H. Overholt, *Strategic Planning and Forecasting* (New York: Wiley Interscience, 1983), pp. 45–46.

42. T. Sowell, *A Conflict of Visions* (New York: William Morrow & Co., 1987).

43. G. Hardin, "The Tragedy of the Commons," *Science*, Dec. 13, 1968, pp. 1243–48.

44. D. Sawada and M. T. Caley, eds., *Mindscapes: The Epistemology of Magoroh Maruyama* (forthcoming, 1993).

45. J. Casti, "The Simply Complex: Trendy Buzzword or Emerging New Science?," *Bulletin of the Santa Fe Institute*, vol. 7, no. 1 (Spring–Summer 1992), pp. 10–13.

46 F. E. Udwadia and T. Agmon, "Trade Deficits: a Look Beyond the Economic View," *Technological Forecasting and Social Change*, vol. 33 (1988), pp. 109–18.

47. H. A. Linstone, *Multiple Perspectives for Decision Making* (New York: North-Holland, 1984), p. 84.

48. Smith, *New Russians,* p. 562.

49. J. F. Kennedy, preface to *Decision in the White House*, by T. Sorensen (New York: Columbia University Press, 1963); quoted in G. T. Allison, *Essence of Decision*, p. vi.

PART 4 INTRODUCTION

1. U. Colombo, "Sustainable Energy Development" (Paper delivered at the IIASA Conference, Laxenburg, Austria, May 12, 1992).

2. N. L. van der Noot and R. van de Berg, referenced in D. H. Meadows et al., *Beyond the Limits* (Post Mills, Vt.: Chelsea Green Publishing Co., 1992).

3. D. H. Meadows et al., *The Limits to Growth* (New York: Universe Books, 1972); Meadows et al., *Beyond the Limits*.

4. H. E. Goeller and A. M. Weinberg, "The Age of Substitutability, or What to Do When the Mercury Runs Out," *Science*, vol. 191, Feb. 20, 1976, pp. 683–89.

5. M. M. Waldrop, *Complexity* (New York: Simon & Schuster, 1992).

CHAPTER 7

6. J. Lears, *No Place of Grace: Antimodernism and the Transformation of American Culture 1880–1920* (New York: Pantheon, 1981).

7. J. Voge, "Information and Information Technologies in Growth and Economic Crisis," *Technological Forecasting and Social Change*, vol. 14 (1979), p. 2.

8. J. F. Coates, "Twenty-twenty Vision," *Technological Forecasting and Social Change*, vol. 30 (1986), pp. 305–12.

9. *Time*, May 4, 1992.

10. *Time*, July 22, 1991; *Los Angeles Times*, Dec. 22, 1992.

11. *Business Week*, May 11, 1992.

12. T. Heymann, *On an Average Day in Japan* (New York: Fawcett, 1992).

13. D. D. Noble, *The Classroom Arsenal* (London: Falmer Press, 1991), p. xi, *New York Times,* Sept. 9, 1993.

14. *New York Times*, Sept. 8, 1992.

15. *Time*, July 20, 1992.

16. *New York Times*, May 23, 1993.

17. *New York Times*, May 28, 1991.

18. *Oregonian*, June 5, 1992.

19. *Wall Street Journal*, Oct. 2, 1991.

20. *Time*, July 30, 1990, p. 48.

21. *International Herald Tribune*, May 14, 1992.

22. *Los Angeles Times*, Mar. 30, 1992.

23. D. Kash, *Perpetual Innovation* (New York: Basic Books, 1989).

24. *New York Times*, June 4, 1992; T. Conover, "Trucking Through the AIDS Belt," *The New Yorker,* Aug. 16, 1993, p. 66.

25. C. Lockwood and C. B. Leinberger, "Los Angeles Comes of Age," *Atlantic Monthly*, January 1988, pp. 31–62.

26. Los Angeles 2000 Committee, "LA 2000: a City for the Future," Final Report (Los Angeles, Nov. 1988).

27. P. Iyer, *Time*, Nov. 18, 1991, p. 109.

28. *New York Times*, July 5, 1992.

29. U. Colombo, "Population Growth of the Third World" (Paper delivered at the IIASA Conference, Laxenburg, Austria, May 12, 1992.

30. Worldwatch Institute, *State of the World 1992* (New York: W. W. Norton & Co., 1992), p. 3.

31. *Time*, Jan. 2, 1989, Apr. 1, 1991; *New York Times*, May 5, 1992.

32. *New York Times*, May 5, 1992.

33. *New York Times*, May 3, 1992.

34. World Development Report 1992, quoted in *New York Times,* May 18, 1992.

35. A. Grübler and Y. Fujii, *Energy*, vol. 16 (1991); T. Kanoh, "Toward Dematerialization and Decarbonization" (Paper delivered at the IIASA Conference, Laxenburg, Austria, May 13, 1992).

36. *New York Times*, Sept. 10, 1991, Sept. 11, 1991.

37. Worldwatch Institute, *State of the World 1992* (New York: W. W. Norton & Co., 1992), p. 3.

38. *Los Angeles Times Magazine*, Apr. 5, 1992.

39. *Los Angeles Times*, Jan. 12, 1992.

40. National Research Council, Bureau of Radioactive Waste Management, "Rethinking High-Level Radioactive Waste Disposal" (1990).

41. J. Lovelock, *The Ages of Gaia* (New York: W. W. Norton & Co., 1988).

42. H. Brooks, "Sustainability and Technology" (Paper delivered at the IIASA Conference, Laxenburg, Austria, May 12, 1992).

43. *Science*, Mar. 1, 1975; *New York Times*, Aug. 14, 1975; *Science*, Dec. 10, 1976.

44. R. Herman, S. A. Ardekani, and J. H. Ausubel, "Dematerialization," *Technological Forecasting and Social Change*, 38 (1990), pp. 333–48; Brooks, "Sustainability and Technology."

45. *New York Times*, May 25, 1992.

46. S. Messner and M. Strubegger, *Potential Effects of Emission Taxes on CO_2 Emissions in the OECD and LDCs* (IIASA Report RR–92–5, Laxenburg, Austria: International Institute of Applied Systems Analysis).

47. *Los Angeles Times*, Mar. 24, 1992.

48. *New York Times*, May 24, 1992.

49. *Science*, June 6, 1992.

50. U. Colombo, "Sustainable Energy Development" (Paper delivered at the IIASA Conference, Laxenburg, Austria, May 12, 1992.

51. J. H. Ausubel, "Does Climate Still Matter?" *Nature*, vol. 350, Apr. 25, 1991, pp. 649–52.

52. Ausubel, "Does Climate Still Matter?"

53. T. Kanoh, "Toward Dematerialization and Decarbonization" (Paper delivered at the IIASA Conference, Laxenburg, Austria, May 13, 1992).

54. U. Colombo, "Sustainable Energy Development."

55. E. U. von Weizsäcker, personal communication, May 1992.

56. H. E. Goeller and A. M. Weinberg, "The Age of Substitutability, or What to Do When the Mercury Runs Out," *Science*, vol. 191, Feb. 20, 1976, pp. 683–89.

57. T. Modis, *Predictions* (New York: Simon & Schuster, 1992).

58. C. Marchetti, "Infrastructures for Movement," *Technological Forecasting and Social Change*, vol. 32 (1987), pp. 373–93; also, "Anthropological Invariants in Travel Behavior," *Technological Forecasting and Social Change,* to be published 1994.

59. J. C. Fisher and R. H. Pry, "A Simple Substitution Model of Technological Change," *Technological Forecasting and Social Change*, vol. 3 (1971), pp. 75–88.

60. Modis, *Predictions*.

61. Modis, *Predictions*.

62. C. Marchetti, "Primary Energy Substitution Models: On the Interaction between Energy and Society," *Technological Forecasting and Social Change*, vol. 10 (1977), pp. 345–56; C. Perez, "Structural Change and Assimilation of New Technologies in the Economic and Social Systems," *Futures*, Oct. 1983, pp. 357–75.

63. Marchetti, "Infrastructures for Movement," also, "Anthropological Invariants in Travel Behavior," *Technological Forecasting and Social Change,* to be published 1994.

64. J. J. van Duijn, *The Long Wave in Economic Life* (London: George Allen & Unwin, 1983).

65. Marchetti, "Primary Energy Substitution Models."

66. T. Modis, *Predictions,* p. 170.

67. W. Strauss and N. Howe, *Generations: The History of America's Future, 1584–2069* (New York: William Morrow & Co., 1991); B. J. L. Berry and H. Kim, "Leadership Generations: A Long-Wave Macrohistory," *Technological Forecasting and Social Change* (forthcoming).

68. *New York Times,* June 5, 1992.

69. *New York Times,* Aug. 25, 1992.

70. C. Sagan, *The Dragons of Eden* (New York: Ballantine Books, 1978), pp. 24–25.

71. G. Gilder, "Into the Telecosm," *Harvard Business Review,* Mar.–Apr. 1991, pp. 150–61.

72. L. J. Perelman, *School's Out: Hyperlearning, the New Technology, and the End of Education* (New York: William Morrow & Co., 1992).

73. J. Coates, unpublished material.

74. C. Marchetti, "Genetic Engineering and the Energy System: How to Make Ends Meet," *Technological Forecasting and Social Change,* vol. 15 (1979), pp. 79–86.

75. A. Toffler, *Powershift* (New York: Bantam Books, 1990).

76. *Time,* Jan. 6, 1992, p. 24; M. Maruyama, "Interwoven and Interactive Heterogeneity in the 21st Century," *Technological Forecasting and Social Change,* vol. 44 (1993).

77. *New York Times,* May 18, 1993; L. Press, "Toward a Truly Global Network," Proceedings of the International Society of System Sciences Annual Meeting, July 12–17, 1992.

78. W. E. Halal, "A Forecast of the Information Technology Revolution" (manuscript, 1991).

79. G. P. Huber, "A Theory of the Effects of Advanced Information Technologies on Organizational Design, Intelligence, and Decision-Making," *Academy of Management Review,* vol. 15, no. 1 (1990), pp. 47–71.

80. Toffler, *Powershift.*

81. V. T. Coates, "Technology and U.S. Stock Markets," *Technological Forecasting and Social Change,* vol. 41 (1992), pp. 1–12.

82. W. E. Tengelsen, personal communication, July 1992.

83. *New York Times,* Oct. 25, 1992; Marchetti (unpublished manuscript), 1992.

84. L. White, Jr., "Technology Assessment from the Stance of a Medieval Historian," *Technological Forecasting and Social Change,* vol. 6 (1974), pp. 359–69.

85. J. Casti, *"Paradigms Lost"* (New York: William Morrow & Co., 1989), pp. 261–339.

86. J. Casti, *Searching for Certainty* (New York: William Morrow & Co., 1990), pp. 404–8.

87. C. Marchetti, "Millenial Cycles in the Dynamics of the Catholic Church: A Systems Analysis." Manuscript, International Institute for Applied Systems Analysis, Laxenburg, Austria, 1993.

88. M. M. Waldrop, *Complexity* (New York: Simon & Schuster, 1992).

CHAPTER 8

1. W. Durant, *The Life of Greece* (New York: Simon & Schuster, 1966), p. 254.

2. G. Modelski and G. Perry III, "Democratization in Long Perspective," *Technological Forecasting and Social Change*, vol. 39 (1991), pp. 23–34.

3. *Los Angeles Times*, Dec. 18, 1992.

4. J. F. Coates, "The Future of War," *Technological Forecasting and Social Change*, vol. 38 (1990), 201–5.

5. R. Ellwood, *The History and Future of Faith* (New York: Crossroad, 1989).

6. J. Miller, "The Islamic Wave", *New York Times Magazine*, May 31, 1992.

7. H. Smith, *The New Russians* (New York: Random House, 1990), p. 400.

8. *Los Angeles Times*, Apr. 8, 1992.

9. P. Schwartz, *The Art of the Long View* (New York: Doubleday, Currency Books, 1991); A. Toffler, *Powershift* (New York: Bantam Books, 1990); G. Friedman and M. LeBard, *The Coming War with Japan* (St. Martin's Press, 1991); H. A. Linstone, "Mediacracy, Mediocracy, or New Democracy," *Technological Forecasting and Social Change*, vol. 36 (1989), pp. 153–69.

10. T. L. Friedman, "Cold War Without End," *New York Times Magazine,* Aug. 22, 1993.

11. Y. Dror, *Crazy States* (Millwood, N.Y.: Kraus reprint, 1980); *New York Times*, Feb. 7, 1993.

12. *Los Angeles Times*, Jan. 6, 1993.

13. R. Barnet, *The New Yorker*, Jan. 1, 1990.

14. *New York Times*, Sept. 2, 1993.

15. L. C. Lewin (widely assumed to be the pen name of J. K. Galbraith), *The Report from Iron Mountain* (New York: Dell Publishers, 1969).

16. F. Pace, Jr., address to the American Bankers' Association, Sept. 1957.

17. K. Morita, J. K. H. Chia, and J. C. Oliga, "The New Japanese 'Intra-Entrepreneur': the Silver Handshake?" Proceedings of the International Society of System Sciences Annual Meeting, Denver, July 12–17, 1992.

18. *Los Angeles Times*, Dec. 22, 1991.

19. *Los Angeles Times*, Feb. 17, 1992.

20. *New York Times,* Apr. 20, 1992, Nov. 14, 1992; *Los Angeles Times*, Jan. 8, 1993.

21. P. A. Corning, "The Power of Information," Proceedings of the International Society of System Sciences Annual Meeting, Denver, July 12–17, 1992.

22. M. T. Brown and R. C. Murphy, "Energy Analysis Perspectives on Ecotourism, Carrying Capacity, and Sustainable Development," Proceedings of the International Society of System Sciences Annual Meeting, Denver, July 12–17, 1992.

23. *Fortune*, Dec. 2, 1991, pp. 102–6.

24. W. Serrin, *Homestead: The Glory and Tragedy of an American Steel Town* (New York: Times Books/Random House, 1992).

25. *Time*, Jan. 20, 1975.

26. C. Lazzareschi, "Under the Big Gray Cloud at IBM," *New York Times Magazine*, Feb. 2, 1992, p. 24.

27. J. Womack, D. P. Jones, and D. Ross, *The Machine that Changed the World* (Cambridge, Mass.: Rawson, 1990); C. Hill, "New Manufacturing Paradigms—New Manufacturing Policies?" Paper delivered at NAE/NAS Manufacturing Forum, Washington, D.C., Aug. 14, 1991.

28. Morita, Chia, and Oliga, "New Japanese 'Intra-Entreperneur.'"

29. Southern California Edison Company, "Planning for Uncertainty: a Case Study," *Technological Forecasting and Social Change* , vol. 33 (1988), pp. 119–48.

30. Womack, Jones, and Ross, *Machine That Changed the World*.

31. *Fortune*, Dec. 30, 1991, p. 59.

32. D. L. Bartlett and J. B. Steele, "America: What Went Wrong?" *Philadelphia Inquirer*, Oct. 20–28, 1991.

33. *New York Times*, June 9, 1992.

34. R. Iwata, "The New Age of Japanese Management," *Look Japan*, Oct. 10, 1985.

35. *Time*, July 15, 1991.

36. F. Cairncross, *Costing the Earth* (London: Economist Books, 1991), p. 238.

37. J. L. Gifford and W. L. Garrison, "Airports and the Air Transportation System: Functional Refinements and Functional Discovery," *Technological Forecasting and Social Change*, vol. 43 (1993), pp. 103–124.

38. L. J. Perelman, *School's Out: Hyperlearning, the New Technology, and the End of Education* (New York: William Morrow & Co., 1992).

39. Bartlett and Steele, "America: What Went Wrong?"

40. J. Kurtzman, *The Death of Money* (New York: Simon & Schuster, 1993); *Los Angeles Times*, Dec. 23, 1991.

41. L. W. Seidman, chairman of the Federal Deposit Insurance Corporation, quoted in *New York Times*, Sept. 15, 1991.

42. G. Gilder, "Into the Telecosm," *Harvard Business Review*, Mar.–Apr. 1991, pp. 150–61.

43. F. A. Rossini, "Transitions: The Synergistic Impacts of Major Technologies in the Twenty-Second Century and Beyond," *Technological Forecasting and Social Change*, vol. 36 (1989), p. 219.

44. E. M. Rogers, *Communication Technology: The New Media in Society* (New York: The Free Press, 1986), p. 181.

45. D. N. Michael, "Too Much of a Good Thing? Dilemmas of an Information Society," *Technological Forecasting and Social Change*, vol. 25 (1984), pp. 347–54.

46. H. Rosenberg, *Los Angeles Times*, Feb. 16, 1991, p. A7.

47. G. Mitchell, "The Campaign of the Century: Upton Sinclair's Race for Governor of California and the Birth of the Media Politics," *New York Times*, May 30, 1992.

48. *Los Angeles Times*, Jan. 6, 1991.

49. *Wall Street Journal*, Oct. 27, 1988, p. A24.

50. *New York Times*, Nov. 9, 1991.

51. H. Smith, *New Russians*, p. 110.

52. Smith, *New Russians*, pp. 119–20.

53. Smith, *New Russians*, p. 161.

54. Smith, *New Russians*, p. 239.

55. *Time*, June 24, 1991.

56. J. Houston, "Prometheus Rebound: An Inquiry into Technological Growth and Psychological Change," *Technological Forecasting and Social Change*, vol. 9 (1976), pp. 251, 253.

57. A. Guillermoprieto, "Obsessed in Rio," *New Yorker*, Aug. 16, 1993, p. 45.

58. D. T. Regan, *For the Record* (New York: St. Martin's Press, 1988), p. 419.

59. *New York Times*, Nov. 5, 1991.

60. I. I. Mitroff and W. Bennis, *The Unreality Industry* (New York: Birch Lane Press, 1990).

61. *Los Angeles Times*, Jan. 22, 1992.

62. *The Federalist*, numbers 10, 14, 18, 20, and 45 (1787).

63. Madison, *Federalist*, number 10.

64. Madison, *Federalist*, number 14.

65. Madison, *Federalist*, number 10.

66. P. Kennedy, *The Rise and Fall of the Great Powers* (New York: Vintage Books, 1989), p. xvi.

67. B. R. Barber, "Jihad vs. McWorld," *Atlantic Monthly*, Mar. 1992, pp. 53–65.

68. J. Kotkin, *Tribes: How Race, Religion, and Family Determine Success in the New Global Economy* (New York: Random House, 1993).

69. *Oregonian*, Sept. 6, 1992; *New York Times*, Sept. 25, 1992.

70. O. Young, "International Environmental Governance: Building Institutions in an Anarchical Society" (Paper delivered at the IIASA Conference, Laxenburg, Austria, May 12, 1992).

71. H. A. Linstone, "Mediacracy, Mediocracy, or New Democracy," *Technological Forecasting and Social Change*, vol. 36 (1989), pp. 153–69; A. M. Rivlin, *Reviving the American Dream: The Economy, the States, and the Federal Government* (Washington: Brookings Institution, 1992).

72. T. Jefferson, letter to J. Madison, Sept. 6, 1789, in P. S. Foner, *Basic Writings of Thomas Jefferson* (Garden City, N.Y.: Halcyon House, 1944), p. 591.

73. A. Jay, *Management and Machiavelli* (New York: Holt, Rinehart, & Winston, 1968), chap. 8.

74. *Los Angeles Times*, Mar. 8, 1992.

75. *Fortune*, May 26, 1986.

76. *New York Times*, Apr. 18, 1992.

77. T. W. Malone and J. F. Rockart, "Computers, Networks, and the Corporation," *Scientific American*, Oct. 1991, pp. 78–85.

78. W. E. Halal et al., *Internal Markets: Bringing the Power of Free Enterprise INSIDE Organizations* (New York: Wiley, 1993).

79. National Research Council, *The Competitive Edge: Research Priorities for U.S. Manufacturing* (Washington, D.C.: National Academy Press, 1991).

80. B. Bowonder, T. Miyake, and H. A. Linstone, "The Japanese Institutional Mechanism for Industrial Growth: an Analysis,"*Technological Forecasting and Social Change*, forthcoming 1994.

81. M. Howland, "Technological Change and the Spatial Restructuring of Data Entry and Processing Services," *Technological Forecasting and Social Change*, vol. 43 (1993), pp. 185–96.

82. D. A. Swyt, "The Workforce of U.S. Manufacturing in the Post-Industrial Era," *Technological Forecasting and Social Change*, vol. 34 (1988), pp. 231–51.

83. R. B. Reich, *The Work of Nations* (New York: Knopf, 1991).

84. M. Thompson and M. Warburton, "Decision Making under Contradictory Certainties: How to Save the Himalayas When You Can't

Find Out What's Wrong With Them," *Journal of Applied Systems Analysis*, vol. 12 (1985), pp. 10, 13, 17.

85. M. Gorbachev, campaigning in Kiev Feb. 20, 1989, quoted in Smith, *New Russians* (New York: Ransom House, 1990), p. 442.

86. H. A. Linstone, J. Fried, W. Yinglin, and S. Hui, *Multiple Perspectives in Cross-Cultural Systems Analysis*, Report 87-2 (Portland: Portland State University, Systems Science Ph.D. Program, 1987), pp. 37–39, 68, 116.

87. W. A. Fischer, "Scientific and Technical Planning in the People's Republic of China," *Technological Forecasting and Social Change*, vol. 25 (1984), pp. 189–207.

CHAPTER 9

1. *New York Times*, July 26, 1992.
2. *New York Times*, Apr. 21, 1992, Aug. 16, 1992.
3. *New York Times*, Sept. 13, 1992.
4. A de Tocqueville, *Democracy in America* (P. F. Collier & Son, 1900), vol. 1, p. 3.
5. G. F. Crystal, "Cracking the Tax Whip on CEO's," *New York Times*, Sept. 23, 1990; *Time*, Apr. 15, 1991.
6. *Los Angeles Times*, Dec. 31, 1991.
7. *New York Times*, Mar. 18, 1991.
8. J. F. Coates, "Immigration: Then, Now, and in the Future," *Technological Forecasting and Social Change*, vol. 39 (1991), pp. 411–16.
9. *New York Times*, Sept. 6, 1992.
10. *New York Times*, May 29, 1992.
11. *New York Times*, Aug. 14, 1991.
12. J. F. Coates, "Democracy in America: a Darkening Future," *Technological Forecasting and Social Change*, vol. 38 (1990), p. 102.
13. *New York Times*, Aug. 18, 1993.
14. J. K. Galbraith, *A Tenured Professor* (Boston: Houghton Miflin Co., 1990), pp. 75–76.
15. *Congressional Quarterly*, vol. 51, no. 15, Apr. 10, 1993, p. 901.
16. *New York Times*, Nov. 12, 1992.
17. *Time*, Apr. 6, 1992, Aug. 17, 1992.
18. *New York Times*, Nov. 29, 1992.
19. A. Bloom, *The Closing of the American Mind* (New York: Simon & Schuster, 1987), pp. 83–85.
20. Plato, *Laws*, in *Works*, Jowett translation, n.d., vol. 8, p. 831.
21. *Los Angeles Times*, Feb. 28, 1992, Mar. 10, 1992; B. J. Stein, *A License to Steal* (New York: Simon & Schuster, 1992).

22. *Time*, Jan. 20, 1992.

23. *New York Times*, Aug. 25, 1992.

24. B. Burrough and J. Helyar, *Barbarians at the Gate* (New York: Harper & Row, 1990), pp. 25, 40, 45, 94–96.

25. J. B. Stewart, *Den of Thieves* (New York: Simon & Schuster, 1991), pp. 83–84.

26. D. L. Bartlett and J. B. Steele, "America: What Went Wrong?" *Philadelphia Inquirer*, Oct. 20–28, 1991.

27. Bartlett and Steele, "America: What Went Wrong?"

28. *Los Angeles Times*, Feb. 27, 1992.

29. *Los Angeles Times*, Apr. 5, 1992; *New York Times*, Sept. 8, 1992.

30. *Time*, May 25, 1987; *Fortune*, Jan. 11, 1993.

31. H. A. Linstone, *Multiple Perspectives for Decision Making* (New York: North-Holland, 1984), p. 93; N. Cousins, *The Pathology of Power* (New York: W. W. Norton & Co., 1987).

32. W. Durant, *The Life of Greece* (New York: Simon & Schuster, 1966), pp. 260, 294.

33. G. A. Geyer, *Oregonian*, Nov. 10, 1991.

34. *New York Times*, Oct. 5, 1992.

35. V. Mayakovsky, quoted in H. Smith, *The New Russians* (New York: Random House, 1990), p. 195.

36. *Time*, Jan. 9, 1989.

37. President Clinton's Address to the U.S. Congress, Feb. 17, 1993.

38. *Los Angeles Times*, Feb. 26, 1993.

39. F. R. Cowell, *Cicero and the Roman Republic* (New York: Chanticleer Press, 1948), p. 276.

40. I. F. Stone, *The Trial of Socrates* (Boston: Little, Brown & Co., 1988), pp. 99–100.

41 Durant, *Life of Greece*, p. 285.

42. L. R. Brown et al., *State of the World 1992* (New York: W. W. Norton & Co., 1992), p. 88; *Los Angeles Times*, Dec. 19, 1991, Jan. 7, 1992, Mar. 1, 1992.

43. A. S. Miller, "'Constitutionalizing' the Corporation," *Technological Forecasting and Social Change*, vol. 22 (1982), p. 96.

44. Miller, "'Constitutionalizaing' the Corporation," p. 101

45. *New York Times*, Nov. 15, 1992.

46. B. Kelly, *Adventures in Porkland* (New York: Villard Books, 1992).

47. *New York Times*, June 7, 1992.

48. *New York Times*, May 16, 1992.

49. *New York Times*, Apr. 29, 1992.

50. *Los Angeles Times*, Dec. 1, 1991.

51. J.-Y. Calvez, "Possibilities of Freedom in Tomorrow's Complex Societies," in *Freedom and Man*, ed. J. C. Murray (New York: P. Kenedy & Sons, 1965), p. 182.

52. *Los Angeles Times*, Mar. 27, 1992.

53. B. Friedan, *The Fountain of Age* (New York: Simon & Schuster, 1993).

54. S. M. Evans and H. C. Boyte, *Free Spaces: The Sources of Democratic Change in America* (Chicago: University of Chicago, 1986).

55. *Time*, May 11, 1992.

56. M. S. Burnett, "Valuing the Future As If It Mattered: The Negative Discount Rate and Sustainable Development," Proceedings of the Annual Meeting of the International Society of System Sciences, Denver, July 12–17, 1992.

57. *Los Angeles Times*, Feb. 12, 1992.

58. *New York Times*, Sept. 2, 1992.

59. J. F. Coates, "Preparing for the Urban Future," *Technological Forecasting and Social Change*, vol. 42 (1992), pp. 309–16.

60. *New York Times*, June 28, 1990, pp. A–1, A–12.

61. *Los Angeles Times*, Jan. 6, 1993.

62. *New York Times*, Sept. 5, 1992, Sept. 16, 1992.

63. E. W. Lawless, *Technology and Social Shock* (New Brunswick: Rutgers University Press, 1977), pp. 208–16.

64. M. L. Cropper and P. R. Portney, *Resources*, Resources for the Future, Summer 1992, No. 108.

65. H. T. Odum, "Simulating Ecological Economic Parameters," Proceedings of the Annual Meeting of the International Society of System Sciences, Denver, July 12–17, 1992.

66. D. Meadows et al., *The Limits to Growth* (New York: Universe Books, 1972), p. 124.

67. J. Rothenberg, *Time Comparisons in Public Policy Analysis of Global Change: an Economic Exploration* (manuscript, 1992, communicated by H. Brooks).

68. J. Dator, "The Dancing Judicial Zen Masters: How Many Judges Does It Take to See the Future?", *Technological Forecasting and Social Change* (forthcoming, vol. 46 [1994])

CHAPTER 10

1. Y. Dror, personal communication, June 1991.

2. M. Smithson, *Ignorance and Uncertainty* (New York: Springer Verlag, 1989), p. 9.

3. A. Tversky and D. Kahneman, "Judgment under Uncertainty: Heuristics and Biases," *Science*, vol. 185, Sept. 27, 1974, pp. 1124–31.

4. *New York Times*, July 21, 1991.

5. M. Caudill and C. Butler, *Naturally Intelligent Systems* (Cambridge, Mass.: MIT Press, A Bradford Book, 1990).

6. J. A. Alic, "Computer-Assisted Everything? Tools and Techniques for Design and Production," *Technological Forecasting and Social Change*, vol. 44 (1993), pp. 359–374.

7. J. Hadamard, *The Psychology of Invention in the Mathematical Field* (Princeton: Princeton University Press, 1945), p. 34.

8. J. Hadamard, *Psychology of Invention*, p. 113.

9. J. Sculley, *Odyssey* (New York: Harper & Row, 1987), pp. 156–58.

10. T. Modis, "Competition and Forecasts for Nobel Prize Awards," *Technological Forecasting and Social Change*, vol. 34 (1988), pp. 95–102.

11. C. Marchetti, "Society as a Learning System: Discovery, Invention, and Innovation Cycles Revisited," *Technological Forecasting and Social Change*, vol. 18 (1990), p. 272.

12. T. J. Gordon and D. Greenspan, "Chaos and Fractals: New Tools for Technological and Social Forecasting," *Technological Forecasting and Social Change*, vol. 34 (1988), p. 14.

13. L. Toma and E. Gheorghe, "Equilibrium and Disorder in Human Decision-Making Processes," *Technological Forecasting and Social Change*, vol. 41 (1992), pp. 401–22.

14. H. Arendt, "Thinking," *New Yorker*, Dec. 5, 1977, p. 212.

15. N. Postman, *Technopoly* (New York: Knopf, 1992).

16. H. A. Linstone and M. Turoff, eds., *The Delphi Method: Techniques and Applications* (Reading, Mass.: Addison-Wesley Publishing Co., 1975).

17. J. F. Coates, "What Do the Tender–Minded Have to Teach the Tough-Minded?" *Technological Forecasting and Social Change* (forthcoming, vol. 45 [1994]).

CHAPTER 11

1. A. Yakovlev, quoted in H. Smith, *The New Russians* (New York: Random House, 1990), pp. 558, 561.

2. H. A. Linstone, "On Discounting the Future," *Technological Forecasting and Social Change*, vol. 4 (1973), pp. 335–38.

3. M. J. Cetron and A. Clayton, "Investigating Potential Value Changes," in *Futures Research: New Directions*, H. A. Linstone and W. H. C. Simmons (Reading, Mass.: Addison-Wesley Publishing Co., 1977), pp. 214–29.

4. W. Bennis and B. Nanus, *Leaders* (New York: Harper & Row, 1985), pp. 87–109.

5. R. N. Bellah, R. Madsen, W. M. Sullivan, A. Swidler, and S. M. Tipton, *The Good Society* (New York: Knopf, 1991), p. 254.

6. B. R. Barber, "Jihad vs. McWorld," *Atlantic Monthly*, Mar. 1992, p. 64.

7. H. E. Goeller and A. M. Weinberg, "The Age of Substitutability, or What to Do When the Mercury Runs Out," *Science*, vol. 191, Feb. 20, 1976, pp. 683–689.

8. *Los Angeles Times*, Jan. 4, 1992.

9. S. Sesser, "A Nation of Contradictions," *The New Yorker*, Jan. 13, 1992, p. 60.

10. *Los Angeles Times*, Jan. 30, 1992.

11. Los Angeles 2000 Committee, "LA 2000: A City for the Future, Final Report (Los Angeles, Nov. 1988).

12. M. J. Kirton, "Adaptors and Innovators—Why New Initiatives Get Blocked," *Long Range Planning*, vol. 17 (1984), pp. 137–43.

13. F. E. Udwadia, "Creativity and Innovation in Organizations: Two Models and Managerial Implications," *Technological Forecasting and Social Change*, vol. 38 (1990), pp. 65–80; J. R. Evans, "Creativity in MS/OR: The Multiple Dimensions of Creativity," *Interfaces*, vol. 23, no. 2 (Mar.–Apr. 1993), pp. 80–83.

14. J. Sculley, *Odyssey* (New York: Harper & Row, 1987), pp. 419, 421.

15. *New York Times*, Apr. 25, 1993.

16. R. U. Ayres, *The Next Industrial Revolution* (Cambridge, Mass.: Ballinger Publishing Co., 1984).

17. *Fortune*, Dec. 2, 1991, p. 56.

18. J. C. Oliga, K. Morita, and J. K. H. Chia, "*Kaizen* Management and the Japanese Gradualist Intrapreneurship," Proceedings of the Annual Meeting of the International Society of System Sciences, Denver, July 12–17, 1992.

19. B. Bowonder, T. Miyake, and H. A. Linstone, "The Japanese Institutional Mechanism for Industrial Growth: an Analysis," *Techno-logical Forecasting and Social Change* (forthcoming, vol. 45 [1994]).

20. B. Bowonder, personal communication, Jan. 1992.

21. M. M. Crow and S. A. Nath, "Technology Strategy Development in Korean Industry: An Assessment of Market and Government Influences," *Technovation*, vol. 12 (1992), pp. 119–36; T. Shin and H. Kim, "Research Foresight Activities and Technological Development in Korea: Science and Technology Policies in National R&D Programs" (Paper delivered at Expert Group Meeting on Technology Assessment, Monitoring, and Forecasting, United Nations, Paris, Jan. 25–28, 1993).

22. M. N. Sharif, "Technological Leapfrogging: Implications for Developing Countries," *Technological Forecasting and Social Change*, vol. 36 (1989), pp. 201–8.

23. S. M. Pollock and K. Chen, "Strive to Conquer the Black Stink: Decision Analysis in the Peoples Republic of China," *Interfaces*, vol. 16, no. 2 (Mar.–Apr. 1986), pp. 31–42.

24. J. J. Bartholdi III, "Operations Research in China," *Interfaces*, vol. 16, no. 2 (Mar.–Apr. 1986), pp. 29–30.

25. *Los Angeles Times*, Apr. 4, 1983.

26. A. Peyrefitte, *The Immobile Empire* (New York: Knopf, 1992).

27. F. Butterfield, *China: Alive in the Bitter Sea* (New York: Times Books, 1982), p. 40.

28. D. Bonavia, *The Chinese* (New York: Lippincott & Crowell Publishers, 1980), p. 45.

29. P. R. Harris and R. T. Moran, *Managing Cultural Differences* (Houston: Gulf Publishing Co., 1987), p. 406.

30. *New York Times*, Aug. 9, 1992.

31. K. Truman, "Organizing the Disorganization in the Aftermath of Central Planning" (manuscript, 1992).

32. *Los Angeles Times*, Mar. 24, 1992.

33. P. Schwartz, *The Art of the Long View* (New York: Doubleday, Currency Books, 1991), pp. 185–91.

34. D. E. Kash, *Perpetual Innovation: The New World of Competition* (New York: Basic Books, 1989).

35. *Fortune*, Dec. 2, 1991, p. 56.

36. *Los Angeles Times*, Feb. 17, 1992.

37. B. Bowonder, T. Miyake, and H. A. Linstone, "The Japanese Institutional Mechanism for Industrial Growth: an Analysis," *Technological Forecasting and Social Change* (forthcoming, vol. 45 [1994]).

38. *Fortune*, Dec. 2, 1991, p. 57.

39. H. A. Linstone, *Multiple Perspectives for Decision Making* (New York: North-Holland, 1984), p. 270.

40. *Time*, July 6, 1992.

41. *The Competitive Strength of U.S. Industrial Science and Technology: Strategic Issues* (National Science Board, 1992).

42. *Fortune*, Dec. 2, 1991, pp. 56–62.

43. *The Competitive Strength of U.S. Industrial Science.*

44. Kash, *Perpetual Innovation.*

45. D. Dimancescu and J. Botkin, *The New Alliance: America's R&D Consortia* (Cambridge, Mass.: Ballinger Publishing Co., 1986).

46. Quoted in Dimancescu and Botkin, *New Alliance*, p. 51.

47. J. Botkin, D. Dimancescu and R. Stata, *Global Stakes: The Future of High Technology in America* (Cambridge, Mass.: Ballinger Publishing Co., 1982); M. M. Waldrop, *Complexity* (New York: Simon & Schuster, 1992), pp. 54–69.

48. Ayres, *Next Industrial Revolution*, p. 84.

49. Ayres, *Next Industrial Revolution*, p. 242.

50. R. U. Ayres, "Technological Protection and Piracy: Some Implications for Policy," *Technological Forecasting and Social Change*, vol. 30 (1986), pp. 5–18.

51. L. C. Seifert and A. Zeisler, "A National Manufacturing Policy: An Industrial Perspective on Promoting Sustained Improvement in U.S. Global Competitiveness," *Technological Forecasting and Social Change*, vol. 35 (1989), pp. 7–9.

52. L. Suarez-Villa and S. A. Hasnath, "The Effect of Infrastructure on Invention: Innovative Capacity and the Dynamics of Public Construction Investment," *Technological Forecasting and Social Change*, vol. 44 (1993), pp. 333–358.

CHAPTER 12

1. J. L. Casti, *Searching for Certainty* (New York: William Morrow & Co., 1990), pp. 77–131.

2. J. G. Miller, *Living Systems* (New York: McGraw-Hill, 1978).

3. H. von Foerster, "Responsibilities of Competence," *Journal of Cybernetics*, vol. 2, no. 2 (1972), pp. 1–6.

4. D. L. Meadows, J. Randers, and W. W. Behrens III, *The Limits to Growth* (New York: Universe Books, 1972).

5. I. I. Mitroff and T. Pauchant, *We're So Big and Powerful That Nothing Can Happen to Us: An Investigation of America's Crisis Prone Organizations* (New York: Birch Lane Press, 1991).

6. C. S. Holling, "The Curious Behavior of Complex Systems: Lessons from Ecology," in *Futures Research: New Directions*, ed. H. A. Linstone and W. H. C. Simmonds (Reading, Mass.: Addison-Wesley Publishing Co., 1977).

7. T. Modis and A. Debecker, "Chaoslike States Can Be Expected Before and After Logistic Growth," *Technological Forecasting and Social Change*, vol. 41 (1992), pp. 111–20.

8. T. J. Gordon and D. Greenspan, "Chaos and Fractals: New Tools for Technological and Social Forecasting," *Technological Forecasting and Social Change*, vol. 34 (1988), pp. 1–25; S. A. Kauffman, "Antichaos and Adaptation," *Scientific American*, Aug. 1991, pp. 78–84.

9. T. J. Gordon, "Chaos in Social Systems," *Technological Forecasting and Social Change*, vol. 42 (1992), pp. 1–16.

10. M. M. Waldrop, *Complexity* (New York: Simon & Schuster, 1992).

11. Waldrop, *Complexity*.

12. P. A. Hansson, "Chaos: Implications for Forecasting," *Futures*, Jan./Feb. 1991, pp. 50–58; Waldrop, *Complexity*.

13. L. Troncale, unpublished work (California State Polytechnic University, Pomona, 1990).

14. B. Bergson, "Mathematical Prediction of Emergent Properties of Systems," Proceedings of the Annual Meeting of the International Society of System Sciences, Denver, July 12–17, 1992.

15. D. Sahal, *Patterns of Technological Innovation* (Reading, Mass. Addison-Wesley Publishing Co., 1981), p. 75.

16. Southern California Edison Company, "Planning for Uncertainty: a Case Study," *Technological Forecasting and Social Change*, vol. 33 (1988), p. 147.

17. C. W. Churchman, *The Design of Inquiring Systems* (New York: Basic Books, 1971).

18. Churchman, *Design of Inquiring Systems*; I. I. Mitroff and H. A. Linstone, *The Unbounded Mind* (New York: Oxford University Press, 1993).

19. M. Crichton, *Jurassic Park* (New York: Ballantine Books, 1991), pp. 306–7.

20. *Time*, June 1, 1992.

21. *Time*, Feb. 10, 1992, Feb. 17, 1992.

22. E. B. Weiss, *In Fairness to Future Generations* (Tokyo: United Nations University, 1989), p. 289.

23. *New York Times*, June 3, 1992.

24. A. M. Schlesinger, Jr., *The Age of Roosevelt: The Politics of Upheaval* (Boston: Houghton-Mifflin, 1960).

25. Waldrop, *Complexity*, p. 198.

26. E. O. Wilson, *The Diversity of Life* (Cambridge: Harvard University Press, Belknap Press, 1992).

27. C. W. Churchman, *Thought and Wisdom* (Seaside, Calif.: Intersystems Publications, 1982), p. 21.

28. M. S. Burnett, "Valuing the Future As If It Mattered: The Negative Discount Rate and Sustainable Development," Proceedings of the Annual Meeting of the International Society of Systems Sciences, Denver, July 12–17, 1992.

29. G. Hardin, "The Case Against Helping the Poor," *Psychology Today*, Sept. 1974, pp. 38–126.

30. Genesis, chap. 1.

31. F. Jansen, "Conforming to Intolerance" (manuscript, 1992).

32. A. R. Vidler, *Essays in Liberality* (London: SCM Press, 1957), pp. 21–22.

33. R. N. Bellah et al., *The Good Society* (New York: Knopf, 1991), p. 284; C. W. Churchman, "A Philosophy for Complexity," in *Futures Research: New Directions*, ed. H. A. Linstone and W. H. C. Simmonds (Reading, Mass.: Addison-Wesley Publishing Co., 1977), p. 90.

34. J. W. Woelfel, *Bonhoeffer's Theology* (Nashville: Abingdon Press, 1970), pp. 22–23.

35. C. W. Churchman, *Thought and Wisdom*, p. 15.

APPENDIX A

1. H. C. Kunreuther et al., *Risk Analysis and Decision Processes* (Berlin: Springer Verlag, 1983), pp. 25–30.

2. H. C. Kunreuther et al., *Risk Analysis and Decision Processes*, p. 236.

3. H. A. Linstone, *Multiple Perspectives for Decision Making* (New York: North-Holland, 1984), pp. 87–94, 322–23.

4. Linstone, *Multiple Perspectives for Decision Making*, pp. 87–94, 322–23.

5. M. Maruyama, "Overcoming the Socialist Aftereffects in East Europe," *Technological Forecasting and Social Change,* vol. 40 (1991), pp. 297–302.

6. R. O. Mason and I. I. Mitroff, *Challenging Strategic Planning Assumptions* (New York: John Wiley, 1981).

APPENDIX B

1. H. A. Linstone, *Multiple Perspectives for Decision Making* (New York: North-Holland, 1984).

2. M. W. Merkhofer, *Decision Science and Social Risk Management* (Dordrecht, Holland: D. Reidel Publishing Co., 1987), pp. 68–69.

3. H. Lorraine, "The California 1980 Medfly Eradication Program: An Analysis of Decision Making under Non-Routine Conditions," *Technological Forecasting and Social Change,* vol. 40 (1991), pp. 1–32.

4. P. Schwartz, *The Art of the Long View* (New York: Doubleday, Currency Books, 1991).

5. T. J. Peters and R. H. Waterman, Jr., *In Search of Excellence* (New York: Harper & Row, 1982), p. 61.

INDEX